Cambridge Imperial and Post-Colonial Studies Series
General Editors: **Megan Vaughan**, Kings' College, Cambridge and **Richard Drayton**, King's College London
This informative series covers the broad span of modern imperial history while also exploring the recent developments in former colonial states where residues of empire can still be found. The books provide in-depth examinations of empires as competing and complementary power structures encouraging the reader to reconsider their understanding of international and world history during recent centuries.

Titles include:

Tony Ballantyne
ORIENTALISM AND RACE
Aryanism in the British Empire

Miguel Bandeira Jerónimo
THE "CIVILISING MISSION" OF PORTUGUESE COLONIALISM, 1870–1930

Miguel Bandeira Jerónimo and António Costa Pinto
THE ENDS OF EUROPEAN COLONIAL EMPIRES
Cases and Comparisons

Peter F. Bang and C. A. Bayly (*editors*)
TRIBUTARY EMPIRES IN GLOBAL HISTORY

Gregory A. Barton
INFORMAL EMPIRE AND THE RISE OF ONE WORLD CULTURE

James Beattie
EMPIRE AND ENVIRONMENTAL ANXIETY, 1800–1920
Health, Aesthetics and Conservation in South Asia and Australasia

Rachel Berger
AYURVEDA MADE MODERN
Political Histories of Indigenous Medicine in North India, 1900–1955

Robert J. Blyth
THE EMPIRE OF THE RAJ
Eastern Africa and the Middle East, 1858–1947

Rachel Bright
CHINESE LABOUR IN SOUTH AFRICA, 1902–10
Race, Violence, and Global Spectacle

Larry Butler and Sarah Stockwell
THE WIND OF CHANGE
Harold Macmillan and British Decolonization

Kit Candlin
THE LAST CARIBBEAN FRONTIER, 1795–1815

Nandini Chatterjee
THE MAKING OF INDIAN SECULARISM
Empire, Law and Christianity, 1830–1960

Esme Cleall
MISSIONARY DISCOURSE
Negotiating Difference in the British Empire, c.1840–95

T. J. Cribb (*editor*)
IMAGINED COMMONWEALTH
Cambridge Essays on Commonwealth and International Literature in English

Bronwen Everill
ABOLITION AND EMPIRE IN SIERRA LEONE AND LIBERIA

Anna Greenwood and Harshad Topiwala
INDIAN DOCTORS IN KENYA, 1895–1940

Róisín Healy and Enrico Dal Lago (*editors*)
THE SHADOW OF COLONIALISM IN EUROPE'S MODERN PAST

B.D. Hopkins
THE MAKING OF MODERN AFGHANISTAN

Ronald Hyam
BRITAIN'S IMPERIAL CENTURY, 1815–1914: A STUDY OF EMPIRE AND EXPANSION
Third Edition

Cambridge Imperial and Post-Colonial Studies Series
Series Standing Order ISBN 978-0-333-91908-8 (Hardback)
978-0-333-91909-5 (Paperback)
(outside North America only)

You can receive future titles in this series as they are published by placing a standing order. Please contact your bookseller or, in case of difficulty, write to us at the address below with your name and address, the title of the series and the ISBN quoted above.

Customer Services Department, Macmillan Distribution Ltd, Houndmills, Basingstoke, Hampshire RG21 6XS, England

---

Indian Doctors in Kenya, 1895–1940

The Forgotten History

Anna Greenwood

and

Harshad Topiwala

First published 2015 by
PALGRAVE MACMILLAN

Palgrave Macmillan in the UK is an imprint of Macmillan Publishers Limited, registered in England, company number 785998, of Houndmills, Basingstoke, Hampshire RG21 6XS.

Palgrave Macmillan in the US is a division of St Martin's Press LLC, 175 Fifth Avenue, New York, NY 10010.

Palgrave Macmillan is the global academic imprint of the above companies and has companies and representatives throughout the world.

Palgrave® and Macmillan® are registered trademarks in the United States, the United Kingdom, Europe and other countries.

ISBN 978-1-349-68412-0   ISBN978-1-137-44053-2 (eBook)

DOI 10.1057/9781137440532

This book is printed on paper suitable for recycling and made from fully managed and sustained forest sources. Logging, pulping and manufacturing processes are expected to conform to the environmental regulations of the country of origin.

A catalogue record for this book is available from the British Library.

Library of Congress Cataloging-in-Publication Data
Greenwood, Anna, 1971–
Indian doctors in Kenya, 1895–1940: the forgotten history / Anna Greenwood and Harshad Topiwala.
    pages cm.—(Cambridge imperial and post-colonial studies series)
    Summary: "This pioneering book offers unique insights into the careers of Indian doctors in colonial Kenya. As such, it deepens and broadens modern understandings of the complex constitution of the British Empire. The British Empire, although ideologically racist, nevertheless relied upon staff of all nationalities and ethnicities. Ideas and practices were imported between various colonial dependencies as much as they evolved responsively to local conditions. The book highlights the complex ambiguities of Empire; advancing modern understandings of the British Empire as a linked, multi-centred global phenomenon, while also providing a case study that enriches local understandings of the practice of medicine in a racially segregated context. Chapters examine in turn the main possible career options for Indian medical graduates as well as setting out the racial and political context of colonial Kenya. An impressively large and varied source base has been consulted throughout resulting in startling new insights into the complex operation of western medicine in this racially segregated world."—Provided by publisher.
    ISBN 978–1–137–44052–5 (hardback)
    1. Medicine—Kenya—History—19th century.   2. Indians—Medicine—History. 3. Traditional medicine—Kenya—History.   4. Doctors—Kenya—History. 5. Ethnopharmacology.
    I. Topiwala, Harshad,   1945– II. Title.
    R464.5.G74 2015
    610.96762—dc23   2015004014

Typeset by MPS Limited, Chennai, India.

Contents

List of Figures and Tables

Figures

Tables

Preface and Acknowledgements

The first seeds of this book were unknowingly sown in Harshad Topiwala's personal experiences. Harshad is the son of one of the Kenyan doctors mentioned in the book: H.T. Topiwala. Growing up in pre-independence Nairobi, Harshad was a child of the segregated colonial world we describe. His Gujarati father was both a private practitioner and a vocal defender of Indian rights in the colony. Although his heated political discussions over supper evidently lodged themselves deeply within Harshad's subconscious, it took the passage of time before Harshad seriously began to reflect upon the delicate balance of challenges and opportunities his father's decision to move to Kenya represented more widely. On one hand, his father was Europeanised in dress, taste, and manners; a keen cricket player; a fastidious advocate of British education; and a staunch believer in the benefits of Western medical science. On the other, he recognised the frustrations of being part of an essentially second-class social group within colonial Nairobi society, felt a strong affinity to his Indian homeland, acted as a benefactor for local Indian associations, and was prepared to stick his neck out with regard to speaking out in the press and other public fora to represent Indian rights. The ambiguities inherent in this particular diasporic identity seemed a potent symbol of a wider topic for study: the forgotten lives of the Indian professionals of Kenya.

It was only after his father's funeral in 1975 that Harshad started thinking about what life had actually been like for him. After the service one of the mourners, Justice Chanan Singh—who himself had played a prominent part in the public life of Kenya—invited Harshad back to his home in Eastlands and showed him his large private collection of papers and books, many of which were on the history of Indians. Harshad had not even realised that the history of the Indian-Kenyan diaspora had been written about and lodged the experience away as something he should one day investigate further.

Five years later, while waiting for a delayed lunch at the Commonwealth Club, Harshad made his way to the library. Browsing the several shelves on East African history he was astonished to see that not a single book had been authored by a non-European. His interest was sparked. Were there really no records left of the non-white Kenyan communities, despite the fact that his personal memory had been that they were a relatively publically articulate sector of colonial society? He himself had experienced a reasonably comfortable upbringing in Kenya. Why had these literate, educated Indians, neglected to tell their own version of their history? This problem stayed at the back of Harshad's mind for several years, until upon his retirement, he started to investigate the history of Indian Kenyan doctors in earnest.

In the early courses of his research Harshad stumbled across a doctoral thesis in the British Library authored by Anna Crozier (now Greenwood). Anna had studied the European Medical Officers of the Colonial Medical Service in East Africa, detailing trends in their recruitment and in-field experiences and arguing for some elements of shared colonial identity. Intriguingly, she did not mention the Indian Surgeons and Sub-Assistant Surgeons who made up the majority doctors of this cohort before 1923.

Harshad telephoned Anna and his personal journey soon became aligned with new strand in Anna's professional one. Anna was fascinated to hear Harshad's critique that she had missed a large part of the colonial history of Western medicine in the region. How could it have been that during four years of doctoral research she had barely heard anything about the Indian cohort of the Colonial Medical Service in Kenya? Using Anna's knowledge of the archives, they went back to colonial records and began to scour them with a fine-toothed comb. It soon became evident that although Indian doctors were recruited in India for the Colonial Medical Service, this policy was rarely spoken about in London by the Colonial Service recruiters, who favoured only recruiting Europeans as the agents of Western medicine in East Africa. Most shockingly, what scant records there were suddenly went dead after 1923, when it appeared that abruptly, yet silently, the majority Indian sector of colonial medical personnel were retrenched. To get to the heart of the drastic shift in colonial strategy they widened their search to the libraries of the India Office, Kenya and Bombay. This brought into relief the parallel histories that existed of non-governmental doctors—those Indians, like Harshad's father, who worked in private practice in Kenya. The result was this history of Indian medical practitioners in the colony before the Second World War. There will undoubtedly be some omissions, generalisations, and unanswered questions, but it is nevertheless the first medical history of Kenya to acknowledge the large Indian contribution.

This book would not have been published without the generous help of many individuals at different stages over a protracted period. At the beginning leads to East African history were provided by Dr Michael Twaddle and Professor David Anderson. Additionally, Professor Sanjoy Bhattacharya, Dr Nandini Bhattacharya, Professor Frederick Cooper, Professor Mark Harrison, Professor Jim Mills, Dr Mridula Ramanna, and Dr Erica Wald provided their invaluable feedback either through informal discussions or during conference or workshop presentations where aspects of the work were presented. They all helped fundamentally to shape a more nuanced and sophisticated articulation of this book's theoretical contributions.

In Kenya Mr Anwar Sheikh, Mr Harshvadan Maroo, Dr Kenneth Ombongi, Dr Andrew Hicks, and Mr Manu Chandaria provided invaluable assistance. Subsequently several individuals who lived (or did live) in Kenya including Mr Benegal Pereira, Mr Kersi Rustomji, Dr Serosh Sohrabji, Dr Manmeet Singh, Dr Brij Sood, Miss Cynthia Salvadori, Ms Zarina Patel, Mrs Judith Aldridge,

Dr Hansa Topiwala, and Dr Sudha Young forwarded very helpful snippets of information about Indian doctors. We also thank Mr Cliff Peireira and Mr Salim Lalani for their assistance. We are very thankful to Mr Lacty de Sousa for forwarding valuable information about his grandparents who were both pioneering doctors and to Dr Tony Jewell for his persistence in tracking Anna down and for showing her the unpublished autobiography of his grandfather, the European Medical Officer, Norman Jewell.

Staff at many archives and libraries in the UK, US, Kenya, and India were always patient in dealing with our queries. While we are really grateful to everyone, particular mention however should go to Lucy McCann, Senior Archivist at Rhodes House Library, Oxford. She went out of her way to dig up relevant papers and photographs and was impressively helpful throughout. We also thank Venita Bryant at the Wellcome Library and Bonnie Ryan at the University of Syracuse Library as well as the staff of the BMA archives London. We are grateful also to Christine Goff for patiently typing up the first draft of this book from a jumbled mess of notes.

In the final stages of revision we are most grateful for the incredibly helpful and considered advice of Professor Pratik Chakrabarti who kindly and patiently read the whole manuscript and encouraged us to submit it for publication. We also thank the Series Editors, as well as the anonymous academic reviewer, at Palgrave Macmillan for their constructive feedback. The editorial and production team at Palgrave were unfailingly encouraging and helpful. We likewise thank Jyoti Dhar for compiling our index. All of these inputs have helped to progressively shape and improve the manuscript. Where we have fallen short of implementing individual recommendations we, of course, fully accept responsibility for our shortcomings.

Above all we are most indebted to our families. We thank our respective partners, Liz Topiwala and Sarb Bajwa, who had to listen to more discussions about the history of Kenya and Indian diaspora identities than is decently mentionable and, practically speaking, took on extra chores and babysitting duties to help us complete the manuscript. Our heartfelt thanks to our children Anya and Kirty Topiwala and Otto Greenwood, who supported us with humour and good will and tried not to roll their eyes too often when ours lit up. Kirty was particularly helpful in providing tangible assistance during the manuscript's metamorphosis into a book. Anna's father, Professor David Greenwood, read most of the chapters and provided invaluable commentary, grammatical discipline as well as his customary encouragement and love. These contributions pooled provided a memorable atmosphere of affection and support that made this book possible and, perhaps most importantly, cemented many of the friendships that will endure well beyond it.

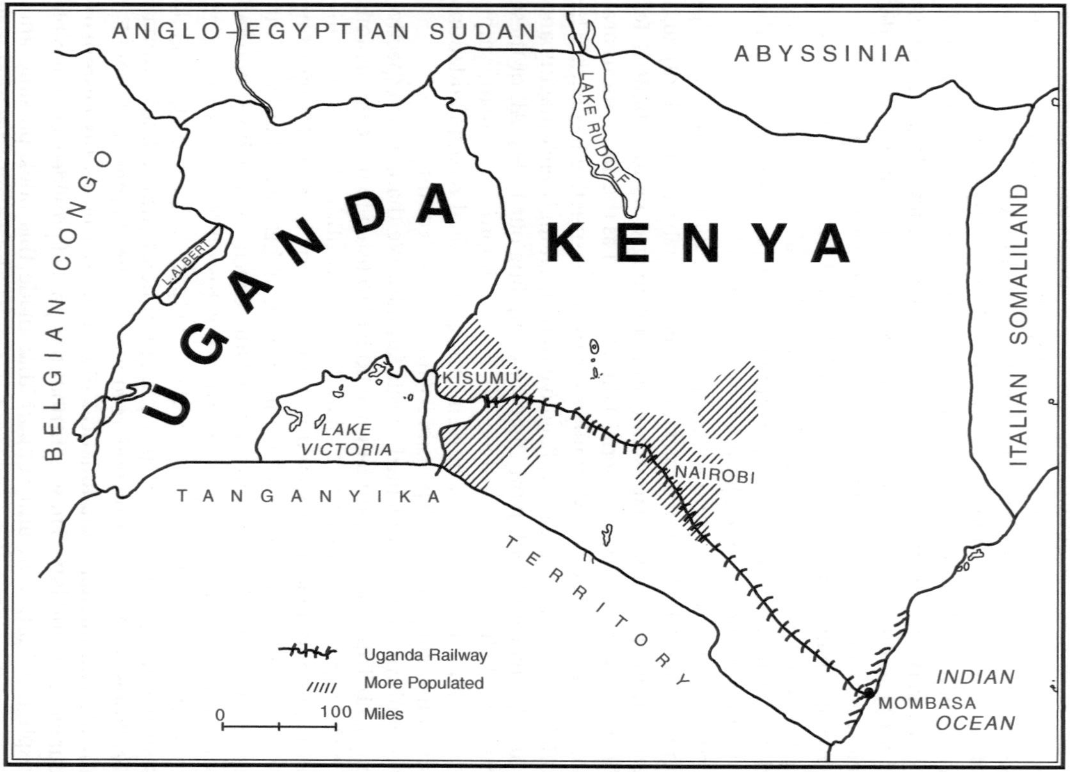

Figure P.1 Kenya, 1925
Credit: Hand-drawn by authors.

1

'The Empire is not white': Indian Doctors in Kenya

> The Empire is not white or English-speaking or Anglo-Saxon
> or British or Christian. It embraces many complexions,
> many languages, many races, many continents, many
> religions.[1]

Indian doctors working in Kenya (formerly East Africa Protectorate) under British rule have been almost entirely written out of the history books. This research gap seems to have passed unremarked by historians of the region, despite various books focussing on the Colonial Medical Service, missionary doctors or the history of Africans entering western medical education.[2] Yet to miss Indian doctors distorts our understanding of the medical history of colonial East Africa—not only in terms of describing the way it imported ideas, medicines, and personnel from the subcontinent—but also by failing to describe a large, diverse and vibrant cross section of the medical community.

Indians should be placed at the very heart of the history of medicine in colonial East Africa with increasing numbers of Indian doctors entering Kenyan ports from the 1890s. Indeed, it would be no exaggeration to claim that Indians were pivotal in the establishment of western medicine in Kenya. They played a large role in the medical care of indentured railway workers in the construction of the Uganda Railway (1896–1901), and cared for troops in the early East African military campaigns of British conquest. During their highpoint at the end of World War One, it is a little known fact that there were actually almost twice as many Indian doctors working for the Colonial Medical Service as Europeans. In fact, if the historian goes back to the colonial medical department files—although Indian names by no means dominated—their presence was nevertheless obvious and consistent until the 1920s, after which time Indian doctors are mostly to be found within the historical records as private practitioners. Collectively, whether working for the railways, the military, the Colonial Service, or as private doctors, Indians provided medical care for hundreds of thousands of

1

Kenyan inhabitants and became the forefathers of medical dynasties— the descendants of which still sometimes work in East Africa today.

Why an entire ethnic group within the colonial medical history of a British colonial dependency has escaped academic scrutiny is open to debate. The explanation lies at least partially in the inherent limitations in researching the history of any cohort who has left few written sources. To be sure, the majority of Indian migrants to East Africa were poorly educated and were preoccupied with earning a livelihood, often in extremely strained conditions. But even Indian traders who regularly produced and maintained accurate business accounts, left few written records in their own vernaculars.[3] Perhaps for the particular professional group of doctors, the persuasive explanation for the lack of published contemporary Indian perspectives is that more educated members of the community, which would have included doctors, consciously avoided publication, fearing political victimisation if they were seen to be too critical.[4] Indians—especially educated and articulate Indians—tended to be silenced by the British authorities, particularly if they were thought to possess views dangerous to the political *status quo*. Although the recipients of British colonial violence were mainly Africans, Indians would have been well aware that they were not exempted from the punitive consequences of critically confronting their European masters.[5]

Whatever the reasons behind the silence of this literate professional group, the overarching aim of the book is to restore the voices of this forgotten cohort of practitioners to their rightful place within history. Indians were more than just useful cogs in the medical administration of empire. In fact, to all intents and purposes, Indian doctors seemed often well liked and appreciated by their European colleagues, at least before the watershed of the Devonshire Declaration of 1923. The publication of this White Paper occurred after extended and high profile debates and endorsed support for white settlers (under the guise of protecting African interests). It was to be the decisive nail in the coffin for Indians in Kenya for the rest of the colonial period, formally curtailing their ambitions for political power and representation.

This sea change in attitudes had the unintentional repercussion of marginalising Indians in historical retellings. After 1923 Indians were rarely mentioned in official medical departmental business and Indian names were omitted in many staff lists. As the book will show, this partly reflected the reality of the situation. After the Devonshire Declaration majority Indian doctors were retrenched from the Colonial Medical Service, but even those that remained employed in this capacity were rarely mentioned in official returns to the Colonial Office. Precisely because they became erased from official records Indian doctors were almost entirely forgotten. The history of Asian professionals in Kenya should be therefore recognised as having fallen between the two stools of the old white triumphalist accounts and the newer focus upon the history of black participation in empire.[6] When the history of medicine in colonial Africa was revisited, it was the history of the Black African doctors that was assumed to be the only crucial missing part of the story.[7] Yet, by reinserting Indian doctors into the social mix of early twentieth century Kenya, a more intriguing, literally more colourful, history of this colonial society emerges. Furthermore, once scrutinised, the relationships that existed between British colonial authorities, Indian doctors and the indigenous population of Kenya are seen to be a complicated array of social and political allegiances with few cohesive agenda, defying any neat generalisations or categorisations of beliefs or loyalties based on skin colour alone. While, for sure, most Europeans accepted racist ideologies prevalent at the time, some can readily be identified as having been remarkably sensitive to both Indian and African rights. Africans in turn can be shown to have both worked in harmony with Indians to their mutual benefit and also, on different occasions, to have been evidently enraged by the lack of solidarity that they felt Indians displayed towards them. To add to the complexity Indians were as likely as Europeans to display racist attitudes towards Africans and individuals within all groups can be shown to have held racial-ideological allegiances that were self-contradictory. Within the Indian communities especially (though Europeans and Africans were not exempt) issues of caste and religious loyalties additionally fractured them into different, often rival, sub-groups, meaning that—although useful as a vague blanket categorisation—it is also crude and inadequate to envisage Kenyan Indians as part of a single, unified, community. Indians were themselves divided as a group and were by no means as homogeneous in their stance towards the British government (or even towards members of their own community). Some Indians were fully collaborative with British government concerns, while others were actively involved in pursuing freedom from the shackles of colonialism. Race, when analysed in this context, then quickly reveals itself as simultaneously both a fragile and a durable concept. Durable because crude, colour based, categorisations of peoples and their dominant behaviours were assumed all the time; fragile because political allegiances, personal loyalties and attitudes, once scratched beyond their surface, cannot always be mapped along ethnic lines with any assurance.

Some of these conflicts and contradictions can be understood as a corollary of the position of Indians in Kenya as members of a group of equivocal status—the middle rank.[8] While often of humble origins, only able to study medicine through the receipt of scholarships—and therefore not 'middle class' in the sense of their class origins—they nevertheless formed a middle *tier* within society: floating in an indistinct middle land between the white ruling elite and the black African 'subalterns'.[9] Although never credited with the so-called civilised status of Europeans, the Indian doctors of Kenya were nevertheless educated, literate, and relatively prosperous. They enjoyed a reasonably high status in their community and, in most cases, embraced Europeanised tastes and standards of living. Mostly, they worked for the colonial state rather than against it, although it is fair to say that even when publically esteemed most Europeans ultimately regarded them as being indecipherable. As John Lonsdale pithily summarised: Indians were the 'unknowable, in-between'. They were at times respected while also at times they were regarded with suspicion. Kenya was a colonial world where all resident communities were struggling for a political and social identity and mutual distrust was a fundamental part of the game.[10] In this context, the Indians who worked for the colonial medical department undertook a variety of official responsibilities that straddled the gap between the lower tiers of junior personnel and the Colonial Medical Officers. The experiences of these middle men (for they were nearly all men), become all the more historically vital, if one thinks of them, as Nancy Rose Hunt has emphasised in her own study of middle ranking hospital aides in Congo, as 'central to processes of translation in a colonial and therapeutic economy'.[11]

The story of Indian doctors therefore offers pertinent insights into the processes that have sometimes swept important components of history under the carpet. It also deepens and broadens modern understandings of the complex constitution of the British Empire. This was an empire that, although ideologically racist, nevertheless relied upon staff of all nationalities and ethnicities. Policies, people and ideas of best practice were imported between regions of empire at least as much as they were dictated from the epicentre of Whitehall.[12] By exploring imperial medical migration as more than just a white phenomenon the study aims to extend diaspora studies that have formerly concentrated upon convicts and labourers, rather than on the movements of professionals. In so doing, this research calls into question ideas of western medicine as a 'tool' of empire emanating principally from the metropolis to the colony.[13] Instead it draws attention to the multiple ways western medicine and its personnel were imported 'sideways' between colonies and will show how the practice of medicine was enacted in ways both collaborative with, and antagonistic to, the colonial government. In this conception of a highly interlinked empire, governors, policy makers, and personnel recruiters were as likely to look towards Bombay as to London for their inspiration.[14]

But this is not to deny the importance of local conditions. While also advancing modern understandings of British Empire as a linked, multi-centred global phenomenon, this research also provides a case study that enriches our local understandings of the *practice* of medicine in a racially segregated context. Colonial Kenya, 'Britain's most troublesome African colony', should be noted as different from its colonial neighbours.[15] It was home to one of the most hard-line racist settler societies of the period—led by Lord Delamere and Ewart Grogan, the settlers of the so-called White Highlands were fierce in their articulation of white supremacy. This provided a local political climate fundamentally antithetical to any enthusiastic accommodation of Indian interests. The loud political voices of the settlers dominated many of the Kenya-related parliamentary debates of the time and have, in turn, influenced the way East African colonial histories have been framed. Barely consciously, subsequent historical descriptions have themselves fallen into the simplistic delineations of white versus non-white that formed the basis of much of the colonial discourse. Yet, as the book that follows will show, any 'them and us' categorisation was rhetorical only and did not begin to describe the intricate social dynamics of either the formal medical administration nor the quotidian community life of this colonial dependency.

Furthermore, by looking at the organisation of medical practice in Kenya, the historical evolution of professional Indian diaspora identity (despite being multiply fractured between kin, family, caste, religion, and political affiliation) is revealed. Even though Indians were not successful in achieving their aims for land rights and enfranchisement in Kenya, they ultimately became the forefathers of a relatively successful community. Over time the ethnic identity of Kenyan Indians began, against the odds, to gain common elements, despite the fact that there was little systemic coherency in the way that different interest groups variously and flexibly deployed their versions of an unpolished generalised identity.[16]

Perhaps most powerfully of all, the reinsertion of the history of Indian middle-men into the story of medicine in Kenya, reminds us of the asymmetries of power. Colonial dominance embodied many forms and was often contradictorily conceived and unequally applied. The British did not speak in one voice, and the way ideologies of racism were deployed were not uniform towards all non-white groups. Indians collaborated as well as resisted, racism was not just the preserve of the British, and local allegiances and colonial policies were all malleably deployed to suit the expediencies of any given situation. In short, this study of the practice of medicine shows not only the colour and diversity of a colonial society, but also reveals (much in the way that John Darwin has shown) that the operation of colonial power should be seen more in terms of the flexible and pragmatic negotiation of deep socio-political asymmetries and networks rather than as the straightforward deployment of dogmatic vertical directives.[17] Local conditions as much as international networks coloured the practice of medicine in the colony—on one hand the Indian subcontinent loomed large as the administrative model for colonial medical administration, on the other hand conditions were formed responsively to home-grown circumstances, particularly under the influence of individuals, such as the Principal Medical Officer (PMO) for much of this period, John Langton Gilks (1880–1971, PMO, Kenya, 1920–33), or the white settler leader, Lord Delamere.

If one were to straightforwardly describe a chronology of the history of Indian doctors in Kenya one would start with the early pioneers who arrived either to attend to the health of the 'coolie' labourers on the Uganda Railway or to provide medical care to the sepoy soldiers who defended East African borders during the establishment of colonial rule. From the 1900s, however, the majority of immigrant doctors came to work for the colonial medical department, although a slow but steady trickle of private doctors also arrived from that time. Especially in their governmental positions, Indians were integral to the establishment of the British administration in the East African region and enjoyed a certain degree of assimilation within colonial departments. Distrust between Europeans and Indians simmered under the surface for almost two decades, but Indians decisively faced a formal barrier to their political ambitions in 1923. The Devonshire Declaration of the same year made a firm statement of policy that halted Indian hopes for equal social recognition and signalled the beginning of a new, much more exclusionist phase in attitudes towards Indians in Kenyan society.

After that point, although it was never officially declared, there was no more active recruitment of Indians to the Colonial Medical Service. In this new context, many of the Indian doctors returned to find jobs in India or initiated a sideways move into private practice. Although this was a turning point, the dramatic cutback to Indian medical recruitment did not come out of the blue. Indians had struggled for greater recognition in Kenya since their arrival and even though they had been regular employees of the government medical department, the colonial authorities refused to acknowledge them as equals, implying that medical qualifications offered in India were inferior to those of the UK. Indian members of the Colonial Medical Service were required to undertake a broad range of medical duties similar to those of their European colleagues, but instead of being called Medical Officers, as British doctors were, Indian doctors were designated Assistant Surgeons, Sub-Assistant Surgeons or (before 1910) Hospital Assistants—titles intended to identify them as occupying a position just (but only just) above non-qualified African medical dressers, medical assistants and orderlies. Terms of employment for Indians within the colonial government were correspondingly designed to reflect their status well below that of their European colleagues. When Indian staffing numbers dramatically dropped by more than a half within a two-year period, almost no justification was given, other than a vague statement of the need to curb expenditure.

But while an increasingly racist staffing policy offers a partial explanation for the historical silence over the role of Indians in the government medical service, it is less persuasive as an explanation for the equally forgotten history of the numerous private Indian doctors practising throughout colonial Kenya. During the 1930s and 40s particularly, Indian private practitioners formed a substantive and relatively influential social group in the urban hubs of Nairobi and Mombasa.

The absence of general practitioners from the historical record can be imaginably explained by a notable failure of this group to write about their experiences; perhaps because the daily demands of general practice were regarded as less interesting than those of colleagues participating in research or government service. It is noteworthy that the same trend is discernible in the lack of histories of European doctors who practised privately, although at least one colourful biography of a pioneering British private practitioner from the same period achieved publication.[18]

Furthermore, although some elements of the experiences of private European doctors can be pieced together through other colonial sources, as well as the records of the proceedings of local institutions such as the East Africa Branch of the British Medical Association (BMA), this was not a place where the voices of Indian practitioners could be found. Indian doctors were denied the possibility of BMA membership until 1935, which undoubtedly also contributed to their relative silence in most historical accounts. Other broader cultural issues, such as not being part of the ruling elite, and the generally racially exclusionist climate of the colony are also likely to have discouraged Indians from telling their own story during, or even after, colonial times.

Plainly the histories of government and private doctors in Kenya are intimately intertwined. They formed one of the elite strata of a diaspora community that although never enjoying parity with their British rulers, had nevertheless been present in the region for much longer than the colonising Europeans and always outnumbered them demographically.

Historiographical background

Histories of colonial medicine

Before independence movements and for about a decade or so thereafter, the colonial history of medicine in Africa was dominated by accounts written by the white participants justifying the colonial project and promoting its beneficent effects.[19] Even if colonialism was admitted to have been fundamentally misguided, even knowingly exploitative, medicine was widely accepted to be one of the few areas in which the colonial project might be justified without shame.[20] After all, had not the British founded hospitals, set up rural health dispensaries, combated epidemics, trained healthcare workers, and established the rudiments of medical education?

This largely positivist view of history dominated until the 1970s and the 1980s when scholars began to question colonial medicine, proposing it to be a much more manipulative tool through which the British had not only established their domination but enacted their presumed racial superiority. Influential proponents of this view included: Daniel Headrick, whose seminal work identified medicine as one of the 'tools' facilitating imperial domination; Macleod and Lewis's volume proposing that medical policy was far from neutral in its commitments; and Megan Vaughan's trenchant exposure of the offensive discourses of power and supremacy that tacitly ran through imperial health policies.[21] These critiques, along with others stemming from the broader post-modernist tendency to question the oppressive and abusive structures inherent to institutional power, set the scene for a more critical evaluation of Britain's colonial past.[22] Rather than be seen as disinterested and benign, assessments of responses to epidemics, public health policies, and imperial recruitment began to describe colonial medicine as, at best, deeply complicit with other colonial concerns or, at worst, actively and subversively damaging to indigenous societies.[23]

This profound about-turn in perspectives, moving from a largely positive assessment to post-modern critiques, coincided with a broadening of spheres of colonial research interests. Whereas histories of empire had once been content with studying the lives of the elites, attention was increasingly reoriented to include the perspectives of the colonised. Influenced by trends within Marxist and Social History (which had been strengthening since the 1960s), and Subaltern Studies championed in the 1980s by writers such as Ranajit Guha, the history of medicine gradually started to additionally take account of the 'small voices of history'.[24] Subaltern Studies particularly refocused attention away from histories written by the so-called thin white line, calling instead for the voice of the common man to be heard, thereby granting agency to voices within history that had traditionally been under-explored. Subaltern Studies also *inter alia* challenged deeply entrenched suppositions that empire was driven only by polices conceived in Whitehall, arguing instead for the formative influences of the actions of the less socially privileged.[25]

A more nuanced picture emerged, one that laid emphasis on the role of non-metropolitan concerns in guiding the aims and enactments of impe-rial policies. This new analysis, moved attention away from the imperial headquarters in London and stressed the British Empire's fundamental reli-ance upon networks, not only of local people (be they collaborating local rulers, or paid servants of the colonial state), but also revealing the vast networks that existed between different parts of empire both officially and unofficially influencing policies and determining the character of imperial cultures.[26] In terms of medical history, David Arnold's study of responses to epidemic diseases in India is widely accepted as having provided the bench-mark for a change of emphasis that powerfully showed the role of protest and resistance in shaping British government actions.[27] In this oft-cited monograph of 1993, *Colonizing the Body: State Medicine and Epidemic Disease in Nineteenth Century India*, Arnold used the three examples of cholera, plague, and smallpox, to show that colonial medical policies in India were shaped by the British government's real, or anticipated, fears of resistance from the Indian people.

Such studies set the scene for a closer analysis of sectors of society that had traditionally been neglected by historians. In accounts of colonial Africa, the roles of black intermediaries, subordinates and clerks, for example, were for the first time identified as central to the day-to-day running of the British Empire.[28] Yet despite this new acknowledgement of the role played by the black Africans in empire, Indians working in the Colonial Medical Service, despite the fact that they were employed in higher positions and with greater clinical responsibilities, were still ignored.

In recent decades colonial historical perspectives have grown from strength to strength. Newer works have provided fresh insights into the development of tropical medicine;[29] the functioning of transnational political relationships;[30] the vagaries of the medical market;[31] and inconsistencies in the implementation of colonial medical initiatives and indigenous responses to them.[32] New themes have also risen to prominence; most notably an exploration of the way colonial medical ideas were inexorably bound up with wider concepts of race, sexuality and political identity.[33] Yet, although the story of Indian doctors in Kenya touched profoundly upon ideas of race, anti-miscegenation, and politics, they were still somehow omitted from medical histories of Africa.

Indeed even histories of the development of western medicine in India are comparatively few and far between in their examination of the role of race in the development and experiences of the Indian medical profession. Some works do touch upon the systemic discrimination faced by graduates of Indian medical schools, or the role of Indian doctors in Indian National Congress to try and gain recognition equitable to their European counterparts.[34] But surprisingly little research touches upon the way racism towards western trained Indian practitioners, in its myriad and various forms, shaded medical practice or government policies. One of the rare exceptions has been Deepak Kumar's insightful 1982 article dissecting the way the Raj systemically discriminated against Indian scientific professionals. This showed that despite the publically lauded benefits of bringing British levels of education and 'civilisation' to India, racial discriminatory practices ultimately held sway and hindered the progression of educated Indians to powerful positions in society.[35] One of the contributions this book hopes to make is to provide a more up-to-date case study of the challenges western trained Indian practitioners faced in delivering their medicine in a racially segregated context.

Histories of East Africa

Old-style narratives of the Kenya concentrated on the adventurous life of the white pioneer amid exotic animals and scenic beauty. Two European authors, Karen Blixen and Elspeth Huxley, played a leading part in shaping these popular perceptions of colonial Kenya. Their highly influential books painted a vivid picture of the gallant white settler combating diverse challenges.[36] Indeed, for many years Huxley's 1935 biography of the settler leader, Lord Delamere, was accepted as the authoritative history of the early colonial period.[37]

As in most fields of colonial history, a more critical evaluation of British social and political interventions in East Africa had to await decolonisation.[38] Within the fresh nationalistic climate, new assessments of colonial interventions gave voice to African experiences of their own country's history, often based on oral rather than written accounts and betraying an inevitable eagerness to demonise their former colonial masters.[39] Most recently, histories of colonial Kenya have emphasised the extreme scale of violence deployed by the British in the conquest, pacification, and administration of the territory, not least in the shocking events of the 1952 Mau Mau uprising.[40] Of course, Indians, as part of the pre-1900 British military efforts, would have also been involved in these deplorable events, although the extent to which they were implicated remains hardly researched. Unsurprisingly Indians in Kenya were not keen to associate themselves, or their ancestors, with the brutality of a regime that later also vilified them.[41]

A relatively small body of reminiscences, unpublished memoirs, and assorted articles also gave insights into the dynamics of medical history of Kenya before 1940.[42] Works such as those published by European colonial doctors John Carman (Kenya, 1926–51) and Henry Bödeker (Kenya, 1902–20), although painting historical pictures of life for a doctor in colonial Kenyan society, failed to discuss Indian doctors. Instead, this group of works tend to be generally ingenuous accounts celebrating the 'triumph of science and sewers over savagery and superstition', although they are still useful in that they provided a flavour of the contemporaneous atmosphere and important insights into colonial attitudes.[43]

After Kenyan independence Ann Beck's history of colonial medical administration was for several decades the only scholarly attempt to be published that tracked British medical interventions in the colony.[44] Although the publication was based upon government medical reports, supplemented with some interviews with European participants, her assessment of the running of the Colonial Medical Service was largely uncritical and did not delve into either doctor or patient experiences.[45] Strikingly, this monograph, and associated articles by the same author, barely mentioned African or Indian perspectives. Indeed, when reading Beck's work, one could be easily forgiven for thinking that only Europeans staffed the Kenya medical service.[46]

It awaited the 1990s before critical views of the medical history of East Africa began to emerge. Innovatively, Dane Kennedy's book on settler society considered medical matters in terms of the way they framed European identity, but it was Megan Vaughan who offered the first really penetrating analysis of medical policy and practice in East Africa in a way that was deeply critical of the British approach and policies.[47] Although still reliant on European source material, Vaughan nevertheless introduced African experiences of illness into the story, using the critical methods popularised by Michel Foucault to scrutinise the British deployment of medical power. Colonial discourses, as described by Vaughan, continually juxtaposed the civility of the European with the primitivism of the African.

Vaughan's book offered valuable new insights, but while her historical investigation was one of the best and most incisive that had been produced by that point, it nevertheless restricted the scrutiny to the tensions between white and black, coloniser and colonised. Vaughan (and most others since) overlooked the ambiguous position of Indians in Kenya, thereby effectively leaving a whole tranche of the racial and political complexity of the colonial medical history of Kenya unaddressed.

Similarly other scholarly works extended the range of topics that were historically considered within East African colonial medical history. John Iliffe's history of African entry into western medical education—*East African Doctors: a History of the Modern Profession*—being a case in point. Not only did this recount the long neglected story of training Africans to be doctors, but it also extended scholarly understandings of the racially exclusionist policies of the British State.[48] Focussing on the polar opposite of the colonial social spectrum, other published research extended analysis of the lives and experiences of the colonial elites. Crozier, for example, demonstrated the strong race-based spirit of solidarity between European doctors in the East African Medical Service[49] and Campbell critically revealed some of the most prominent European doctors in Kenya to be active in local eugenics movements.[50]

Although these accounts have deepened and extended our understanding of the dynamics of colonial medicine in Kenya, the role and experiences of Indian doctors in East Africa remained hidden between the cracks. In short, medical historical work extended academic understands of both black and white history, but somehow forgot about the brown races that also played a key role in the introduction of western medicine into East Africa. Indeed, the only mentions of these Indian professionals that have been found in European accounts are confined to a few pen portraits of Indian general practitioners within Robert Gregory's study of Asians in East Africa and a small collection of reminiscences from interviews with surviving relatives of Indian medical families conducted by Cynthia Salvadori during the 1990s.[51] Interesting though these rare insights are, they offered neither critical assessment of the experiences of these doctors, nor analysis of their contribution to the establishment of western medicine in Kenya. Most tellingly, a 1997 compilation, *Heroes of Healthcare in Africa,* failed to mention a single Indian doctor.[52]

Diaspora studies

A great deal of scholarly work on more general aspects of the Indian diaspora to East Africa has been undertaken however. The tensions surrounding the 'Indian question' which dominated Kenyan politics from the end of World War One until the Devonshire Declaration have been the topic of extensive investigation, and a number of other published works have explored in detail the importance of Indians in any history of Kenya.[53] Various publications focused predominately on the political and social experiences of East African Indians. Some of these with the aim to shed further light on the complex race relations in Kenya, with others aimed to extend our understanding of the contributions of Indians to the working of the British Empire more generally.[54] Since independence, considerable efforts have been made to address the hitherto untold stories of Indian traders,[55] industrialists,[56] politicians,[57] and the intricacies of workings of the complex Indian society in Kenya.[58] Authors such as M.G. Vassanji added to post-colonial interest, by fictionally retelling the history of the Indians in East Africa, focussing particularly on the development of their diasporic identities forged through the complex circumstances of their migration, and their concomitant negotiations for citizenship and ethnic identity.[59] As part of an inevitable backlash since independence perhaps, other more polemical works also started to emerge deeply critical of Indian influences on East African society.[60] So fertile is this field for exploration that a bibliography published in 2002 listed over six hundred books, articles, and reports about the history of Indians in East Africa.[61] Yet, even amid the rapidly advancing historiography, doctors are among the last (and arguably the most important) social groups to remain unexplored.[62]

Scope of the study

The area dealt with in this study is confined to the modern borders of Kenya, although these borders only came into place in 1920, when Kenya Colony was established in place of the somewhat smaller East African Protectorate.[63] The country provides an interesting site for study, not only because of its popularity as a destination for Indian emigration but also because, visualised differently (both locally and internationally) to its East African neighbours as a settler colony, it was also the location for particularly hard racist attitudes. These forcefully racially exclusionist ideologies were most obviously manifested in the infamous White Highlands Policy and also demonstrated via the history of the eugenically motivated *Kenyan Society for the Study of Race Improvement* (KSSRI), founded in 1933, although—while undoubtedly operating under the assumptions of their own innate superiority—not all settlers necessarily actively espoused these scientific racist ideas.[64]

The study starts shortly after the period of the official chartering of the Imperial British East Africa Company (IBEAC) in 1888 until immediately after the outbreak of the Second World War in 1939. The year 1888 signalled the official commencement of British interest in the region, although Indian doctors did not begin arriving to the mainland of East Africa until 1895. By 1940 a second generation of Indians were a firm fixture within the social landscape of Kenya, although recruitment to the government medical service became limited after 1923, the number of private practitioners continued to expand in service of this growing community until independence in 1963. The period before the Second World War is interesting because it was characterised by the establishment and consolidation of colonial power and was the period in which the 'Indian Question' was most hotly debated. It also represents a pioneering era of research in tropical medicine and predates the widespread availability of antibacterial drugs that radically changed the ability of doctors to treat many common infections.[65] Additionally, within medicine the period after 1940 was one in which the nature of medical practice began to change. The post-war period saw an increased reliance on medical specialists, rather than generalist 'all rounders' in the medical sphere. The first half-century of colonialism was also the time when racial animosity was most pronounced within the Kenya. After 1940—particularly as a result of British acknowledgement of Indian and African support in the international war effort—racial tensions subsided somewhat.[66] In short this book would be a different study, of a distinctive social, medical, and political world, if it also dealt with the period after 1940.

This history of the Indian medical community has had to be pieced together from the available source material, which—in the tradition of Subaltern Studies—has meant reading the colonial sources as much as possible 'against the grain'.[67] Although the bulk of published primary sources came from the pens of Europeans, a few of them acknowledged the presence of Indian doctors. Interestingly, the Indian doctor who gets most mention in European accounts is Rozendo Ribeiro (Kenya, 1899–1951), a Goan, who was one of the first private practitioners in Nairobi. Clearly an eccentric as well as likable figure, Ribeiro was much admired by the senior administrators and European doctors.[68]

Aside from the odd mention of Ribeiro, some fuller biographical accounts do exist, however. For example, a publication based on the personal recollections of, Dr K.V. Adalja, a distinguished private doctor who practiced in Nairobi between 1926 and 1964 and two papers on the clinical challenges of maternity care by Dr S.D. Karve (1922–1983) that are additionally revealing of the social and political climate of the colony.[69] Such published work was supplemented by material from a large number of newspapers. Of particular value was the Indian newspaper, the *Daily Chronicle*, published in Kenya until 1922.[70] Two other Kenyan newspapers, *The East African Standard* and *The Leader* were also consulted. Although these newspapers were well-known for their espousal of settler causes, they provide wide historical coverage of medical developments and also include numerous letters to the editor from Indian doctors presenting their side of various debates. These newspaper sources provide strong evidence of how Indians were commonly targeted as a key source of corruption and political malcontent in Kenya. Indian leaders in Kenya often complained precisely of this, feeling that the European residents actively utilised Kenyan media for 'spreading mischievous propaganda' about the Indian menace.[71] To this has been added a number of personal interviews with those who had Indian medical ancestors in Kenya and a few private papers that were in possession of the authors.[72]

A rich source of material from which to piece together this history was the National Archives in Kew. References to the activities of Indian doctors employed by State before 1922 abound in colonial dispatches to London, displaying a constant demand from the Kenya medical department for Indian Assistant Surgeons and Sub-Assistant Surgeons. Indian doctors can also be clearly identified in the British Library's India Office files dealing with the construction of the Uganda Railway and the early military camp-aigns of the colony.[73] The *Annual Medical Reports* and the *Official Gazettes* of the East Africa Protectorate usefully listed the names of Indian doctors prac-tising in the colony. Although the Medical Reports tend to be rather bland and official (especially on controversial matters), their silence on Indian doctors after the 1920s is in itself revealing of a perceived need to quietly avoid mention of this troublesome group.

As well as the Colonial Office records housed at the National Archive, and the India Office Records at the British Library, information was obtained from minutes of Legislative and Municipal Committees, BMA archives, the Rhodes House Archive, and minutes of Indian National Associations. Kenya National Archive was also visited, although little was found that was not also available on microfiches at Syracuse University Library. The India Office files in the British Library contain one of the best sources of data.

Some crucial records were lost during the exodus of Asians from Kenya between 1962 and 1970, but nevertheless, the relatively large amount of source material that has been uncovered makes the historical silence about Indian doctors all the more perplexing.[74]

A broadly chronological survey has been adopted as the framework to this book whilst all the time returning to the central intention to describe the practice of medicine in a segregated society. Some groups of doctors came and went, while others were present throughout the period under consideration. As such, railway doctors and military doctors are presented first in the narrative as they were recruited specifically as part of the foun-dational establishment of the colony. In contrast government doctors and private doctors after 1900 were a continuous feature of the daily medical infrastructure and therefore the chapters that describe them deal with them together to keep thematic unity. Although Indians were initially welcomed in Kenya, to the point where their immigration was actively encouraged, in the period after the turn of the century (but particularly after 1923) the political landscape of colonial Kenya changed significantly, silently but forcibly pushing Indians out of any places of influence or esteem. To indicate the backdrop to these changes, contextual chapters (Chapters 2 and 4) relating the broader aspect of immigrant profiles and the situation that greeted them in Kenya are also inserted as an essential part of the story.

The scene is initially set through a description of the history of Indian medical migration to Kenya, to allow readers to understand the sorts of influences that motivated Indians to make the journey to Kenya, and to additionally describe the kind of expectations these early pioneers brought with them, and the challenges they faced (Chapter 2). During this period the model of colonial medical administration of Kenya was almost entirely imported from the British Raj, so the problems that occurred in transposing one system to another social and political context form an important part of the discussion.

A detailed description of the earliest Indian doctors who came to work as practitioners in the Protectorate then follows. This third chapter predominantly examines the individuals who came to medically oversee the Indian labourers in the construction of the Uganda Railway, but also pieces together a picture of the Indian doctors who worked for the British military in early East African campaigns. Although records for this early period before 1900 are sparse, and individuals initially only came on secondments, it is nevertheless clear that, despite dominant racist attitudes, Indian doctors were relied upon by the British as an early constituent part of government medical personnel.[75] Indeed, if anything, the period is characterised by repeated calls for more Indian doctors to come to Kenya to participate in the establishment of British colonial rule.

The narrative then moves, in Chapter 4, to a description of the social context of doctoring in the racially charged atmosphere of the colony: here the practice of medicine is set in its appropriate setting, describing the segregated housing areas, shopping markets, schools and hospitals and the policies used to justify the separate employment structures that were put in place for Europeans, Indians, and Africans. In this chapter particularly, our intention is not to give an impression of firm and absolute boundaries between people of different ethnic origins. Communities did experience common types of discrimination, but equally circumstances can be found where Europeans worked cooperatively with Indians and Indians did much—despite their supposed lack of sympathy towards African peoples—to forward and improve African causes. The way racial politicking on the ground played itself out was, we contend, subject to common themes, but nevertheless was impossible to neatly categorise, with exceptions always breaking the rules. In presenting the complicated dynamic, this chapter provides the broader ideological context for the events that were to occur in the 1920s in the Colonial Medical Service, when the Indian cohort was significantly diminished.

The next two chapters (5 and 6) then specifically explore the history of Indians in the Kenya service of the Colonial Medical Service, tracing the early contributions that Indian doctors made, and their typical career experiences. Attention is drawn to the major, but largely unpublicised, turnabout in medical recruitment policy implemented by the Principal Medical Officer John Gilks in the early 1920s—decisions that led to the loss of approximately half the active Indian medical staff within 18 months under the guise of making financial economies and the pretext of focusing on Africanising medical personnel.

Finally, private doctors are considered, with Chapters 7 and 8 outlining what is known about their lives and also some of the political struggles in which they were engaged. Although there was some limited professional and social fluidity between government and private doctors, there was also an atmosphere of rivalry between the two groups, the possible reasons for which are also examined. In this section the diverse experiences of being an Indian doctor in Kenya before World War Two are discussed, showing overall however that this group—although far from united—was nevertheless a successful and relatively prosperous middle ranking tier of medical society.

Overall the book aims to enhance our understanding of certain key questions: what were the experiences of Indian doctors? Why has their story been forgotten for so long? And last, but not least, how does their story enrich our understanding of the practices that both surrounded and constituted western medicine in the settler colony of Kenya?

***

Indians in Kenya lived in their own large communities, and though they were segregated from both the Africans and the Europeans, they actually had greater daily contact with Africans than the British did. Their position as 'middle men' put them in the ambiguous position of being both valuable intermediaries, but also potentially rather suspicious collaborators. Indians were readily lumped together with Africans as the sites of filth, disease, and corruption, but were simultaneously seen (especially when holding such a view was politically or socially expedient) as one tier above the black majority and as relatively useful members of the community and of the colonial civil service in particular. They received more privileges than were typically granted to Africans, but this position also meant that they were the targets of African resentment.

Unsurprisingly, the experiences of Indian doctors reflect these complex ambiguities of social standing and their political, national, and ethnic affiliations. On one hand, this was a group trained in western medicine, providing essential preventative and curative services, yet it could never be fully trusted by a British elite that saw itself as inherently superior.[76]

The history that follows is one that reinvigorates the previously hidden story of the Indian doctors of Kenya. It is one that sheds light on the complex reality of empire. It was an empire of many ethnicities, and with constantly shifting allegiances and rivalries and numerous ideological contradictions. Between 1895 and 1940 it can be seen that Indian doctors variously negotiated this intricate cultural landscape in their search for a social identity within the British colonial structure. To miss their contribution to both the implementation of colonial medical governance as well as to the various facets of Kenyan society is to deny some of the vital diversity of the socio-medical world of the African empire. Finally, acknowledging the participation of three, rather than two, ethnic groups in the history of the establishment and growth of western medicine in Kenya takes us for the first time beyond the studies of Crozier, Iliffe, Kennedy, and Vaughan. This case study shows above all else that empire was concurrently subject to both local and international influences and was asymmetrically complex and infinitely flexible in its networks of power and privilege. Hopefully this study will challenge as many knee-jerk reactions about race and social identity in the colonial period as it confirms.[77] It is a story the telling of which is long overdue.

2
Indians, Migration, and Medicine

> East Africa is, and should be, from every point of view, the
> America of the Hindu.[1]

The Indian population of Kenya, formerly known as the East African
Protectorate, increased from under 5,000 in 1895 to over 90,000 in 1948.
These immigrants accounted for more than three quarters of the population in 1921 and half in 1948.[2] Most of these Indians were traders, artisans,
or lower level administrators, but an increasing number of professionals
such a doctors joined the migratory cohort. The trend for Indians to seek
opportunities in East Africa was far from new. Centuries before, Indians and
Arabs had established strong trading links with the East African coastline
and a small Indian community was already present in East Africa when the
British arrived. This established migratory precedent, coupled with the fact
that the British also ruled nearby India, meant that it was natural for the
British to look to India for subordinate staff to help establish and run its
newly-acquired East African possessions. However, the policy was not to last.
After an initial period of promoting Indian settlement, the tide decisively
turned in the early 1920s when the British government vetoed ideas for
the establishment of an Indian enclave in the region and curtailed Indian
political representation. Perhaps surprisingly, despite the waning of official
support, Indian immigration remained remarkably active and, by the time
of independence, a substantial and varied group of Indians were a firmly
established demographic component of Kenya.

Indian Migration to East Africa

Early navigational handbooks called *periplus* reveal that trade between East
Africa and the Middle East occurred with some regularity from around
120AD. These sources also indicate that the trading initiatives were accompanied by smaller movements of people between the coast of East Africa and
India, South East Asia and Indochina though the evidence concerning the

early activities is not conclusive.[3] There was a further boost of commercial activity in the eighth century, when the Omani Arabs established their first permanent trading posts in Africa.[4] By the thirteenth century, it has been suggested, trading excursions by both Indians and Arabs to the East African coast were commonplace with dhows regularly moving between the shores of India, the Gulf States and East Africa using the seasonal winds. By the time Portuguese navigator and explorer, Vasco da Gama landed on East African shores in 1497, a small but significant Indian settler community existed. Notably, Vasco da Gama himself evidently saw the benefits of tapping into local Indian knowledge, with accounts describing how he picked up a Gujarati speaking Indian pilot on this African coast to guide his Portuguese fleet on the last leg of his voyage to India.[5]

Yet although the coast of present-day Kenya undoubtedly received early attention from Indian trading populations, it is Zanzibar that provides the key to the origin of the Indian community in East Africa.[6] For several centuries, Zanzibar was famously the hub of the slave trade between Africa and Oman and India. Although this trade was chiefly in the hands of the wealthy Omani Arab elite, the role of the less politically powerful Indian community should not be underestimated. As important traders and middle-men, Indians were both owners and vital intermediaries in the principal Omani trades of slaves, ivory and spices, and were prominent fixtures in the East African commercial landscape.

From the early sixteenth century until the late seventeenth century the Portuguese ruled Zanzibar. In 1698 power moved to the Sultanate of Oman, where it remained until the islands were formally declared a British Protectorate in 1890. Throughout this period, the Indian community managed to navigate their various Portuguese and Omani (and, later, their British) masters with relative equanimity, effortlessly succeeding in ingratiating themselves with all of them. The Portuguese, as defenders of Christendom against Islam, were relatively hostile to interaction with Muslims and favoured the Hindu Vanias for developing their regional trade. But the Omanis too saw the advantage of working closely with their Indian associates. The Indians knew the geography of the interior well and could act as sponsors for potentially treacherous caravan expeditions, they were also prepared to act as important financiers in trading initiatives to both India and the Americas. The Indian community thus played a balancing role between the Portuguese and the Omanis, becoming key participators in trading activities in Portuguese Mozambique as well as the Omani coastal possession of Kilwa in what is now Tanzania.[7]

Indian models of business organisation became highly esteemed by both the Portuguese and the Arabs, particularly the idea of the 'joint family' organisation: a Hindu legal concept of the family ownership of assets linking ancestors and descendants.[8] In this model, members of joint families had pooled financial resources and high levels of mutual trust, despite being dispersed at different locations as their businesses expanded. Several prominent Indian trading families including the Sewji, Dhamji, Paroo and Topan successfully expanded around this culturally orientated organisational model.[9] Tharia Topan (1823–1891) is perhaps the most famous example of an Indian rising to prominence in Zanzibar, and managing to simultaneously win favour with both the Omani Sultan and the British colonial authorities. Described by Henry Morton Stanley as 'one of the richest merchants in town', Topan was made honorary Prime Minister of Zanzibar by Sultan Sayed Bargash, and later knighted by Queen Victoria.[10] The place of Indians as well-established, vital cogs in the early commercial success of Zanzibar is illustrated most powerfully through the words of the third Aga Khan in his 1918 book, *Indian in Transition*:

> Commerce in every branch, the development of agriculture, the supervision of works of public utility, the higher forms of skilled labour, the exercise of no insignificant share of political influence amongst the chieftains—all these were in Indian hands for many a decade before Europeans began the work of thorough and scientific exploration of the East African mainland. And the pioneers of this great enterprise, Stanley, Kirk and others, were indebted to Indians, such as the late Sir Tharia Topan, for the organisation of their expeditions into the interior.[11]

Plainly, his opinions hold more than a touch of bias, but this extract nevertheless, emphasizes the undoubted importance of the Indian community, not only in Omani trade, but in the establishment of British power.

British interests in East Africa took a big step forward after 1888, when William Mackinnon's Imperial British East Africa Company (IBEAC) was granted a royal charter to trade in the region. In 1895 the company folded and the administration of the East Africa Protectorate was formally transferred, first to the Foreign office and in 1905 to the Colonial Office. In the period during which British domination was established, Indian army regiments were brought in to impose control over the new territory. This was to be the first of many such initiatives to use Indians as key personnel in the establishment and consolidation of British power in the region.[12]

The directors of IBEAC had envisaged a crucial role for Indian immigrants in the imperial project to colonise and develop the country.

> The question of immigration from India appears to the Directors to be of great importance, with a view to colonisation by trained agriculturalists of the unoccupied districts of the Company's territory, more notably at Witu, in the country between the Tana and Juba Rivers and in the Sabaki Valley where the climate, soil and general conditions are particularly favourable to their settlement. The entire trade of East Africa has long been carried on by wealthy resident British Indian merchants, themselves

large plantation owners, who would greatly welcome and encourage their countrymen to settle in Africa. The Directors have under consideration the expediency of initiating the movement by offering grants of unoccupied lands to approved families. With such support and encouragement Africa may in future become to the natives of India what America and the British colonies have proved to the mother country and Europe.[13]

This idea for the 'Indianisation' of East Africa was supported by a number of other colonial administrators, not only those working within the top echelons of the IBEAC, but also respected colonial experts such as Sir Bartle Frere (who stated 'hardly a loan can be negotiated, or mortgage effected, or a bill cashed without Indian agency')[14], John Ainsworth[15] and Harry Johnston, the special commissioner for Uganda.[16] Compelling reasons for supporting Indian settlement were put forward by Johnston and famous colonial administrator Frederick Lugard, including arguments that immigration schemes would help ease population pressures in India and encourage the economic growth of Africa.[17] Furthermore, the historical legacy of established Indian trade and banking in the regions was seen as providing familiar networks into which prospective Indian migrants could integrate. The British admired the professional capability of Indians serving as advisors to the Sultan of Zanzibar; these privileged few appeared to enjoy their status and execute their public offices with pride, making them ideal employees of the colonial state. As Lloyd William Matthews, first Minister to the Zanzibar government succinctly put it: '[i]n Zanzibar Indians are very well treated and when in charge of departments or offices allowed to feel they are someone and gentlemen'.[18] Added to this favourable impression was the cooperative way Indian troops had helped to consolidate British colonial expansion in East Africa.[19] Ironically in the light of later developments, the potential contribution Indians could make to colonial social cohesiveness was also cited as being a desirable corollary of Indian settlement, with Johnston going to far as to surmise that the presence of 'the Indian, liked by Black and White, will serve as a link between these two divergent races'.[20] Lugard reiterated these views in his statement that the arrival of Indians in East Africa would bring a civilising effect as '[t]he African ... is extremely imitative.'[21] Given these interconnected factors, it is unsurprising that early blueprints for the development of colonial Kenya placed Indian immigration firmly at centre stage.[22]

The policy before 1923 of staffing of the bottom and middle rungs of the IBEAC, and later the Colonial Service, with Indians and also of actively supporting Indian immigration to East Africa should be seen in the context of broader political dynamics from the middle of nineteenth century, when the British in India began to exert increasing pressure on Zanzibar, to 1920 when the East African Protectorate was made a colony. In this period the British Indian administration in Bombay was perceived as the natural sub-imperial headquarters of the East African Region.[23] The proximity and presence of experienced colonial administrators made India a natural power centre from which to extend rule to the island and the mainland. Accordingly, until 1883 the British agent in Zanzibar reported to Bombay rather than to London, the Indian silver rupee was introduced as the colonial currency and the legal codes and systems of colonial administration were exported directly from the British Raj.[24] When British influence extended to the East African mainland this trend was simply replicated. The words of Sir Henry Bartle Frere, former governor of Bombay, show how the East African regions were viewed: 'all our interests in Persia, Arabia and EA were primarily Indian interests.'[25] With the bustling bazaars, mosques, and its cosmopolitan society, Zanzibar and the East African coast was effectively a 'huddled, unplanned block of Asia'.[26]

The issue of having two administrative headquarters, however, brought with it an unnecessary amount of bureaucratic complexity and soon became a headache for the colonial administration. For many years, the British agent stationed in Zanzibar received his orders from the Governor of Bombay as well as the Foreign Office in London and both sets of correspondence were copied to India Office in London.[27] This periodically caused conflict between the departments in London and India, their different perspectives and priorities naturally meaning that they sometimes espoused divergent ideas over the way policies should be regionally implemented; for example, how revenues from taxation should be utilised.[28] To overcome this cumbersome administration, Zanzibar was transferred, to the Colonial Office in 1913, just as East Africa had been in 1905, as part of moves by the Colonial Secretary, Joseph Chamberlain, to centralise all of the colonial administration. But, despite this streamlining of affairs, the precedent of East Africa to look to India had already been set and the character of the East African coastline was profoundly bound up with India. As Michael Fisher has shown, the origins of both East and West African systems of indirect rule were to be found in British India more than a century before.[29]

With this climate of shared administration it is not surprising that the British looked to India when they embarked on their first major construction project in East Africa. This was the Uganda Railway (1896–1902), a project to link the coastal city of Mombasa to the edges of inland Lake Victoria (the line was later extended to Kampala) and thereby to facilitate colonial ambitions to make the interior of the region more accessible. To expedite the construction of the line the British government recruited, over six years, about 30,000 indentured labourers from India, finding them to be the most reliable work force in the project. Indeed, their preference was even promoted as a kind of social justice. The employment of Indians in the construction of the Uganda Railway was justified as a visionary policy on the grounds that it would reduce demographic pressures in India by moving populations away from the more congested and overpopulated districts of their homeland.[30]

Although the absorption of these labourers temporarily represented a large rise in Indian immigration, more than two thirds of them returned to India during the construction or once the railway was completed, either because their contracts had expired or because they were repatriated as invalids.[31] Less than 1000 Indians continued to be recruited for service in the colonial Public Works Department, to maintain the line, build new branches, and to act as surveyors, clerks and carpenters as well as labourers. As the next chapter will describe, a number of Indian doctors (12 Assistant Surgeons and 13 Hospital Assistants) were also recruited during the railway project to supervise the health of those Indians involved in the administration and construction of the railway.

The most substantial permanent migration of Indians into East Africa took place after the construction of the Uganda Railway and was of subtly different profile to the earlier waves of migration. Traders who had come to the region to make their fortunes in the preceding centuries were now joined by a wider variety of middle-ranking professionals, such as administrators, clerks, teachers, skilled workers and a few university trained doctors, engineers and accountants. These were to be the early colonial servants of the East African administration, imported directly from the Colonial Service in India to staff all elements of the early tiers of the new administration. Johnston himself recognised that '[t]he intermediate role played by the Indian sepoy, non commissioned officer, surveyor, clerk, surgeon, botanical collector, trader and horticulturalist in all East Africa' was a significant contribution towards securing British success in the region. He expressed his particular gratitude for the support of 'Indian troops, Indian doctors and Indian clerks' as vital aides in the establishment of British rule.[32]

Despite these positive successes, a waning of enthusiasm in British policy towards Indian settlement occurred after the railway was completed in 1902, after Sir Charles Eliot took over from Sir Arthur Hardinge as the Commissioner of East Africa Protectorate. Eliot had quite different views from his predecessor and was keen to promote white settler interests above any concern for the Indian community. The issue came to a head after World War One when debates over whether Indians should be rewarded land grants for their considerable contribution to the British war efforts in East Africa reached a climax (although this proposal had been raised a couple of times before). As stated by Lonsdale: 'Indian nationalists saw Kenya as a test of Britain's word' to grant them equality of status following their war efforts.[33] To this end, a campaign from within Indian circles (led by the East African-based entrepreneur Alibhai Mulla Jeevanjee) petitioned to reserve, at least part of, the newly-won East African territories for Indian settlement. The issue was taken to the British parliament, but was ultimately defeated in a campaign led by Secretary of State Lord Curzon and Viscount Walter Long. Both expressed their strenuous concerns that backing such a policy would alienate the increasingly powerful European settler community in Kenya and might be seen as a dangerous softening of the all-important racial boundaries of colonial rule. Indian hopes were finally overturned between 1919 and 1923. This change in sentiment towards the Indian community in East Africa undoubtedly paved the way for the moves to enshrine African 'paramountcy' in the 1923 Devonshire Declaration and curtail Indian rights.[34]

***

The character and reputation of the Indian community

Immigrant population statistics

Estimates of the number of Indians living in Zanzibar and East Africa before the British arrived are difficult to make, as no reliable census was taken. It seems likely that the total Indian population in East Africa in 1860 was somewhere between 4,000 and 7,000 with approximately half of the group located on the island of Zanzibar and the rest spread over 20 coastal areas including Bagamoyo, Pangani, Tanga, Mombasa, Malindi, Kilwa, Lamu and Kismayu.[35] To put this into perspective, statistics for 1870 estimate the number of Europeans living in Zanzibar to be just 66.[36]

In what is now Kenya, the Indian population grew relatively rapidly from barely 2,000 in 1895 to more than 50,000 by 1940 (Figure 2.1). Over this period, Indian population distribution diversified considerably. Whereas at first, Indians mostly lived along the coastal strip, by the 1940s they had also spread to the interior, although mostly remaining in commercial hubs.

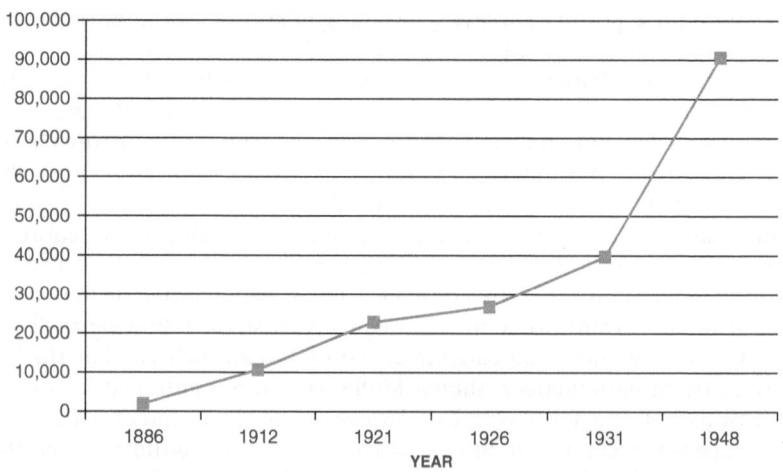

Figure 2.1   Indian population of Kenya

The steady movement of people to East Africa was part of the wider influx of Indian nationals that was occurring throughout the British colonial territories, largely to fulfil British needs to supply indentured labour to its new colonies. By 1920 over a million and half Indians were residents in regions of the British Empire outside of India.[37] It has been argued, however, that the relative proximity of Kenya to the motherland meant that Indians were less likely to assimilate with local customs in ways that many did in destinations further afield, such as Natal, Mauritius, Fiji, Guiana, Jamaica, Trinidad, Panama and Australia.[38] The proximity to home meant that ties to home traditions were strong. No doubt the possibility of returning to see family, or returning to be married, would not have seemed so remote as in more distant lands and the long history of trade between India and East Africa would have altered perceptions of relations with local populations and their way of life.

Some estimates have placed the number of Indians in Kenya in 1948 as over 90,000.[39] Local Indian births increased and families established new generations as both traders and skilled workers. Migration continued despite the increasingly unwelcome atmosphere of colonial Kenya, principally because although hardly rosy in Kenya, economic conditions were perceived as being even worse in India.

Cultural subgroups

The social history of Indians in Kenya is essential to understanding their medical practice and therefore the broader context of the social history of medicine in Kenya. Culturally the migrant population was extremely heterogeneous, reflecting the wide variety of regions, castes, sects, religions and languages of the Indian homeland. In addition to the Sikh and Muslim indentured labourers from Punjab, two other communities in particular were important in the history of the region, albeit in rather different capacities: the Gujaratis and the Goans.

The Gujaratis comprised a large number of the traders drawn from the Western Indian port province of Gujarat (including Kutch) who had had long historical trading links with Zanzibar. Their presence meant that from very early on Zanzibar became demographically colonised by a religiously and ethnically diverse Gujarati speaking population of Ismali Khojas, Bhoras, Suni Memons, Hindu Vanias and the Parsis.[40] Frustratingly for historians this group of immigrants left few records, but it is widely agreed that they were the ancestors of the communities of Indian traders that the British encountered when they arrived in Zanzibar and the East Coast of Africa at the end of the nineteenth century.[41] Despite the lack of cultural homogeneity within the Gujarati community, a substantive majority were closely associated with trade, although some had wider interests. With characteristic cultural clumsiness, however, the British tended to label all Indian merchants together as 'Indian banyans'.[42]

The Goan community also was important largely because Goans had a long historical association with Zanzibar and with the mainland. After the annexation of Goa by the Portuguese in 1510 the local population was particularly receptive to absorbing European cultural influences, including language, religion (most Goans converted to Christianity), clothing and social habits. This apparent penchant for European values led the Portuguese in Goa and the British in India to preferentially employ Goans for government administered services; a trend which was mirrored in East Africa once it was annexed by the British at the end of the nineteenth century. Numerous examples of the high regard that British officers held for Goans can be found within the literature. The British regarded them as the 'best' kind of Indian and consequently often employed them (although they were rarely appointed to the Colonial Medical Service as anything higher rank than compounders or dressers). An unusually high percentage of the clerical staff of the IBEAC and the Uganda Railway was drawn from the Goan community.[43] However, the perception of Goan professionals depended upon where they obtained their qualifications. Within India graduates of the Goa medical school were regarded as very much second rate in terms of their medical training.[44]

Apart from Gujaratis and Goans various culturally distinct Indians from Kutch, Punjab, Gujarat, Bengal, and Kerala sought a permanent home for themselves in Kenya, each bringing with them the distinctive cultural identities of their homelands. Culturally sensitive accounts written by Richard Burton and James Christie[45] of the diverse Indian communities that resided in East Africa during the nineteenth century identified the Bohras (Muslim traders) Khojas (Shia Muslims), Parsis (Zoroastrians) and Vanias (Hindu traders) as the most easily observable sub groups of the Indian population.[46] To these groups can certainly be added the Lohanas, Punjabi Muslims, Oshwahlis, Patels, Sikhs, Bhattias, Kutchis and Punjabi Hindus, the histories of which have been investigated by a number of scholars.[47] Each of these fragmented and non-cohesive groups of immigrants had different traditions, diets, ways of life, social and commercial networks and employment patterns. Although social and business interactions were not uncommon, inter-community integration remained fundamentally superficial in character, with each guarding their own ethnic and religious purity. Intermarriage was seldom permitted. Even among communities that were culturally relatively close, such as the Hindu Gujarati speaking Patels and Lohanas, exogamous marriages were extremely rare. Indeed it was regarded as preferable that Indians found appropriate marriage partners in India, rather than resort to a culturally mixed one in East Africa. Marriage to an African or European was almost entirely unheard of before 1940, aside from truly exceptional examples.[48]

Furthermore, the historiography on the issue of inter-racial cooperation within Kenya is divided. Historians such as Robert Gregory, Keith Kyle and

Michael Twaddle argued that Indians and Africans found few common causes against the European colonisers.[49] Recently this accepted view has been challenged by Sana Aiyar, who has instead made much of the cooperation that went on between Indian and African nationalist leaders, particularly in the period immediately after the First World War. Aiyar's argument needs to be taken into account and it is clear that the nationalist discourses of Kikuyu Harry Thuku rhetorically evoked the potential meeting points between Indian and African struggles in Kenya.[50] Nevertheless, it is difficult to extrapolate that these select political discourses represented evidence of a widespread mood of interracial cooperation throughout Kenyan society. Indeed, individual Indian responses to African people were confusingly varied, in some ways unpredictable. Nevertheless, it is hard to argue against the dominant impression gleaned from the sources that describes a colonial society where Indians, Africans and Europeans all remained within their own social and cultural community enclaves in Kenya. Which is not to say, however, that *necessarily* all relations were bad, that Indian nationalist leaders did not find common causes with African ones, nor that the Europeans did not use a language of Indo-African antagonism as justification for their policies when it suited them. As Megan Vaughan has usefully shown, Europeans selectively used racialised discourses to justify their attitudes and policy decisions.[51] They did this no less frequently in the way they generalised Indians for their own rhetorical ends as they did Africans. The untidy conclusion being that although notable exceptions existed, ultimately each sub-community predominately looked inward rather than outward for support and validation.

Within Indian society the situation was further complicated by internal divisions, particularly fractures which existed between members of different castes. This meant that even if people shared the same origins and religion, group members who were lower castes were forbidden to associate with those of the high castes. For example, the predominantly Christian Goans of Nairobi had within their own community three separate social organisations, each with a different caste composition carefully organised on a rigid hierarchy of social rank.[52] These internal divisions were characteristic of most Indian communities, causing a lack of ethnic cohesiveness that may have hindered Indians when they needed to present themselves socially and politically as a united front. As will be touched upon in later chapters, the inclination for Kenyan Indians to be socially fragmented and inward looking, was to prove a very real hindrance in mobilising a coordinated Indian defence against the white settler community and the increasingly restrictive policies of the colonial government.[53]

Lastly, the tendency of the colonial Indian community of East Africa to remain confined to their own communities needs to be understood in a repressive context where physical violence was an integral and characteristic part of European domination'.[54] Aside from a few notable exceptions,

Indians were reticent to publicly complain about their treatment in the hands of their British masters, preferring rather to remain silent on colonial political matters. Several reasons, apart from coercion, have been put forward for the community's acquiescence to the colonial system in Kenya, though between 1880 and 1940, numerous examples can be found within the archives of Indians being punished for questioning, or at least being perceived as challenging, the British colonial regime.[55] The reality of physical punishment was very real and would have been observable to all residents. As early as 1908, settlers of the Nakuru Association passed a resolution which declared 'physical pressure will be resorted to so far as being required to show persuasion being exercised is intended' in this same resolution, it was stipulated that Indians who resisted settler demands should be moved away to the coast.[56]

Trade and commerce

The dominating image of the Indian settler in Kenya is perhaps that of the Indian small scale, but prosperous, shop owner (dukawalla). Although this group was certainly extremely important in the economic landscape of Kenya, shop keeping and trading were by no means the only ways that Kenyan Indians made a living under British colonialism. Significant numbers of Indian migrants were also involved in construction work, or—if more educated—they were employed in the middle and lower tiers of the colonial administration or within banking or cooperative enterprises.

Since independence, Indians have been seen as a representative of some of the more wealthy and prosperous communities in Kenya, but this was not always the case and the picture of their economic status before World War Two was significantly bleaker.[57] Although the importance of small retailers is indisputable (an official Government report published in 1936 confirmed 'the assets of Kenya are almost entirely agricultural or pastoral and production is in the hands of Europeans and Natives, the Indian community being engaged in marketing and in distribution'[58]), in reality it seems that the wealth and status of this community of traders were more modest than prosperous. For sure, a few exceptional examples of thriving Indian commercial families could be found (Visram, Jeevanjee, Mehta, Chandaria, Madvahni, Manji, Khimasisa) in the period before 1940, but most Indians engaged in commercial activities were small dukawallas (shopkeepers) who worked long hours and subsisted on small profits.[59]

The publication in 1936 of the report of the Commission appointed to inquire into the financial position and system of taxation in Kenya confirmed the relatively low income of the Indian, but failed to document the reason for this state of affairs.[60] In general Indians had meagre accumulated capital to invest, but they were deliberately confined to less lucrative business segments through a series of restrictive policies and practices implemented by the imperial administration. For example, during the colonial period, restrictions were placed on the ability of Indian companies to import and export numerous goods. Licensing regulations effectively excluded Indians from forming their own markets for exports and imports compelling them instead to buy and sell through European-owned concerns such as The British East Africa Corporation, Smith Mackenzie and Gailey and Roberts (now Unilever). Indians were also excluded from the most attractive agricultural activities by limitations imposed upon their right to own land under the Land Ordinance of 1915.[61] Furthermore, private Indian companies had great difficulties in obtaining credit facilities from European banks.[62]

The situation improved after 1945 when the Bank of India and the Bank of Baroda established branches in Kenya, but before that period the colonial Government Gazettes are full of examples of Indian companies being forced into liquidation.[63] It was only after 1939, that general trends appeared to change for the better, and more examples can be found of Indians accumulating enough capital to make significant moves into secondary processing and manufacturing.[64] Given the restrictive context of conducting business under the British colonial government, the preponderance of Indians involved in small-scale retail becomes more understandable. This was one of the few areas where the community could operate within relative freedom and independence without state imposed impediments.

Education and Status

Most of the Kenyan Indians were not highly educated, though doctors and lawyers were, clearly, an exception to the rule. According to a survey of Nairobi in 1915, a quarter of the Indians resident in the city were literate in English though many more may have been able to read their own vernaculars, such as Gujarati, Punjabi and Urdu.[65] Other estimates put the range of literacy in English between 10 and 40%.[66]

Thus, most Indians who came to Kenya had not received secondary education, but there was a small but significant number of literate and educated—at least to secondary school level—Indians particularly from the Goan, Parsi and Patel communities. It was members of these groups who typically staffed the middle and lower ranks of government administration, railways, banks and commercial organisations.[67] Poorly educated traders and labourers certainly constituted the majority of Indian immigrants in the period before 1940, but the demographic picture was more diverse than the stereotype of a community of 'coolies' that many settlers alluded to in the Kenyan press.

Very little has been written on the role of Indians in the British Colonial Service in Africa, although it is known that Indians were restricted to the lower status posts. On average an Indian employee cost one third to one half of the equivalent European, so there was a strong economic incentive to follow this sort of staffing policy.[68] From the Indian perspective a Colonial Service career offered relatively good job security and marginally enhanced status within the colonial regime. Although there have been studies of the Colonial Service in both British India and Africa, these have failed to describe or quantify the Indian participation in the governing bodies of empire in any detail.[69] A few straws of information have occasionally come to light, for example one estimate of the number of Indian employees in government departments in Kenya referred to 1447 individuals being employed in 1920 and 1220 in 1935.[70] The same report went on to conclude that: 'on the whole the recent tendency has been to replace Asians by Europeans and to some extent by Africans, but up to the present the African share has been a very small one.'[71] Yet, even if they were rarely formally acknowledged or enumerated, a glance at Colonial Office records, provides indisputable evidence that a great many Indians staffed the lower and middle echelons of the Colonial Service; whether as clerks, police officers, teachers or medical assistants.[72]

It would be no exaggeration to say that during the period in question the Indian community became increasingly disliked in Kenya by segments of both the European and African communities. The long history of Indian involvement in the slave trade had left deep scars on the Indian image for many Africans and Indian participation in the European military campaigns of colonial conquest, as well as their close involvement with the day to day running of the colonial bureaucracy, very much categorised Indians, to most African eyes, as untrustworthy colonial collaborators.[73] Furthermore, Indian society embodied a long-standing cultural tradition of ascribing superiority to fair skin, and subsequently they were frequently accused as being racist and superior in their attitudes towards Africans.[74] Indians frequently called the Africans 'kala lok' (black people) in a derogatory manner and, those who could afford it, also employed Africans in subservient roles, as Europeans did. Unsurprisingly, after independence, some African politicians continued to propagate these pejorative images as part of their new articulations of independence.[75]

But effective political hostility to the Indian community came mainly from within the white sectors of society. Despite the fact that most early European colonials were happier to work with Indians than Africans, Indians were ruthlessly rebuffed by legislation when they appeared to make too many demands, or raised their heads above the parapet in ways perceived to threaten European hegemony. They were vehemently abused in the local English language press and routinely characterised as active exploiters, unscrupulous financiers (the 'Jew of Africa'), and purveyors of filth and disease.[76] In recent years historians refuted the veracity of many of these allegations, but some of the mud that was thrown by the white settler community against Indians, and their supposed habits, inevitably continued to stick for a considerable time.[77] Images associating Indians with the spread of unhygienic practices, and therefore being the origin of much bad health and disease, had been in increased circulation from the beginning of the nineteenth century, but became particularly vehemently expressed in the light of the great cholera pandemics that spread to Europe (from India) from the 1830s. Indians, because of their habits of pilgrimage and customs to bathe in the unclean water of the Ganges, were directly associated with the spread of diseases. The need for quarantine to guard against these risks was hotly debated from the first International Sanitary Conference of 1851, putting the Indian culpability for global health panics at the centre of international debates.[78]

***

Medical Migration

The earliest accounts of Indian doctors in the East African region, unsurprisingly come from Zanzibar where Indian doctors trained in western biomedicine were present in the Sultan of Zanzibar's regime from mid nineteenth century. One of the earliest references is to a Parsi, Dr Nariman, who was recorded as having arrived in Zanzibar in 1883 where he acted as the Sultan's physician.[79] By the 1890s, there seem to have been at least three Parsi and two Goan doctors, in addition to three Europeans, also practising in Zanzibar, outside the court circle, although little is known about these doctors.[80]

The earliest mention of Indian doctors on the African mainland are found among the records of the Uganda Railway and the Indian Regiments which were deployed in East Africa. The case for the use of Indian doctors was made, amongst others, by the IBEAC Medical Officer, Dr Mackinnon, stating that they were not only cheaper than European doctors to employ, but also that they were more culturally familiar both to the workers on the railway line and the Indian troops.[81] The Indian doctors who were recruited to serve the Railways and the Army were predominantly Punjabi Muslims, Anglo-Indians and Christians, though Parsis, Hindus and Sikhs were also represented (see Appendix 1). Most of the early private doctors appear to have been Goans who were favoured in the British administration for their Christianity and adoption of European habits (*vide supra*).[82] Goans and Parsis were often categorised by the British as superior to other Indians in East Africa. Their disproportionate representation among the early private medical community of Kenya reflects, unsurprisingly, the profile of students who attended Grant Medical College in Bombay, the institution that supplied most of the Indian doctors who came to work in Kenya.[83] Later, as Indian western medical education expanded and became universally accepted, the profile of immigrant Indian doctors diversified to include members of other Indian communities; although, Hindus were consistently underrepresented during the very early colonial period.[84]

The small trickle of Indian doctors before 1900, turned into a steady stream once the British consolidated their systems of administration after the Uganda Railway was constructed. In short, a policy of employing

Indian medical personnel was pursued consistently and enthusiastically in Kenya until the end of the First World War, though certainly never publically proclaimed. Almost as soon as the British arrived in the Protectorate they employed Indian doctors to accompany the army regiments that were instrumental in establishing colonial control and they immediately recruited a number of Indian Assistant Surgeons and Hospital Assistants (renamed Sub-Assistant Surgeons after 1910) to serve the Indian labour force imported for the construction of the Uganda Railway. Indeed, despatches from the Colonial Medical Department in Kenya before the First World War were full of pleas to recruit more Indian medical staff, with the early heads of the colonial medical department making no bones about their perceived necessity within the workings of the new colony.[85]

A career in the Colonial Medical Service was the most obvious destination for Indian medical migrants before the 1920s. It was only after the Indian population had increased substantially and recruitment of Assistant and Sub-Assistant Surgeons was abruptly curtailed during the 1920s that private practice became the more viable option for doctors seeking a career in East Africa. Which was not to say that there were no private doctors working in Kenya before 1920. On the contrary, doctors such as Dr Luis Lobo, Dr Rozendo Ribeiro, Dr L.A. da Gama, and Dr Edward Dias were seemingly able to make a reasonable living from their private practices established in the first years of the twentieth century (see Chapter 7).

Just as the legal, administrative and political structures of East Africa were 'in the early decades of colonialism, almost an extension of India itself', replicated also were many of the imported medical infrastructures and beliefs.[86] The medical practices of East Africa reflected those used within the wider Indian Ocean region. This involved the interaction of four main systems: the Unani (or Quranic the chief Islamic system), the Ayurvedic (the traditional Indian system developed chiefly from within the Hindu and Buddhist traditions), the Western (after the nineteenth century, allopathic and biomedical) and the various African indigenous practices.[87] While traditional African medicine, based around herbalism and the use of diviners as healers, was always the dominant form of pre-colonial medicine in Africa, the conversion to Islam of large numbers of people on the Swahili coast during the centuries before 1850 meant that Islamic traditional medicine—itself traditionally open and flexible with regard to absorbing external influences—was already a regular feature of the medical landscape before the British arrived.

As the Indian settler community grew throughout the nineteenth century Unani medical methods from Oman were extended to include Unani practices from India as well as Ayurvedic techniques. Western medicine was a relatively late arrival. Although, the first Portuguese invasions at the end of the fifteenth century had ensured a limited exposure of western medicine in East Africa, it did not really become a significant influence until European missionaries arrived in the mid-nineteenth century.[88] Similarly Indians educated as western medical practitioners did not come to Africa until the second half of the nineteenth century, though it is likely that traditional medical practitioners existed among the early Kenyan Indian settler communities, as immigrants would have sought recourse to treatments similar to those they could have received in their home countries.[89] Almost without doubt, Indian traditional healers existed, but as there was no official bureaucracy to register these early practitioners, it is difficult to ascertain how many of these doctors were practising in the region, or where they were to be found. Most Ayurvedic and Unani medicine was customarily practised within the home, with knowledge being inherited through the generations and medical advice informally sought from respected elders within the Indian community. Although an assessment of the extent of Ayurveda within colonial Kenya is beyond the scope of this study, it is worth noting that business records from the 1920s, when record keeping had become more systematic in East Africa, consistently showed a number of Indian wholesalers vending Ayurvedic and Unani medicines successfully along the coast.[90]

Although the reality of medical experience in colonial Kenya would have remained closely tied to the traditional medical systems of Africa, Asia and Oman for most poor Africans or Indians seeking medical care in their communities, the gradual availability and uptake of western medicine by these groups was nevertheless one of the most profound changes of colonial contact.[91] It is often forgotten that Indians were essential conduits for the dissemination of western medicine in Africa, illustrating the multiple and shrewd ways western ideas came to enter non-western cultures. In short, western medicine spread 'sideways' via empire to Africa from India, as well as along the traditionally acknowledged linear route from Europe to Africa. This was accomplished through Britain's long colonial relationship with India, where a tradition of absorbing, adapting and amalgamating western medical methods had existed since the arrival of the British East India Company in 1600. Over time western medicine had gained cultural prestige in influential Indian circles, with several Indian princes having enthusiastically welcomed European practitioners into their courts by the 1800s.[92]

The way the British managed the medical care of its large Indian armies and its first Indian colonial employees had further acted as a means of gradually introducing western medical methods into the lives of everyday Indians. For sure, western medicine was always just one part of a larger medical market in India; and yes, examples can be found of ways it was actively resisted, or selectively appropriated; but nevertheless by the 1850s it was accepted as a viable system (among other viable systems) by most educated Indians.[93] Given this relatively receptive context, it was natural that Britain should look to India to help staff its new Colonial Medical Services in East Africa. Simply speaking, Indians provided a cheap, and also an appropriately educated, medical labour force.

***

Western medical education in India

The Indian-based training of medical assistants for the British colonial administration really started to get underway systematically from the middle of the nineteenth century. However, the establishment of medical colleges in India with a view to educate Indians in the theories and methods of western medicine were not the first training schemes designed to train non-European staff. In fact, one of the earliest projects involved the educating of former African slaves as low-level medical aides. In the mid-nineteenth century a group of freed African slaves in India, which became known as the 'Bombay Africans', were given a rudimentary training in Bombay and then sent back to Africa to provide various services to European expeditions and missionaries.[94] German missionary, Johann Ludwig Krapf, as well as British explorers such as Richard Burton, John Hanning Speke and David Livingstone were all said to have made good use of these aides and certainly they seemed, for a time, quite popular. Their training, knowledge of local geography and customs and the ability to communicate in different languages were helpful attributes. However their impact should not be exaggerated, although approximately 3,000 Bombay Africans were sent to East Africa between 1846 and 1880, most came to work with missions operating in the coastal towns of Freretown and Rabai in non-medical capacities. Only a tiny handful were medical aides.[95]

Aside from the exceptional initiative to train 'Bombay Africans', Africans were not regularly employed with any systemic regularity by the Kenyan colonial medical department before World War One. Indians were recruited instead, largely because they were deemed to have had the appropriate medical training in institutions regarded as comparable to—although not equal to—those which Europeans had access to in their home countries. A comprehensive history of the development of western medical education in India remains to be written, but some regional accounts of medical professionalization provide useful insights.[96] Although the Portuguese had brought western medicine to India in the sixteenth century, it was the British who established the first hospital in Madras in 1664 to serve the health of European traders and soldiers who travelled to India.[97] Later, the first British-sponsored training facility was founded in Calcutta in 1822.[98] The establishment of this first medical school (Calcutta Medical College) in 1835 was stimulated directly by the needs of the British government. It had become increasingly obvious that the swelling European presence in India needed a health service, so to this end the Indian Medical Service (IMS) was established in 1763.[99] Naturally, the higher echelons of this service were staffed by European doctors, but it was realised, quite pragmatically, that a subordinate staff of native doctors, apothecaries, compounders, dressers, aides and was also needed to provide the infrastructural support fitting for a government medical service. In 1822 it was decided that a Native Medical Institution (NMI) should be established in Calcutta with the aim of training Indians as subordinate medical assistants specifically for the government-run medical service. This vocational training was initially offered in the vernacular, and included training courses in Ayurvedic and Unani medicine, although it also instructed its students through translations of European medical texts and provided (although dissections were not performed) some western-style clinical and anatomical experience.

For a short period, this must have seemed to have been the perfect amalgamation of different systems of medicine, appropriate to the spirit of intercultural exchange that many of the early European Orientalists—sympathetic to Indian history, languages, and culture.[100] However, this open policy was not to last and in the 1830s a committee headed by the Governor General of India, William Bentinck, was set up to enquire into the state of medical education in India. By 1834 the results favoured the termination of the teaching of traditional indigenous medicine, stipulating instead that an institution should be set up to provide medical education in English and more closely along the lines prevailing in Europe.[101] Consequently the NMI was abolished, officially marking the end British government patronage of native medical learning. In its place and in the spirit of the new emphasis, Calcutta Medical College opened its doors in February 1835 as the first institution explicitly designed to train young Indian boys, irrespective of caste, in the principles of western medicine in English. Entry was by competitive examination; students were offered a stipend by the British government and were expected to undertake training from between four and six years, with lectures delivered by a variety of European medical luminaries. By 1846 the qualification was formally recognised by the London Royal College of Surgeons and by 1857, the year in which the institution was confirmed as having full university status, Licentiate in Medicine and Surgery (LMS), Bachelor in Medicine (MB), and Doctor of Medicine (MD) were all offered, in line with the medical degrees available in the UK.[102]

The patterns established for how the western medical education of Indians should be delivered, accredited and officially perceived were soon copied elsewhere in the country. Madras Medical College admitted its first Indian students in 1842 and Lahore Medical School (also called King Edward Medical School) was established in 1860.[103]

The most important training ground for Indian medical migration to Kenya, however, was Grant Medical College in Bombay, affiliated to the University of Bombay in 1860. Indeed, more than two thirds of the private Indian doctors who registered to practice in Kenya between the Medical Ordinance of 1910 and 1940 had received their medical education there.[104] This is not entirely surprising, as the British Bombay Presidency, had, since 1818, been the de facto headquarters for the large amounts of trade that passed between the city and East African ports. Furthermore, Bombay was Britain's principal naval base from which it safeguarded its Indian Ocean interests and it was a hub for railways, industry and international departures and arrivals to the Indian subcontinent. Understandably, it was also a popular destination for internal migration, with many Parsis and many Christians, especially Goans, moving to the city because of its relatively attractive educational and employment prospects. Robert Grant, the Bombay Governor, first raised the idea for a medical college for Indian students in Bombay in 1834. Despite much initial opposition, and thanks to a generous grant from the Parsi philanthropist, Sir Jamshedji Jeejibhoy, the college finally opened its doors to its first students in 1845.

Following the lead of Calcutta Medical College, the first students that entered Grant were between the ages of 16 and 20 and were selected on set standards of vernacular language, arithmetic and English.[105] The first group of students admitted to the college included nine Christians (including Goans) and two Parsis. In the early years Parsis provided 40% of the intake and formed the largest contingent, though by 1886 Hindus from the higher castes came to represent almost 30% of the students in the college.[106] As was typical of these colonial institutions, the first five Professors of the college were British graduates. Two levels of instruction were offered: Indians could undertake a course to qualify as doctors, with the title Assistant Surgeons, or take a shorter course to allow them to practice as medical subordinates, e.g., Sub-Assistant Surgeons, Hospital Assistants, and Sanitary Inspectors, for the British government services.[107] Initially the main medical degree offered to those taking the longer programme was Graduate of Grant Medical College, but the course was gradually streamlined in accordance with UK standards, so that by 1862 the first Licentiate of Medicine (LM) could be offered, later joined by MD and LMS degrees.[108] As the medical colleges, including the Grant Medical college, became affiliated to Universities the training of the subordinate medical staff was transferred to the more numerous medical schools established at regional centres such as Poona, Ahmadabad, Hyderabad and many others in the different regions of India. One estimate within the *British Medical Journal* suggested that there were ten university medical colleges and 27 medical schools in India by 1939.[109] These schools, which issued diplomas and certificates as opposed to medical colleges that issued degrees, also provided the bulk of the Indian medical staff for the Colonial Medical Service in Kenya. The two tiers of medical training that were offered (i.e. to become either licenced or registered) did nothing to advance internationally the image of medical degrees gained in India.[110] Overall the levels of medical education and training were thought to be inferior to that which could be received in the UK or Europe and, undoubtedly, this perception contributed to the long-term omission of the role of Indian doctors in the medical histories of the region.

Job Titles

One confusing issue complicates this study and that is to do with the fact that, upon qualification from a medical school in India, there was little systematisation in terms of how these newly-qualified practitioners were titled. Deep rooted ideas of racial superiority in the Colonial Service meant that Indian colleagues in Kenya, even if in possession of LM, MBBS or MD degrees, were never allowed the job title of Medical Officer, which was reserved solely for Europeans.[111] Instead, Indian medical practitioners were variously called Native doctors, Assistant Surgeons, Sub-Assistant Surgeons (known as Hospital Assistants before 1910) once they took up posts with the British government services. The variety of job titles means that before the Medical Registration Ordinance of 1910 in the Protectorate it is difficult to tell from the job titles alone whether Hospital Assistants and Sub-Assistant Surgeons had similar licenced medical qualifications, or whether Hospital Assistants were actually less well qualified subordinates, who had under-taken a shorter curriculum of training, more akin to a training that would be expected of medical aides. Certainly the titles Hospital Assistant and Sub-Assistant Surgeon designated a lower status than Assistant. For example, Hospital Assistants undertook a shorter subordinate training course at Grant Medical College after 1861 though the duration and content of the of course improved significantly subsequently.[112] Furthermore, because there was no systemisation of the medical titles awarded (neither throughout India, nor between India and East Africa) different contexts named Indian doctors differently, almost indiscriminately before 1910. Indeed, in some sources, the designation Hospital Assistant and Sub-Assistant Surgeon were used interchangeably as if they described exactly the same thing, while in others the title Hospital Assistant seemed to precede the rank Sub-Assistant Surgeon.[113] Historically this is hard to dissect, particularly when analysing a context that anyway had a tendency to lump all racial minorities together with little differentiation between people of different sub-community, caste, or—as in this case—professional rank. To complicate matters Africans, with a minimum of education or medical training, who were utilised as dressers in Kenya after 1923 were sometimes also misleadingly referred to as Hospital Assistants. This study aims to describe Indians formally qualified and legally entitled to practice western medicine, but there is always a slight lingering doubt in the period before the standardisation of the nomenclature, whether those described as Hospital Assistants had achieved formal medical qualifications or not. Although we cannot know definitively, it seems most likely that the first early mentions of Hospital Assistants accompanying the British military and also the construction of the Uganda Railway were doctors who, after 1910 in Kenya, became known as Sub-Assistant Surgeons.[114]

Numbers

Estimates of the number of Indian doctors in Kenya between 1880 and 1940 have to be pieced together from a variety of sources. Even then, it is diffi-cult to ascertain verifiably accurate statistics. Some of the most substantial waves of early medical migration, such as the Sub-Assistant Surgeons and Hospital Assistants, brought in to supervise the labourers on the Uganda Railway, were doctors recruited for specific assignments, and were never permanent residents in Africa. Moreover, in the period before the 1910 Medical Practitioners and Dentists Ordinance in Kenya, the listings (for example, in the 'Blue Books' after 1901) did not always differentiate between doctors with a formal medical degree and those with lesser qualifications. Additionally the Blue book was not published for ten years after 1914.[115] A source of frustration with the official Medical Register published in the *Official Gazette* after 1910 is that no systemic provision was made (before 1935) to remove the names of those who had left the country or had ceased to practice, so that it often contained out-of-date information.[116] Another problem with the information within the *Official Gazette* was that though it listed Government-employed Indian doctors, it seldom gave indication of their qualifications. Furthermore, the *Gazettes* did not distinguish between doctors working for the government and those who were in the private sector or working for missions. The names of Indian doctors can also be partially ascertained from the Annual Medical Reports (regretfully not all) produced by the Colonial Medical Department, but these are useful only for capturing the names of government staff, and then only up to 1923 after which, Indian names were infrequently included in descriptions of the Colonial Medical Service. Imprecision also extends, though to a lesser degree, to the number of European doctors because of the lack of clarity in the *Annual Medical Reports* between sanctioned, filled and loaned medical staff. Additionally there was also doubt about whether doctors from the Sanitary Division and highest echelons of the Service were always included in the official counts of staff numbers.

Cross-checking these sources with the *Medical Directory* (produced annually in the United Kingdom) was only of limited use, since the information is wholly dependent upon doctors voluntarily filling in an annual return and many doctors who worked abroad in the far-flung colonial dependencies unsurprisingly failed to register. To illustrate this point, between 1900 and 1935, more than half of the Indian doctors listed in the *East African Government Gazette* are not listed in the relevant *Medical Directory*.

Best estimates of the numbers of doctors practising in Kenya between 1897 and 1940 are provided in Table 2.1. Bearing in mind the caveats outlined above, the number of qualified medical practitioners rose from a modest group of approximately 30 at the start of the century to a much more substantive community of around 200 people in 1940.

***

It is clear that the history of Kenyan Indian medical practitioners trained in western medicine is intimately bound up with the evolution of western medicine in the Indian sub continent, but a word of caution is advised. To use the

Table 2.1 Registered and licensed medical practitioners in Kenya

Year	1903	1910	1920	1923	1928	1930	1932	1937	1940
CMS									
European	13	18	39	31	44	72	54	52	60
Indian	13	26	72	32	27	25	25	26	25
Priv. Prac (or**)									
European	3	6	36	36	Not available	Not available	66	Not available	121
Indian	3	4	6	6	Not available	Not available	40	Not available	57

Note: **Potentially this group could include doctors working in the army, as missionaries or within private companies.

words of Deepak Kumar, medical '[t]ransmission was a complex process; it was never linear.'[117] Although the enthusiastic way Indians embraced western medicine could (should) be seen as example of the victory of western cultural imperialism, it should also be seen as part of a dynamically interactive tradition of the medical exchange of ideas between east and west which—at least before the biomedical revolution—also saw plentiful examples of Europeans learning from Indian practices. Indeed, part of the reason for British success in India initially was their tolerant, and actively enquiring, attitude towards many aspects of Indian traditional medicine.[118] Notably, the British administration in India positively encouraged the practice of Unani and Ayurvedic medicine for a considerable period before withdrawing government patronage and there was a surprising permeability of the boundaries of medical knowledge. Indians and Europeans residing under the pre-twentieth century Raj existed in a medically pluralistic context, picking and choosing remedies from various traditions and sometimes using them simultaneously, or adapting them to their own needs.[119] This is not to deny that people naturally gravitated towards their more familiar cultural traditions, but rather to stress the mixed medical environment in which some Indians chose to become western practitioners, particularly before 1900. As Deepak Kumar put it: 'From the Indian point of view, the mid-nineteenth century was a period of looking for fresh opportunities and acquiring new knowledge. Syncretism, not revivalism, was the agenda'.[120] It was only after the Sepoy Revolt of 1857, which strengthened divisions between the British and their Indian subjects, and after the bacteriological revolution of the subsequent 50 years, that boundaries between 'eastern' and 'western' realms of knowledge gradually strengthened. After this point Indian support for western medicine, western education and western business initiatives, became mobilised in different ways as part of the Indian nationalist agenda.[121]

By the 1930s and 1940s western medicine was so embedded in certain sectors of Indian society that it was not generally viewed as an oppressive symbol of western hegemony; in fact for some time it was seen as a means of achieving social mobility and of breaking through the old caste monopolies of the key professions.[122] Despite Gandhi's vocal criticisms of western medicine, its meaning and utility had changed for most of the Indian doctors trained in its methods.[123] This was not necessarily to be perceived as an example of western superiority, but rather as a pragmatic means of working towards a new, modern, independent India.[124] The relationship of educated Indians to western medicine was incredibly diverse and nuanced and certainly cannot be characterised simply as imperialistic or scientific. After 1920, it increasingly became to some a means of demonstrating modernity and national self-confidence.[125] The Indian doctors that came to work in Kenya in this period were representative of this delicate change in perceptions. Their support of western medicine by no means suggested necessary sympathy with the British colonial project.[126] Indeed some of the Indian doctors in Kenya politically positioned themselves as active Indian nationalists, despite (or maybe because of) making their medical livings under the British Kenyan colonial regime.[127]

The life of an Indian doctor in Kenya before 1940 was not an easy one, although their prosperity was plainly better than that of indentured labourers or smaller shopkeepers. For the Indian doctor who made his home in Kenya colony, life was full of challenges and perils. Once the new immigrants arrived in Kenya they would have found themselves facing new diseases, sub-standard housing, poor sanitation and a lack of political power. Indians engaged in professional occupations, which required the possession of a high-end education, represented a minority of less than 1% of all Kenya immigrant Indians.[128] Most of the Indian population of Kenya consisted of poor migrants who came to Kenya principally for economic gain. The openings and social mobility created by the colleges of western medical education established in British India, most of which offered stipends and bursaries for their students, allowed a large proportion of first-generation Indian doctors from modest backgrounds to seek employment. In short, becoming a doctor would have meant social advancement for most Indian medical school entrants, as the majority were the first generation to achieve such professional credentials. Ultimately, the primary motivation of the Indian doctor seeking employment in East Africa was money and greater opportunities to earn it than at home. Factors such as the lust for adventure, a sense of Christian missionary zeal, or a social duty to enact reform, such as have been identified as being among the prime motivators for European doctors seeking a life in Africa, were not a strong binding element of the collective identity of the Indian medical cohort.[129] Kenya provided a place of potential prosperity, not too far from home, with familiar colonising structures and an already existing history of Indian settlement. For many it must have seemed a convenient, if not natural, choice of location in which to further their career.

3
Indians, Western Medicine, and the Establishment of the Protectorate

[Indians are] a potent factor in the process of civilising the African[1]

Despite the perceived advantages of Indian immigration as outlined in the Hamilton Report quoted above, the role of Indians in the East Africa Protectorate was one that was hotly debated. Even when their formidable early role was acknowledged, medical contributions were rarely seen as part of the equation. As Winston Churchill reminisced:

It was the Sikh soldier who bore an honourable part in the conquest and pacification of these East African Countries. It is the Indian trader, who, penetrating and maintaining himself in all sorts of places to which no white man would go or in which no white man could earn a living, has more than anyone else developed the early beginnings of trade and opened up the first slender means of communications. It was by Indian labour that the one vital railway on which everything else depends was constructed.[2]

Military and political gains represented the glory of British might. A combination of determined coercive strength and tactful diplomacy were fêted as the cornerstones of British colonial success. Although this was undoubtedly an important part of the story, historians in the past 25 years have nevertheless supplemented this picture by concentrating on other rationales behind colonial victories. Most relevantly in terms of the subject matter of this book, science, technology and medicine were put forward as fundamental in providing the necessary infrastructural prerequisites for colonial domination.[3] In terms of technology, how could military and political success be assured if the colonising forces did not have the necessary ships, navigational aides, maps, logistical methods and armoury? In terms of medicine, how could long-term success even be contemplated if the dominators were not able to master the health of their new environments? Effective medicine not only ensured that the colonisers themselves were able to successfully settle, but also, vitally contributed to British economic ambitions in helping to ensure (at least) the minimum requirements for a productive indigenous workforce. In setting up a new colonial possession the physically gruelling, and therefore physiologically debilitating, undertakings of military conquest campaigns, construction and land reclamation were the infrastructural priorities of the British government. Relatively healthy bodies were needed to undertake this work.

Despite scant source material, a little is known about the earliest Indian doctors in Kenya. These medical men arrived along with the earliest European pioneers, as part of the first contingents to scope out the economic possibilities of the territory and shore up its defences. In the earliest days of British colonial rule thousands of Indians were imported, either as indentured labour to help in the infrastructural development of the colony, or as sepoys to serve in the military. Indian doctors, felt to be more culturally relevant to treating the new immigrant population, were recruited too and therefore had a presence in the colony since the 1890s.

The early recognition that the management of health had to be a priority if colonial rule was to be successful was not surprising. Arguably, the urgency to master local health problems was nowhere as pressing as it was in Africa. The so-called Dark Continent was not only impenetrable in terrain and climate, but was feared as being the site of horrifying diseases as had never been seen before in Europe.[4] Furthermore, the shocking mortality and morbidity figures for members of the European military stationed in West Africa readily confirmed the health reputation of the continent as 'White Man's Grave'. As, Philip Curtin has shown through his detailed statistical analyses: soldiers, missionaries and colonial administrators of the nineteenth century were warranted in their fears that a posting to Africa might result in encountering disease, even death.[5] Although the disease environment of East Africa was regarded as being slightly healthier than the West, the same rhetorical brush of negativity also tarred its reputation. The majority of Victorian discourses about Africa failed to make any differentiation between different regions of the continent, preferring instead to lump everything together in a generalised myth of mystery, pathology and violence.[6] Given this context, it is no surprise that the successful management of health was a key priority in the colonial government's mind if it was going to make its new African ventures viable. In Kenya the case for extending medical services to the African reserves was justified to keep the African labour needed for the European estates, in a reasonable state of health.[7]

Fortunately the final years of the nineteenth century, when the British wanted to establish themselves in Africa, were also the years when tropical medicine solidified as a new discipline and offered for the first time very real possibilities for taming, if not entirely conquering, tropical diseases. After centuries of caution, for the first time a tangible confidence accompanied debates about the European conquest.[8] To this end, the new tropical medical discoveries of the founding fathers of the sub-discipline, Patrick Manson, Robert Koch and Ronald Ross, cannot be doubted as crucial stepping-stones for colonial expansionist ambitions in Africa.[9] However, both the creation and movement of medico-scientific knowledge were more nuanced than is contained within the traditional narrative of an undeviating 'western' victory, lineally emanating from the efforts of white men working in the European mother-ship. Two points should be made in this regard. First, with regards to the formation of so-called 'western knowledge', it should be noted that both Manson and Ross (and to a lesser extent Koch), although Europeans by ethnicity, both made many of their most famous discoveries outside of Europe, in the so-called periphery of Empire.[10] Second, with regards to the dissemination of 'western' knowledge, the way western medicine was spread in Africa (and indeed also in other parts of Empire) was not purely a western phenomenon. Rather, both medical knowledge and medical personnel were sourced from the outposts of Empire. Facts were assembled from experiments conducted outside of Europe, in rudimentary laboratories built by locals, and large teams of indigenous laboratory assistants and microscopists facilitated research projects. As shown in the previous chapter, India from the 1840s became increasingly important as a training ground for western medicine. In short, once analysed, the bottom line, unadulterated 'western-ness' of western biomedicine needs to be subtly reassessed.

With regards to the specifics of this book's subject matter, although Indian medical migration to East Africa was initially small, it was nevertheless a fundamental cog in the wheels of the bigger, more publicly celebrated, processes of the formation and enactment of British colonial power more generally and the establishment of biomedicine specifically. In the dynamics of military conquest, in the projects of infrastructural construction and in the establishment of government bureaucracies Indians were increasingly employed specifically to assist in the management of health. In this period, two strategic priorities, occurring roughly simultaneously, dominated the British government agenda in East Africa. First the firming up of borders and the consolidation of colonial rule via military operations of conquest and second, the construction of the Uganda Railway as a major infrastructural initiative providing a way into the interior of the region from the coast. In both cases Indians were recruited in large numbers to establish and extend British administration in East Africa. What is less known is that, along with the mass immigration of workers and sepoys, in each case Indian doctors were also brought in (as well as European doctors) to manage the health of these new groups of personnel. Although the group was initially not a large contingent, it nevertheless marked the substantive systemic beginnings of Indian medical migration to the region.[11]

***

Indian Doctors in the Army

Although the distribution of European colonial influence in Africa had been decided through the 1884 Berlin Conference, the actual establishment of territorial boundaries locally, was far from a straightforward process, especially in a place characterised by numerous ethnic rivalries such as Africa presented. In this context, when it became quickly clear that the locals did not welcome British colonial ambitions, the British, under the auspices of the Imperial British East Africa Company (IBEAC) brought in troops to help forcibly establish colonial power. The communications of 1893 between officers of the Foreign Office, and the British Consul in Zanzibar, are indicative of the hostility of most indigenous people to European colonial expansion. The colonial authorities bemoaned the difficulties in taming the 'intractable Wakikuyu' and realised that the possibility of combating the 'standing menace of the Masai' in the interests of Empire was going to be a formidable, if not an impossible, task.[12] Given the local resistance to British expansionist ambitions, it was strenuously advised that administrative control could only be exercised by crushing all resistance by a highly organised military force, which the Company then set up to establish.[13] The Foreign Office was approached by the IBEAC in 1889 for permission to recruit 200 Indian troops but the proposal was rejected by Lord Salisbury (although 100 retired volunteer Indian soldiers were employed to undertake police duties in the colony).[14] The first military contingent that was assembled in 1894 consisted of a mixture of Swahili and Sudanese solders, but these Africans were said to have been unsatisfactory for tackling the native revolts, such as the burning of the town of Malindi, that were occurring throughout the East African coastline. In response, the IBEAC reinforced the contingent with additional soldiers from Sultan of Zanzibar's mercenaries but, regrettably, the new force also proved undisciplined and inadequate.[15] It was at this stage of desperation, when it was felt that the British might not be able to effectively establish themselves in the region that another request was made by the Foreign Secretary to the Secretary of State for India to urgently provide an Indian infantry regiment to Mombasa.[16] This contingent, who arrived in March 1896, was to mark the beginning of the short, but nevertheless, effective history of using Indian soldiers from a number of different regiments, particularly those from the Punjab, in the British pacification of the East African region.[17]

With the new Indian force in place, the first sorties were against the Mazuris (1896) and the Wakamba (1896) and subsequently, building on these successes, the Uganda Mutiny (1897) was put down. The Commissioner of the

Protectorate commended the actions of the Indian contingent during these early campaigns, describing their efforts as 'exemplary.'[18] Undeniably, the casualty figures for the Indians were higher than those of the African soldiers confirming their deployment at the sharp end of the fighting. The importance of the Indian regiments in the success of a number of the early punitive expeditions was so lauded by the British authorities that several Indian officers were granted military honours for their distinguished service.[19] Once the early penetrative phase of conquest was complete, the new Protectorate's army was reformulated as the East African Rifles (EAR), a force comprised of Sudanese and Swahili as well as Punjabis.[20] Finally, in 1902 the Indian contingent became separately organised as the 5th King's African Rifles (KAR), but the job of conquest had been achieved and the battalion was disbanded in 1904, by when most of the Indian soldiers had returned home.[21]

In all cases, Indian Hospital Assistants and (or) Sub-Assistant Surgeons accompanied the troops and attended to the much-needed medical requirements of the regiments. Their organisation was based on a model that had become well established in the nineteenth century in India. A British doctor was usually, but not always, the Senior Medical Officer of the Indian regiment with under his direction several Indian doctors or lower rank who saw to the day-to-day healthcare provision for the troops. Similarly, regarding campaigns outside of British India, Indian subordinate doctors made up the bulk of the medical capacity of the overseas expeditions although there were also cases when an Indian doctor headed the medical contingent (see case of Masani, below).[22]

Because of the limited evidence available for this period, regrettably, not a great deal is known about all the specific names, precise qualifications and activities of Indian medical men working for the British military in East Africa. Nevertheless, some positive comments over Indian medical contributions can be found, occasionally naming names. The British army officer Lloyd-Jones, for example, described the gallant activities of a number of Indian medical men, singling out Bannerji, Maula Baksh and Parvati as having made valuable contributions. He made particular comment about Parvati in his descriptions, referring to him as 'remarkably clever and kind' in reference to a story where this doctor went out of his way to make a 'hazardous journey in the remote areas around Marsabeit' to save the Lloyd-Jones' life.[23] Lloyd-Jones provides one of the rare accounts of the role of Indian doctors in the early East African campaigns, and although his voice is one of only a few, the time he spends praising the 'sterling work by sub assistant surgeons' within his more general reminiscences is nevertheless striking.[24] Maula Baksh, serving under the European Medical Officer at that time, Dr Turner, also gets particular mention in another work by Lloyd-Jones, where he was described as 'first of his countryman who had done yeoman work of this sort, literally and actually "off the map" in

Equatorial Africa.'[25] Similarly, European Medical Officer, Norman Parsons Jewell in his recently come to light autobiography, fondly remembered a Sikh Sub-Assistant Surgeon, Zorawar Singh, who served with the 3rd East African Feld Ambulance. Singh, described by Jewell as 'a fine man' and 'a popular and respected figure with the Indian, African and European troops' travelled hundreds of miles from Dar es Salaam to Nakuru specifically to say goodbye to his friend Jewell once he had heard he was leaving Africa to return to his Indian home.[26]

The career of Surgeon-Major Harmasji Dadabhai Masani, the Parsi IMS officer in charge of the medical team in the military expedition of 1896, provides some bones of a life story for an Indian individual working in medicine in the East African campaigns of military conquest. Despite the preference to recruit Europeans into the senior positions of authority within the army, Masani was designated Principal Medical Officer to the Mombasa Field Force and was in charge of the first Indian Regiment (24th Bombay) sent to quell the East African rebellions in 1896.[27]

From his biographical entry in the Roll of the Indian Medical Service, it is known that Masani was born in 1850, attended Grant Medical College, Bombay for his medical training, receiving L.M.S. in 1873. Thereafter he worked as a Civil Assistant Surgeon in Bombay between 1874 and 1876. At that point, or sometime before, he appears to have spent some time in London as in 1876 he is listed as having achieved the M.R.C.S. and L.R.C.P. Lond. On return to India he joined the Indian Medical Service, in which capacity he spent time working on campaigns in both Afghanistan and East Africa. During his time in Afghanistan he won a medal for his services in 1879, to which he added an additional medal (1896) for his work in the East African campaigns against the Mazuri rebels. Clearly an exceptional individual, Masani rose through the ranks quickly, from his initial appointment as a Surgeon in 1877 to that of the senior rank of Surgeon-Major in 1897. He finally retired from service in 1898, after an IMS career of nearly 20 years.[28]

Some details of Masani's duties in East Africa and the conditions he encountered are to be found in a report he submitted after his 1896 Mombasa Expedition.[29] Masani was accompanied on his first expeditionary venture in the region by five Indian Hospital Assistants: Balkrisna Kashinath, Maula Baksh, Rahim Baksh, Sheikh Ahmed and Niyamtulla (no forename given). All were later awarded the Ashanti Medal for their services in East Africa.[30] From Masani's account it is clear that he felt that he was understaffed considering the high incidences of ill health (especially diarrhoea and dysentery) that accompanied the battalions. Consequently, he quickly saw the need for more hospital facilities for the care of sick and wounded members of military.[31] In the description given by Masani a sense of the medical organisation of a typical military contingent can be pieced together. A Hospital Assistant was attached to each separate detachment, each of whom were given 'suitable drugs and surgical equipments' (*sic.*).[32] Because of the poor conditions, restrictive diet and excessive heat, however, diarrhoea and dysentery were said to be prevalent amongst the troops, with six deaths on the campaign from disease as well as eight cases of invaliding.[33] Although each detachment brought with it field hospital, mules, attendants and doctors, it was clear that the medical teams were struggling. On the basis of his experience, Masani made a number of recommendations in his final report for future East African campaigns by the Indian army. Importantly, he specifically stressed the need to recruit 'a good many more medical subordinates' and better basic provisions for the troops (blankets, rations, clothing, and tents) as well as the need for a separate hospital for the 'coolies'.[34]

There were a number of advantages of utilising Indian medical staff in the military. The low salaries of the Indians compared to Europeans naturally provided a major economic incentive and, furthermore, Indian doctors had first-hand knowledge of tropical diseases and their experience of which was thought to be of particular value in the equally warm climate of East Africa. Another significant advantage cited at the time in terms of utilising Indians, was that Indians subsisted on simpler diets and also had much lower porterage requirements, making them again more economically efficient than their European counterparts.[35] A senior officer had explained the situation aptly:

... a supply of medical necessities is always one of the essential loads ... and services of skilled men desirable. European doctors are scarce and none are as a rule attached to military forces; besides an extra European means extra loads and therefore a larger caravan. In these circumstances the gap was often filled by Indian Assistant Surgeons trained upon European lines and might well have been these devoted men lived up to the tradition of their profession.[36]

The ability to subsist on inferior rations, was, however strange as it may seem now, a key money-saving factor in favour of the employment of the Indian Sub-Assistant Surgeon, over the European Medical Officer as well as being, presumably something in favour of the employment of Indians in all sectors of the administration.[37]

By 1900, the pacification of East Africa was largely complete and the number of troops brought into the country substantively dropped. This was not to quite mark the end the history of Indian troops in the region, however, and almost 20 years later Indians were drafted in again to partake in some of the East African battles of the First World War that raged over the Tanganyika borders. The incredibly high death toll on the campaign has been described as 'a war of attrition and extermination which [was] without parallel in modern times'.[38] In total these battles have been recounted as

having cost nearly 100,000 lives, the majority of which were African porters (the Carrier Corps) and soldiers, although many Indians (and a smaller
number of Europeans) also lost their lives.[39]

The successes of the World War One campaigns, however, were not as
easily won as those that had been fought for at the end of the nineteenth
century. Alien fighting conditions, poor military command and a lack of
centrally coordinated strategy to combat the guerrilla tactics of the Germans
proved to be disastrous for Indian regiments in terms of casualties.[40]

Furthermore, a disturbing range of diseases along with poor facilities for
the treatment of the wounded and the continued to colour campaigns. A
1917 report by Major General T.E. Scott on the East African First World
War campaigns sheds valuable light on the health conditions that existed
at this time.[41] Scott listed the top five causes of 'abnormal wastage among
Indian units' as, malaria, jigger fleas, tick fever, dysentery and 'sun fever'.
He also commented that food, clothing and supplies were still lacking in
terms of what he felt was essential for the Indian battalion, thus reiterating many of the same concerns outlined in Masani's report after the 1896
campaign more than 20 years before.[42] In terms of medical staff Scott made
specific mention of the 'shortage of Sub-Assistant Surgeons, complaining
that their number was 'quite inadequate for the needs of the force'.[43] He
also noted, as Masani had done before him, that hospital facilities were
lacking.[44]

In essence, the role and position of Indian Sub-Assistant Surgeons in the
British army remained the same as it had been 20 years previously. But there
was one important discernible change, which went beyond formal policy.
A shift in attitudes can be seen in the language deployed by Europeans
to describe their Indian colleagues. Whereas, before 1900, comments
(although admittedly infrequently documented) about Indian doctors were
almost uniformly positive, after 1910 more overly racist attitudes were
beginning to become evident. This was reflective of the increasingly charged
atmosphere surrounding racial politics in the colony, which came to a head
in the early 1920s. By the beginnings of the First World War both Indian
and African Nationalists were, slowly but surely, starting to assert their views
and gain support.[45] A particular milestone occurred in 1914 when the East
African National Congress was founded by Jevanjee and specifically started
defending Indian rights against settler demands to deny Indians access to
the highlands and enfranchisement.

The history of Indian involvement in the British military in East Africa
was not a long one, but it nevertheless further regularised expectations of
using Indians as a substantive section of the medical personnel of the East
African British Empire. Indians were also important, however, in another
great imperial project of that time—one that went hand-in-hand with
territorial conquest—the construction of the Uganda Railway.

***

Indian Doctors in the Uganda Railway

The construction of the Uganda Railway—by far the most expensive and ambitious initiative the British undertook in East Africa—continues to provoke historical imaginations and has been the subject of a large number of monographs.[46] Commitments in terms of both finance and manpower for this project were on a scale that had few parallels in the history of colonial Africa. Furthermore, its impact was known to be profound at the time, with Governor Edward Grigg going so far as to declare that Kenya was 'not conquered by force of arms but by the Railway'.[47] A report by Commissioner Charles Eliot of 1901 reiterated this view through stating that the railway construction was without a doubt 'the most important event in the history of East Africa'.[48] Grigg went so far as to claim that the railway marked 'the beginning of all history in Kenya.'[49]

Despite the difficulties presented by this ambitious problem in terms of finance, staffing and engineering, the reason why the project was doggedly pursued was clear. Having a railway running from coastal Mombasa to the shores of Lake Victoria opened up the less-accessible interior of the continent for commerce, and colonial exploitation. The railway was said also by some to strategically provide increased British control of the Nile delta. In public at least, humanitarian aims were also cited as an uppermost reason for its construction. Not only would the railway be a tangible means of improving communication and transport, it would help the British to police any violations of the abolition of the slave trade by giving them greater access to previously inaccessible districts.[50] But even if there could be little doubt as to its strategic and infrastructural benefits, the project (sometimes referred to as the 'lunatic railway') still attracted much controversy within British parliament because of the huge costs associated with it.[51] In the years when the possibility of the scheme was most debated, the subject was sufficiently contentious to warrant the publication of a handbook in 1892 specifically dedicated to spelling out the arguments in favour of and against the project.[52] In many ways this caution was justified, the cost of the railway imposed a crippling burden of debt on Kenya that had long lasting implications. Even after the railway was opened, the colonial government had to continuously tackle the substantive debt issues it had directly caused and made strenuous efforts to make the railway as financially profitable as possible.[53]

Staff from India

Given worries about its cost, it is no surprise that the government looked to India for staffing its construction where it knew it had a source of relatively reliable, resilient and cheap labour. Furthermore, both East African Commissioner Arthur Hardinge and George Whitehouse, the Chief Railway Engineer, had extensive experience of India so were amenable to build their plans and technological and personnel choices on Indian precedents.[54] Yet, although in retrospect, the decision to use Indians on the railway was perhaps predictable, the decision was nevertheless not finally taken until some experiments using local African labour had been tried and had failed. Sir John Kirk Chairman of Uganda Railway told the Sanderson Committee many years later: '[w]e began by trying native labour ... then came time of famine and labour was impossible to get. Then we appealed to the India Government'.[55] But it does not seem that famine was the only reason why Africans were not employed in this capacity. Rather their alleged inability to work was also cited:

> Strenuous endeavours have been made, are still being made, to induce natives of the country to come forward and work on the line. The idea of organised labour is utterly foreign to most of the tribesmen, and the country is under the rule of numerous petty Chiefs who only possess authority over a limited following. The native has a strong home instinct and dislikes work at any great distance from his own district.[56]

Given the dominant attitudes, the employment of African labour throughout the first three years of the construction of the railway line was limited to porterage. Even though the Indian railway labourers cost 14d a day (as opposed to 4d a day for African labourers), the Indian predisposition for the work made them purportedly more suitable.[57] In addition, Indian Assistant Surgeons and Hospital Assistants were considered to be more useful in helping Indian 'coolies', it was even admitted more so 'than European Medical Officers ... who would be unable to converse with or understand the language of their patients.' (Figure 3.1)[58]

Once the decision to opt for Indian indentured labourers had been made, immigration numbers rose quickly. From the original first batch of 350 Indian labourers who embarked for Mombasa in January 1896, the number rose dramatically by 1897–4,000; by the end of the same year this had increased to 6,000 and finally by 1898 to over 13,000.[59] At the peak of construction just over 20,000 indentured Indians were working for the project.[60] Although the majority of Indians returned back to their homeland after their labour contracts expired or were invalided back to India, their presence utterly changed the East African landscape, making it into what Sir Harry Johnson called a long 'strip of India in Africa'.[61]

The beginnings of bureaucracy

Although much has been written about this Indian labour force on the railways, no scholarly work has examined the Indian doctors who were also employed by the Railway Medical Department to undertake the day-to-day medical care, under the supervision of a handful of European Medical Officers.[62] This trend of relying on Indian medical staff was apparent from

Figure 3.1   Indian medical staff, Uganda Railways, Nairobi, 1900
Credit: Humphrey Winterton Collection of East African Photographs, Melville J Herskovits Library of African Studies, Northwestern University.

the very first caravan expedition of 1891 that surveyed the route of the proposed railway. On this first exploratory mission, only one Indian doctor (Hospital Assistant) was recorded as having accompanied the group as the only medical resource dedicated to the 389 person strong caravan after efforts to obtain two Indian medical men had been unsuccessful.[63] The caravan party consisted of 7 Europeans including four army officers and a surveyor, 41 Indians including a surveyor and two draughtsmen and 341 Africans of whom 270 were porters.[64] While the official report of this expedition thanked a number of European doctors for their organisation of the medical arrangements, it made no mention of the part played by the Indian Hospital Assistant.[65] In many ways this was typical of the climate of the time, which prioritised European contributions, although it is notable that the account of Macdonald's later expedition to Uganda by Major Austin of his experiences did make passing mention in of the 'able work' of hospital assistant Molah Bux.[66]

After the results of the survey had been accepted, the British knew that the railway would not be built without a relatively healthy workforce, but provision was not immediate. Despite the fact that concerns had been voiced from India about the hasty export and welfare of indentured labourers,[67] Whitehouse's first brief upon appointment as Chief Engineer to the project in December 1895 made no specific recommendation, among the seven it listed, as to how to manage health of the railway workers.[68] The selection and terms of remuneration for one European Medical Officer, two Indian Assistant Surgeons and six Hospital Assistants were still being finalised in March 1896. It was to take until the middle of 1896—and much internal haggling (especially about salaries) between the Railway Committee, The Indian Office and the Foreign Office—before the final establishment of the Railway Medical Department.[69] A tardiness which shows not only the bureaucratic difficulties in coordinating three separate organising bodies, but also one that shows that, although health provision was seen as a necessary part of the infrastructure, it was not the first priority and concern of the authorities (Figure 3.2).[70]

The Railway Medical Department operated separately from the main government medical department, until the two departments were amalgamated in 1901.[71] During its short life, from 1896 to 1901, four Principal Medical Officers (PMOs) were in charge of the railway medical service. The first appointment was a transfer from the IMS, Dr Turner (served 1896–1897), who left after a year's service to join British military campaigns in East Africa. He was briefly succeeded by Dr Oliver (7 months of 1898), who himself was even more briefly succeeded by the temporary appointment of Dr Brock (less

Figure 3.2 Indian workers on Uganda Railways, Londiani, 1902
Credit: Bodleian Library, Oxford, MSS.Afr.s.530, Londiani.

than three months of 1898). Eventually Dr Sieveking was appointed to the position in 1898 (serving until 1903). In any case, Sieveking's appointment marked the end of three years of changes to the Railways medical leadership, arguably occurring during the crucial period of the project when the health of the Indian workers was highly vulnerable and a large percentage of imported labourers were sent back to India.[72] Exactly as in other colonial territories, the PMO was always a European who was given administrative charge for overseeing the workings of the medical department and held ultimate accountability for the health of all the staff employed on the railway. In the case of overseeing the medical needs of the railway, this in effect meant responsibility for the health of a few Europeans, hundreds of non-European solders and thousands of Indian labourers.

Recruitment and Numbers

The systematic recruitment of Indian subordinates to staff the medical service started almost immediately. When Captain Turner took up his position at the head of the Railway Medical department in 1896, a number of Indian Sub-Assistant Surgeons (2) and Hospital Assistants (6) were reassigned with him to form the early medical department.[73] Turner's first tasks were to supervise the building of a small hospital for railway workers at Kilindini and obtain the sanctioned number of medical staff from India: evidently, according to his complaints contained within his annual medical reports on the progress of the railway works, a difficult task.[74] In 1896 the establishment comprised only one European MO, two Assistant Surgeons, six hospital Assistants (Sub-Assistant Surgeons), four dressers and nine compounders.[75] An unexpected challenge was the 1897 revolt of the Sudanese troops in Uganda which diverted many of the European doctors to the front line including Turner himself.[76] The first mention of an actual increase in European MOs for the Railways project was recorded in the 1900 Progress Report which indicated the number had increased to four. The medical establishment had eventually increased to six European MOs, twelve Assistant Surgeons and six Hospital Assistants but difficulties continued with recruitment of staff to the sanctioned numbers.

Although it is difficult to identify all the Indian individuals by name, a near contemporary account for Dr Sieveking's period as PMO listed his medical staff as consisting of six Europeans: 'Brock, Rose, McCulloch, Waters, Carne and Paget' as well as much larger contingent of Indian medical staff.[77] The Indian subordinate staff was judged to be 'very superior and intelligent' despite the fact that facilities were rudimentary and diseases amongst the railway workers were common.[78] It should be noted that a few reports countered this largely favourable representation of the Indian medical workforce, though Guildford Molesworth in his official report presented to parliament offered no comment on the adequacy of the medical staff.[79] Notwithstanding the odd complaint and their need to work in harrowing conditions, however, Indians remained the backbone of the Railway Medical Service throughout its existence, mostly operating with little supervision from their European superiors.[80] In total it is estimated that 21 Indians worked as either Assistant Surgeons or Hospital Assistants, supplemented by 5 Anglo-Indians of the same rank (see Appendix 1). In short, there were roughly three times as many Indians as Europeans conducting medicine along the railway line in the five years of the Railway Medical Department's existence.

However, the number of Indian doctors was always felt to be inadequate and numerous references can be found within the archives of calls from the Railway Medical Department between 1897 and 1901 to substantively increase Indian recruitment.[81] Interventions supporting this policy came from the highest echelons including Lord Elgin, (Viceroy of India, 1894–99, later also Secretary of State, 1905–1908), Lord Hamilton (Secretary of State for India) and Marquis of Salisbury (Sec of State for Foreign Office).[82] The instructions of the Marquis of Salisbury are particularly illuminating in this regard. In many instances he made explicit pleas for retention of Indian Medical staff. His November 1898 appeal to the Secretary of State for India, directly resulted in the two year extension of the periods of service of five Indian medical men: Messers Orman, Ram Saran, Rahmat Ali, Ahmed Husseni and Nehal Chand (compounder).[83] But this was just one of many examples and special arrangements for the recruitment of medical staff from India continued throughout 1899 and 1900 as demonstrated by the continuing correspondence between the Foreign Office and the India Office. For example, Surgeon W. Hussey, Hospital Assistant Muhammad Ibrahim and Hospital Assistant Muhammad Ali Khan were all transferred from India specifically for employment in the Uganda Railway.[84]

Experiences

The terms and conditions of service for Indian medical employees were much more meagre than that which were offered to European Medical Officers in the Railway Medical Department. While European Medical Officers were paid R600 a month, Indian Assistant Surgeons with less prestigious qualifications were paid half as much, receiving R300 per month. Hospital Assistants, although often possessing qualifications from medical institutions in India not drastically inferior to those appointed Assistant Surgeon, were even less favourably remunerated, being granted only R100 a month, a figure which was then renegotiated by the India Office to just R70 a month.[85] Contracts were issued for three years 'terminable if services proved unsatisfactory'. Free rations, quarters or tents were supplied and travel costs incurred in taking up the employment were covered. Indian medical staff received pension and leave allowances in line with those that were offered for Indian government employees in the Indian Civil Service.[86]

By combining different types of source material from the India Office Records at the British Library, it has been possible to confirm a list of the 21 Indians working either as Assistant Surgeons or Hospital Assistants within the Railway Medical Department (see Appendix 1). To find out more about their experiences, however, has been more challenging. Several accounts that have survived from this period concur that Indians provided the bulk of medical services, and were deemed to be particularly relevant as they spoke the languages of their patients and were familiar with their ailments. The high incidences of disease, the continued calls for supplementary staffing and the descriptions of the poor treatment facilities suggest that most Indian doctors were in high demand and were likely to have felt pressured by the heavy workload.[87] The majority of accounts tended to lump the Indian medical staff together as one group without personalisation of the individuals involved. Nevertheless, a couple of individuals were singled out for praise, for example Dr Orman was said to have given 'skill, learning and compassion' in the service of the Uganda Railway.[88] Another notable exception is that of Rahmat Ali, an Indian doctor who left a remarkable account of his work during this period in the form of an appeal for promotion, supported by the Principal Medical Officer of that time, Dr A.R. Sieveking, which has provided some unique insights into what the experience of being an Indian doctor in East Africa at that time entailed.[89]

Rahmat Ali had qualified in 1895 as a Hospital Assistant, it is not known from which institution he graduated, but he was listed as having won the 'Burton Brown Medal' for his achievements in his final medical examinations. On arrival in East Africa in 1896 the young Rahmat Ali was put in charge of Uganda Railway's Headquarters hospital at Kilindini, where he found himself immediately in a position of substantive responsibility. He recounted how onerous he found the work, not least, as he had to routinely carry out:

> [T]he duties of an Assistant Surgeon, Hospital Assistant, Storekeeper, clerk and compounder; also the duties of Medical Officer when Dr Turner used to go up the line, where he used to stop for weeks and weeks and during his absence I used to have the full charge of Officers, Subordinates and Employees etc.[90]

While Ali's efforts impressed his superiors, they nevertheless took a toll on his health and in 1897, on the advice of the local acting Medical Officer— Dr Carre—he was sent back to India for three months of rest and recuperation. On his return things did not get much better. Ali was immediately posted to the Voi hospital to take charge of matters there (the previous incumbent of the job, Assistant Surgeon Desmond, having deceased). After a short period at Voi, Ali was recalled back to work at Kilindini in 1898, from where he wrote his petition for promotion from the rank of Hospital Assistant (3rd grade) to a position three steps higher up the salary scale to Senior Hospital Assistant.[91]

Ali's experience of work at Kilindini was evidently as stressful as he had experienced at Voi, for, after the transfer of Assistant Surgeon Orman he found himself elected 'to take up charge of headquarters and hospitals both the European and native'. During this time he had to also treat many Europeans within the colonial administration, including the railway chief engineer, Mr Whitehouse. Evidently with some success as he recounted in his plea for a salary increase that:

> My hard work, careful attention and familiarity have so much won the hearts of employees that each and every one from Kilindini to Railhead prefers to be treated by me.[92]

In both hospitals Ali's work and diligence were recognised by his superiors, to the extent that he was awarded a special discretionary salary bonus. Yet, however well he was accepted and despite the fact that his work was highly praised by his European peers, his position as a subordinate was ultimately confirmed when his request for three step promotion was limited to a single step by the Secretary of States in both India and London, despite the favourable endorsements his proposal had from his PMO, Dr Sieveking and from Whitehouse, the Chief Engineer.[93] As the next chapter will show, broader racial ideas that increasingly imbued all levels of colonial life in Kenya were to also limit the professional and social mobility of East African Indian doctors.

Health on the Railways

Throughout its short history, it was not unusual to find up to 10–15% of the railway workforce hospitalised and a greater percentage incapacitated at any one time.[94] And this was not the worst. A strikingly poor month occurred in January 1897 when it was estimated that 50% of the entire labour force was on the sick list.[95] One report prepared for Whitehall in 1897 is indicative of the health situation:

> the health of the staff and the labourers was very bad. The effect of turning up soil, ... resulted in great increase in malaria.... Ulcers have been very prevalent amongst the Indian coolies and over 50% of them have been down with malarious fever.[96]

Not only did disease spread quickly down the line with labourers living and working together in cramped and insanitary conditions, but none of the reports and correspondence address the need for basic necessities, such as the provision of extra clothing and blankets during the colder months, which took time to appear. 1898 saw a particularly high number of repatriations to India because of ill health. It was said that the exceptionally wet weather of that year had resulted in 'much sickness and mortality', particularly from malaria, diarrhoea, dysentery, liver complaints, scurvy and ulcers.[97] Another report about health along the railway construction site, by Guildford Molesworth in 1899, complained of the 'waves of fever' that were common amongst the workforce and bemoaned the constant problems of jiggers that were said to 'infest the coolie camps'.[98] This recurrent pest, which was almost unheard of in India, had incapacitated hundreds of labourers, most of whom (as the photographs that survive from that period attest) had to perform their duties barefoot without protective footwear.[99] If dealt with quickly, jiggers are not particularly dangerous, but the delays in dealing with them among railway staff frequently made their presence worse. It was noted by the authorities that the lack of attention paid to jigger infestations led to greater, and more costly, medical problems, such as ulcers but also sometimes even 'the amputation of one or more toes'[100] Tellingly, Miller's history of the railway observed: 'rare was the Indian worker or European supervisor who did not carry at least one fist sized blob or livid purification on his arms, legs or trunk.' The same author went on to vividly describe the 'common sight' of seeing Indian labourers 'drop their rail length, stumble down embankments frantically lower their trousers or dhotis as amoebic dysentery opened their bowels'.[101] Exacerbated by seasonal climatic variations and the long hours expected from labourers, the majority of the accounts that survive from this period made repeated reference to the poor health endemic among the construction workers. Estimates for the overall morbidity and mortality incidences vary, but, conservatively, it seems that about 20% of all Indians employed on the railway were repatriated as invalids, while 8% of the Indian workforce were said to have died on duty.[102] Furthermore, even if no longer fit to work on the line, the fate of the Indian labourer was likely to be unfortunate. In accordance with economic considerations, the official policy of the railway medical department was to repatriate the sick who could not be cured locally back to India rather than to provide them with longer-term local restorative care.[103] Unsurprisingly, given the meagre medical resources available, most of the workers who returned to India were said to have 'obtained little benefit from their exile', and arrived back to their homeland in a far worse physical state than in which they had left it.[104] The opinion of the Port Health Officer who inspected a returning ship in Karachi in 1900 is indicative: of the 279 labourers it carried, he remarked, 75% of them were 'more or less broken in health'.[105]

One of the problems common to all aspects of colonial medical provision during the early years of British rule in Africa was the lack of good hospital facilities. Plans to build hospitals were advanced after the first batch of Indian workers had landed in East Africa, but these mostly never came to fruition.[106] Detailed descriptions from this early period are rare, but the overwhelming image of hospital provision for the railway workers is that it was ad hoc and makeshift. The only permanent hospital before 1899 was established in 1896 at Kilindini, Mombasa, and the other hospitals built along the line, such as at Voi and Makindu, were impermanent encampments of a very poor quality, usually just consisting of a few 'fly infested' tents 'full of refuse and stench of sickness'[107]. The conditions existing in the temporary hospital at mile 483 were described in similarly stark terms in an official report commissioned by the Foreign Office:

> In the hospital for the earthwork and plate laying coolies at mile 483, the medical officer had been absent for 12 days and patients were in charge of a native of India hospital assistant. There were 225 patients in all with thirteen in each tent, which was not sufficient to hold more than 8. The sick were lying on dried grass on the ground, which had not been raised, and, in consequence, a good deal of rainwater had entered in places.[108]

Even European officials, a group who usually tended to put a positive spin on things, occasionally let down their guard and admitted these temporary facilities (and the lack of management by the senior European medical officers) were 'improper' for the health needs along the line.[109] For the most part however, official declarations about health along the railway remained upbeat and optimistic. Sir Guilford Molesworth—who visited East Africa between 1898–99 to compile an official report on the railway project—declared the death rate of about 20 per thousand per annum as a 'wonderfully low average considering the climate and conditions'.[110] In a similar vein, some other descriptions portrayed the workings of the railway medical department as 'most efficiently carried out', although it is clear from the FO commissioned reports by both Gracey and Molesworth, as well as the final report of the Uganda Railway Committee of 1904, that the medical department was overworked and under-resourced for the level of health problems that they experienced.[111] The table in Appendix 1 compiled from the information supplied in the final 1904 Report provides the stark figures concerning the poor condition of the health of the Indian labourers. It lists the cumulative number of imported workers, those repatriated, invalided, as well as those who died.[112]

Several factors were responsible for the poor state of the health of the Indian labour force. First, retrospectively at least, it is not hard to conclude that health issues simply was not prioritised enough. Speed of construction overrode consideration of the health of the labourers.[113] Shocking to the modern reader, heavy mortality amongst non-Europeans was accepted as an inevitable 'accompaniment of big engineering works in the tropics.'[114] But also, intervening troubles, such as the Sudanese Revolt of 1897 'swallowed up all the available staff in either Protectorate' and meant that European MOs were diverted away from the railway project.[115] As a consequence, when the needs of the railway workers were becoming really urgent (1897–99) the medical staff was at its most depleted. The year 1900 registered the highest number of deaths at over 900 when the average number of staff employed was 19,000. The official report for the year offered little explanation of the exceptionally high mortality. It merely commented 'the medical staff of four European doctors, assisted by qualified assistants lent by the Indian Government, have been in charge, and, with adequate hospital accommodation at their disposal, have been able to cope with the diseases as well as to carry out the special examination of imported labourers and out-going invalids and time-expired men.[116]

Predictably, news of the poor conditions travelled back to India and seems to have made an impact on the image of East Africa as a favourable destination for Indian Sub-Assistant Surgeons. Certainly by 1897, the problem of Indian medical recruitment was regarded as sufficiently urgent to prompt high-level exchanges between Foreign Office and the India office.[117] As one commentator put it, securing, 'equally qualified men' was proving increasingly difficult.[118] One idea raised was to increase the salaries of Indian doctors who came to Kenya as a means of providing more of an incentive to entice suitable talent. Suggestions to improve the pay and pension of the Indian contingent met with some resistance from the Foreign Office—partly due to budget considerations surrounding the railway projects and partly because it was deemed inappropriate to further widen the gulf between pay of doctors who stayed in India and those who immigrated to Kenya.[119]

***

A short postscript to the history of health on the Uganda Railway would be to identify a positive legacy. The authorities eventually noted the unacceptably high incidences of morbidity and mortality among Indians and steps were made for a similar situation to be avoided in future.[120] In the words of Christopher Wilson writing in the *Kenya Medical Journal* in 1925: 'the day has passed for heavy mortality to be considered a necessary accompaniment of big engineering works in the tropics'.[121] Correspondingly, the programme to extend the railway network to Thika, Nyeri and Yala undertaken between 1924 and 1929 was better planned, with more careful and systemic attention paid to the medical aspects.[122] The project this time was dependent upon an African workforce, in the hope that utilising such labour might create a domino effect locally if satisfied labourers returned to their homes as unwitting apostles to 'the gospel of work and an effective agent in recruiting other workers from his district.'[123] The new railway project administrators paid more active attention to public health considerations such as the planning of the location of camps, latrines and dietary requirements. The results were satisfying, and relative invaliding and mortality rates were significantly lower than they had been for the construction of the Uganda Railway.[124]

The management of health was necessary for the conquest and establishment of the British Empire and in East Africa certainly Indian medical recruits played an important supportive role in this task. By the time the railway was completed in 1901 healthcare facilities in the East Africa Protectorate were still underdeveloped but nevertheless the presence of Indian doctors working with the British government had become customary. As the positive comments about their work and the calls for their accelerated recruitment attest, Indian Assistant Surgeons and Hospital Assistants were becoming not only a familiar and accepted part, but also a welcomed part, of the colonial medical hierarchy.

As much as Indian doctors contributed to the spread of western medicine in Kenya the British (even if for self-serving reasons) also contributed to the internationalisation of western trained Indian practitioners. The migration of Indians into Africa, hastened by the military and construction projects described in this chapter, suggested new routes abroad in which Indian medical expertise could be utilised. The work was hard and the conditions poor, but for many this would have seemed a way to escape even worse conditions at home—perhaps even to gain a bit of social prestige through working for the British colonial government. Moreover, in this early, formative period of colonial rule in East Africa at least, from the British perspective the comparatively civilised Indian was seen to be a good influence on colonial Africa.[125] As the events of the years after 1900 were to show, these findings were to be deeply ironic in the context of Kenya. Whereas once Indians were comparatively welcomed, even sometimes applauded, within 20 years the European attitude towards this immigrant group was to become profoundly more negative, at least within public and political debates. As British rule solidified and gained confidence, developments within Kenya and beyond, particularly in the realms of biological racism, started to influence both white settler mentalities, and government policies towards Indians. As the next chapter will show, after only a short period of positive relations between Europeans and Indians, during which British power was established in the region, racist ideologies eventually overtook pragmatics in terms of judging who the best servants of empire should be.

4
Race and Medicine

> Physically, the Indian is not a wholesome influence because of his incurable repugnance to sanitation and hygiene.... The moral depravity of the Indian is equally damaging to the African, who in his natural state is at least innocent of the worst vices of the East. The Indian is the inciter to crime as well as vice,.... The presence of the Indian in this country is inimical to the moral and physical welfare and the economic advancement of the native.[1]

By the time the main line of the Uganda Railway had reached the shores of Lake Victoria in 1901, the Indian population of Kenya was not only an accepted part of colonial society, but one that was publicly esteemed for its positive influence. Remarkably, within 20 years this image had almost totally reversed, with the 1923 Devonshire Declaration effectively putting a formal limit on fully participative Indian social integration in Kenya. While the events surrounding the 'Indian Question' that resulted in Devonshire are well described within the scholarly literature, the indirect impacts of this policy on medical recruitment and retention have never been explored.[2] Yet within both the political and medical realms anti-Indian discourses did not appear out of the blue and were the culmination of mutually reinforcing events and attitudes that subtly transformed Kenyan society since the early 1900s. On one hand, changes in perceptions towards the Indian community can be traced to the allegiances and priorities of a number of key individuals. In other ways, attitudinal changes reflected in policy decisions can be linked to broader discourses surrounding race that were solidifying in East Africa at the time. Various political milestones can be identified in the hardening of attitudes to Indians in Kenya, while simultaneously medical ideas evolved in such a way as to specifically target the Indian population. These events make the dramatic pruning of Assistant and Sub-Assistant Surgeons from the Colonial Medical Service after 1923 comprehensible. To fully understand the lives and experiences of Indian doctors, it is necessary to first grasp the dominant attitudes towards race and segregation that existed in Kenya. These standpoints can be seen to have toughened over time and were often supported directly by medical opinion, and sometimes by medical policies.

Political Landscape

One of the most dramatic shifts in attitude towards the Indian community in Kenya can be pinpointed to the change in Commissionership (Governorship after 1906) in 1900 from Sir Arthur Hardinge to Sir Charles Eliot. This marked the concerted beginnings of the 'White Highlands Policy' that favoured the white settlement of the area located to the North of Nairobi and the West of Mount Kenya, which was thought to have the most fertile land and the most agreeable climate.

During his time as Commissioner of the East Africa Protectorate (1895–1900) Hardinge had established a model of colonial administration that drew heavily on Indian personnel, not only as indentured labourers, but also as staff in positions of responsibility. In 1895 he had sought appointments of senior Indians as the heads of the Works and Transport Departments.[3] While Hardinge agreed that the Highlands should be reserved for white settlement, he nevertheless accepted Indian landownership as part of the cultural landscape of the East Africa Protectorate.[4] Hardinge was not alone in this thinking and a number of Colonial officials with experience of Africa, such as Sir Bartle Frere, Sir John Kirk, John Ainsworth and Sir Harry Johnston were also keen to find space for permanent Indian settlement.[5]

When Eliot succeeded to the headship of the protectorate it was quite clear that his sympathies were elsewhere. He saw the key to the economic development in the encouragement of British and South African immigration and the prioritisation of the rights of these settlers. Immediately upon his investiture Eliot drove a private bargain with the leader of the 'white frontiersmen' of Kenya, Lord Delamere, and invited White South Africans, who had suffered from the aftermath of the Boer War, to migrate.[6] In stark contrast to the more circumspect Hardinge, Eliot was vocal in broadcasting his negative views towards the indigenous peoples of Africa. He complained that most of the inhabitants of East Africa were characterised by their 'low intelligence, their natural timidity and mistrustfulness and their utter ignorance of everything a mile or two beyond their own village'.[7] He similarly had little evident respect for the local Indian population, squarely blaming them for the high incidences of plague within the protectorate.[8] Indeed, the only non-white group he seemed to harbour any respect for were the intelligent and 'warlike' Masai.[9]

Eliot's vision for the settlement of the East Africa Protectorate aroused some concerns among senior administrators, such as John Ainsworth, about the wisdom of trying to replicate South African social systems in Kenya.[10] Sir Harry Johnston warned specifically of the tendency of the colonialists, if left unchecked, to 're-establish slavery in their own fashion'.[11] Even Winston Churchill commented, after his visit to the country while still a junior minister in 1907, that he perceived that ' "COLOUR" is already the dominant question at Nairobi'.[12] With Eliot's support, however, vocal settler leaders such as Lord Delamere, Lord Francis Scott and Colonel Ewart Grogan gained political influence beyond the numerical strength of the settler community, or it's economic contribution. Settler objectives, often expressed within the pages of the two prominent settler newspapers, the *East African Standard* and *The Leader*, were commonly explicit. The settlers were desirous to establish a white aristocracy with a host of Africans coerced to work for them.[13] White settler demands included: European majority on decision-making bodies, franchises to be based on communal allocation, reservation of the Highlands for Europeans, the exclusion of non-Europeans from senior Government jobs, and the ability to command African labour by imposition of a taxation system with a heavy dependence on hut tax.[14] Although many of these conditions focused upon the social and economic separation of whites from Africans, the place of the Indian within this vision of a settler-ruled Kenya was also clearly designated:

> The Indian here takes the place of the European as the journeyman in skilled trades and however useful the Indian may be and we would fare ill without him at this early stage of our settlement, his capacity and stamina may not be deemed to approach that of the European of a similar trade and occupation... The European workman and tradesman in British East Africa fill the role of the overseer and master man having natives or Indians under him and to assist him.[15]

The Colonial Office in London did little to meaningfully hinder Kenyan settler ambitions. Between 1919 and 1939, complaints voiced in Parliament mainly through information supplied by parties in India and two prominent colonial administrators who had had first-hand experience of Kenya— Norman Leys and W. McGregor Ross—had helped to 'prevent the creation of a Great White Dominion in East Africa', but did not succeed in entirely curtailing settler aspirations.[16] Although other prominent members of the Colonial Office during the same period including W. Churchill, W.C. Bottomley, E.F.L Wood and Sir James Masterton Smith were similarly prepared to argue for the fair treatment of non-Europeans, a tacit preference to refrain from interference in local Kenyan politics underscored most debates.[17] The actions of the former High Commissioner for South Africa, Viscount Milner, who served as Secretary of State from 1919–1921, seem fairly typical: although Milner publicly criticised the 'abhorrent actions and tirades against Indians and Africans' of the settlers in Kenya, he actually offered the settlers significant support.[18] In short, although Whitehall occasionally stepped in to temper local policy decisions, in general settler interests were rarely curbed and as a result racially exclusionist policies were allowed to be sustained, even to flourish, in Kenya.[19] Eliot was to be only the first of a series of governors to fall 'willingly into settler clutches'.[20] Many of his successors including, Sir Edouard Girourard (1909–1912), Sir Henry Belfield (1912–1917), Sir Charles Bowring (1917–1919) and Sir Edward Northey (1919–1922) all displayed sympathy for the settler position and, increasingly, an accompanying disregard for the Indian community of East Africa.[21]

Given the comparative strength of settler power in the region and their racially exclusionist aims, it is unsurprising that Indians found themselves increasingly side-lined, often grouped together with Africans in public discourses as the non-white majority in service to the minority ruling whites. There was little possibility of widespread African-Indian collaboration to counteract this perception. Indians in general felt no special affinity with their African peers, and though examples of collaboration between these two broad ethnic groups, can be cited they were not a common feature.[22] Most Indians, with few exceptions, because of their numerous historical and contemporary links to business and trade in East Africa resented being grouped together with the majority African poor and preferred instead to see themselves as deserving parity with European settlers, befitting the contributions they had made to colonial development, not least in terms of military service.[23] These expectations had some external support. Sporadically the issue of whether Indians should be awarded land grants in East Africa as a formal recognition of their contribution to the infrastructural developments and war efforts of the colony was seriously discussed, first upon completion of the Uganda Railway project and second towards the end of World War One.[24] Nevertheless, attitudes towards Indians became explicitly more restrictive between 1919 and 1925 when they became increasingly recast as the most problematic non-white group of East Africa. Paradoxically, it was during this same period that attitudes towards Africans gradually became more open and expansive.

The key to this change in approach lay partially in the economic threat that the Indian community was perceived to embody. Most portrayals of Indians made allusion to their thrift and business acumen, which were increasingly seen not as a quality to be praised, but as a characteristic that other members of the colonial community should be suspicious of. Indeed, the lack of an Indian coordinated response to encroaching settler power was explained in terms of the amount of time and attention the community committed to their business affairs. As one senior administrator, C.W. Hobely, said in 1905: 'one hears little of their grievances because they are nearly all making money and they indirectly contribute very considerably to the revenue'.[25] Similarly, Winston Churchill, no great friend of the Indian community in East Africa, summarised their threat concisely: 'In every single employment of this class, his power of subsisting upon a few shillings a month, his thrift, his sharp business aptitudes give him the economic superiority...'[26].

In 1920, when the issue of the rights of the Indian community in Kenya was hotly debated in the House of Lords, Lord Islington, expressed his concerns that the importance of the Indians of East Africa was being overlooked. He reminded his audience that the Indian community had:

> [I]n the country for, I think, at least three centuries ..., outnumber the European population by something like four to one .... The development of the industry and the wealth of British East Africa have uniquely advanced due to these Indian traders and settlers ... who pay the larger proportion of the taxation.'[27]

Indians' apparent desire to make money and to use their economic contribution to the colony, as a bargaining chip for reforms in their favour, clearly grated with the Europeans in Kenya. One of the turning points, in terms of the strengthening of racial attitudes, appears to have occurred in 1914, when Indians participated in a mass strike against the enforcement of non-African poll tax.[28] To the manager of the railways this strike very much demonstrated the danger of being dependent on people over whom 'we have no control'.[29] Whatever their financial contribution to the colonial economy, the Indian community had shown itself to be worryingly demanding and had a potential articulacy that raised white colonialists' hackles in defence.

The most prominent crystallisation of economic animosity towards the Indian community came in the findings of The Economic Commission of 1919. The Commission was set up by Governor Belfield with the specific aim to report on a 'sustainable economic future of the colony'.[30] The conclusions of the final report of this commission were unambiguous. Africans, it stated were exposed 'to the antagonistic influence of Asiatic as distinct from the European philosophy' which needed to be countered 'to avoid a breach of trust'. There was also the recurrent theme that Africans were threatened with 'economic stagnation' through the presence of the ambitious and money orientated Indian.[31]

Final recommendations of the report explicitly stated that senior posts in Government, railway, municipalities and European firms should be reserved exclusively for Europeans. Among its recommendations was a complete halt to Indian immigration and a call that all Government departments 'should, as quickly as possible, replace Indian employees by Europeans in the higher grade and Africans in the lower'.[32] Indians were described as morally depraved, displaying an 'incurable' repugnance to sanitation and hygiene and with disproportionately large numbers of their community associated with crime and vice.[33] Unsurprisingly, public responses from Indians and their sympathisers to this intemperate statement of racial dislike, in both London and Kenya, were fierce; so much so that the Colonial Secretary of State, Viscount Milner (generally supportive of the segregation initiatives in Kenya), felt the need to issue a public condemnation of this official document for its extreme anti-Indian pronouncements and language, In 1920 Milner publically denounced the report's contents as 'deplorable' and reprimanded the individuals involved in its authorship.[34]

Following the 1914 strike, and the findings of the 1919 Economic Commission, anti-Indian sentiment within East Africa palpably strengthened. Calls started to be made by both settlers and members of colonial government demanding the cessation of Indian immigration and transfer of Indian held jobs in government to Europeans and Africans in line with the Commission's recommendations.[35] This new, exclusionist, line of argumentation reached its high point with the findings of the 1922 Bowring Committee (officially called the Economic and Financial Committee). This committee, which had been set up to evaluate the need for measures to rebalance the economy such as expenditure cuts and protective tariffs in the colony, proposed in its on-going recommendations one sixth reduction in all Asiatic civil servants working for the British Government in East Africa.[36] With seemingly no sense of contradiction, this projected staff reduction was justified on grounds of ameliorating the deteriorating financial situation of the colony. Indian leaders were quick to point out the illogicality of removing Indian staff as 'the salaries of some of the Asiatic staff are at present less than a quarter of the minimum salaries drawn by European staff' who were being retained, but this reasoning appeared to carry no weight in a colonial administration determined to reduce the influence of the Indian community.[37]

These unequivocal statements against Indians, particularly solidified through the reports of the 1919 Economic Commission and the 1922 Bowring Commission, were accompanied by moves to disallow Indian community representatives a voice in East African colonial politics. The disappointed and bitterness was evident in an editorial of the *East African Chronicle* 'the bulk of revenue for sanitary purposes comes from Indians, who get nothing but plague in return....'[38] Governor Edward Northey was particularly instrumental blocking Indian interests; within weeks of his arrival in the colony in 1919 he had formally ruled against the demands of the so-called Indian 'agitators' who were asking for equal representation on the Legislative Council of East Africa Protectorate (known colloquially as Legco). As can be seen, by the 1920s the scene was already set for the pronouncements that were to be enshrined in the 1923 Devonshire Declaration.

What is most startling about these moves to counter Indian financial and political power was how out of step it was with the reality of the economic status of the majority of Indians residing in Kenya. While the settler leadership included the powerful personalities of Lord Delamere, Colonel Grogan, Lord Francis Scott and Lord McMillan who all hailed from affluent, well-connected families in the UK, Indian leaders, in contrast ultimately relied on relatively small local donations from the impoverished local community as well as the support of a few prosperous merchants.[39] Although they attempted to defend their interests, this lack of access to political and economic power in an imperial context that displayed increasingly blatant racism towards them, meant that their chances of achieving their aims actually lessened, rather than increased, during the first half of the twentieth century.

Indian Responses

The Indian community was led at various times by a number of prominent individuals, including L.M. Salve, A. Jeevanjee, S. Thakur, Manilal Desai, Ismail Sham-su-deen, B.S. Varma, U.K. Oza, Isher Dass and J.B.Pandya. This was by no means an unambiguously radical, anti-imperialist, group—at least not at first. Most of the Indian leaders had originally been loyal to British imperial authority. So much so, that the transformation of allegiance of many Indians from a loyalist position to one of hostility towards the British provides clear testimony of the high levels of dissatisfaction they increasingly came to feel under the British East African regime.[40]

In this regard Jeevanjee's journey was a particularly significant one. A successful entrepreneur, Jeevanjee had been involved in many business transactions with the fledgling British administration, procuring labour and goods for the Uganda Railway. His contributions were so esteemed that he was nominated in 1910 by Governor Percy Girourard as the first Indian to serve on the Legislative Council.[41] However, despite his initial eagerness to represent Indians within the colonial administration, by World War One Jeevanjee had adopted a position of outright hostility to the British. This change in allegiance caused him much personal anguish, but nevertheless was one that he felt to be his only political option given imperial hostility to the community he represented.[42] Gradually moving from his pro-British stance since 1910, by 1920 he pointed out that, despite much goodwill on the parts of Indians, their rights had been trampled by a 'series of odious measures' such as the 1915 Land Ordinance, the 1918 Segregation of Races Act, the 1919 Municipal Corporation Ordinance and the 1919 Electoral Representation Ordinance, which cumulatively excluded Indians from rights granted to Europeans.[43] Jeevanjee was by no means alone in changing his attitudes to the British authorities; other Indian leaders became progressively disillusioned and migrated from one position to another over time.[44]

The history of Indian collective action in East Africa can be dated to 1906.[45] Realising that their interests were being increasingly side-lined, a mass meeting, chaired by Jeevanjee, was held in that year in Mombasa during which an Indian delegation was charged with the task of travelling to London to represent the community's grievances directly to the imperial government.[46] The meeting demanded an end to discrimination and challenged the issue that 'foreigners have preference in public matters over British Indians in a British Protectorate'. In 1912 an Indian statement of purpose was published in a booklet entitled *An Appeal on Behalf of Indians of East Africa.*[47]

By 1914 the need of the Indian community of East Africa to become more formally organised, resulted in the foundation, with Jeevanjee as its first president, of a representative body to defend Indian rights in the form of the East African Indian National Congress (EAINC). The funds raised by this body, although numerous, were individually small, and could in no manner match the financial resources at the disposal of the Europeans.[48] Furthermore, the leadership of EAINC remained contested between moderates and radicals, namely those who felt it was more effective to work within the framework of colonial structure or those who preferred to take a position of outright opposition to cooperation with the state. In the latter group, radicals such as M.A Desai, Isher Dass and U.K. Oza supported boycotts and argued vehemently against sacrificing the principles of common roll and racial equality, just to ensure some minimal representation. The differences between these two models of political action came to a head several times. Examples include a rowdy and heated meeting in Nairobi on 18 January 1925 that broke up in an acrimonious manner and later when I. Sham-su-deen and Isher Dass both contested the position of the General Secretary of EAINC in 1931.[49]

The colonial state was not averse to encouraging the rift between the Indian leaders, exploiting it whenever possible and dispatches between Kenya and London were quick to identify potential splits among the Indian leaders.[50] Differences that developed between the Hindu, Muslim and Sikh sections of the community were publically declared to hinder 'the cooperation in the work of the Legislative Council'.[51] Settler newspapers likewise revelled in giving front line coverage to the apparent lack of cohesion between Indian community leaders, and frequently published letters about the issue.[52] The Indian press, such as the *East African Chronicle*, edited by Desai and published in Kenya between 1919 and 1922, was not allowed such freedom of expression in return, however. Articles questioning European authority were considered dangerous and subversive, and the paper was charged with sedition and libel, an accusation that directly led to the newspaper's bankruptcy.[53]

Race and Medicine

Alongside the progressively restrictive political climate in Kenya, quasi-scientific racial attitudes within Kenya also gained popular currency between the 1910s and 1940s. Much of this discriminatory discourse was unambiguously aimed towards the African community; for example the research conducted by Colonial Medical Officers, Henry Laing Gordon and Frederick Vint, in the 1930s aimed at proving, principally through comparative skull measurement, the mental inferiority of the African.[54] Enthusiasm for this line of research reached a peak with the establishment of the *Kenyan Society for the Study of Race Improvement* (KSSRI) in 1933. This society, which had an overtly eugenics based agenda, hosted a series of lectures, popular with the settler community, implicitly arguing for restrictive policies towards Africans because of their intrinsic lack of suitability for, so-called, 'western' standards of civilised life.

Although Indians were not the focus of attention within the proceedings of the KSSRI or the research of Gordon and Vint, this racist discourse reinforced colonial attitudes towards Indians. Indians were lumped together as one homogenous community, one characterised in general terms as a public health risk to both themselves and to other members of the colonial society. No attempt was made to distinguish between different socio-economic groups or communities. In this regard colonial attitudes seem to have regressed rather than moved forward.

This had been going on for some time. As early as 1876, James Christie in his book on cholera epidemics in East Africa published his erudite observations on Indian social customs as a potential determinant in the spread and severity of cholera.[55] Christie was intrigued by Richard Burton's earlier observation that Banyans and Muslims experienced different levels of mortality in cholera outbreaks. To test these claims, Christie examined the lives of separate communities analysing their rituals of diets, sanitation and burial.[56] He concluded that the Hindus, who followed strict rituals of washings, observed restrictions on who touched their food and their person, did not contract the disease as frequently as members of other Indian sub-communities.[57]

Yet, despite Christie's findings, discourses about health still tended to gravitate towards mass pronouncements about 'Indians'.[58] In part this arose as a natural extension of the nineteenth century ideal of the 'civilising mission', which construed the British role as being to educate and to reform indigenous populations. But there were also clearly some wider motives underlying the need to differentiate so strongly between white and non-white groups. Medical historians are mostly agreed that this phenomenon can be understood as indicative of broader racial anxieties. In health discourses particularly, as Warwick Anderson has argued, whites portrayed non-whites as the 'pathological other', highlighting what were regarded to be uncivilised behaviour, as evidence to confirm their deep-seated beliefs in their own racial superiority.[59]

Within the colonial hierarchy, Indians were placed a little higher than Africans. In Kenya, an Indian doctor, K.V. Adalja, recalled:

> The Europeans even in the early days had a high standard of living, good civic sense and reasonable sanitation whilst the Africans were poor and backward, ignorant of the most elementary principles of sanitation and personal cleanliness and were catered for medically almost entirely by the

Government. Between these two extremes lay the Asians amongst whom I practiced almost exclusively.[60]

The few personal accounts or memoirs that remain from European (but also, if one views the quotation from Adalja above, Indian) medical men in East Africa support this dominant tendency to categorise Indians and Africans as one homogenous group, and endorsed health policies that protected white interests. Henry Albert Bödeker, one of the earliest European government medical officers in Kenya (served 1902–1913; 1914–1918, 1920), held indicative views.[61] He was scathing about the unsanitary habits of Indian shopkeepers and, irrespective of the damage it would do to family economies, favoured the radical measure of eviction against them.[62] Similarly, the views of the Irish private doctor, Roland Burkitt (uncle of Denis Burkitt, (1911–1993) of Burkitt's lymphoma fame) who arrived in Kenya in 1911, seem little different.[63] Burkitt, President of Kenya Branch of the British Medical Associate (BMA) in 1924 and 1932 was famously eccentric and extremely vocal about his opinions. Although two sources claim that he was not averse to treating private Indian patients—was even well-liked by them—his violently discriminatory opinions expressed in other circles about the Indian community of Kenya are nevertheless striking.[64]

At a mass meeting at the Theatre Royal, Nairobi in 1921 Burkitt provocatively asserted that 'God especially created the English race to rule if not the whole world at least that portion containing dark skinned races' and 'Christ's teaching of Brotherhood is nonsense'.[65] At the same meeting at the Theatre Royal in Nairobi in 1921 Burkitt spoke at length on the differences of races seen from a doctor's point of view. He ended his speech with the words 'why even in the simple matter of urination the difference between races is profound. The European male stands to urinate but the female sits, whilst with the Indians the reverse is true'. The meeting was reported to have 'dissolved into unrestrained laughter'.[66] In fact, Burkitt's frequent public outbursts against Indians were so out of step with the norm (which mostly preferred to hold these views tacitly, rather than explicitly before World War One), that several residents of Kenya publicly voiced their opposition to his views. The *East African Chronicle* in 1921 described Burkitt's views as a 'doctrine of racial hatred in the name of Christianity'.[67] A European resident of the colony, Mr O'Toole, was also sufficiently incensed to publish a letter to the editor of the *East African Standard* criticising Burkitt's contributions as 'ill advised and thoughtless'.[68]

Time and time again Indians were cited collectively as sources of diseases and ill health or uncongenial behaviours. But, although racial tensions undoubtedly underscored these discourses, the reality was more blurred than a simple dichotomy of white versus non-white might indicate. Indeed, some Indians can be seen to have spoken out against their own people, feeling that the lack of hygienic practices displayed by certain sectors of their populations warranted their bad reputation. One of the earliest recorded

Indian accounts of health conditions in Kenya from 1902 squarely blamed uneducated Indians for 'having no idea of cleanliness' and being responsible for 'white man's revulsion for dark people'.[69] Similarly, Indian doctors were shocked and frustrated by the habits of some segments of the Indian population in the township bazaars, declaring them to be 'ignorant as well as being prejudiced against change'.[70] Indian leaders, sensitive to Europeans scapegoating the whole community, were vociferous in demands for the severest punishment to Indians who were prosecuted for breach of sanitary regulations.[71] Again, Indian disunity contributed to the differences of opinion typified by these sorts of comments. Higher castes, felt themselves superior in both attitude and behaviour to members of the lower castes, and members of some religions displayed distinct biases against others.

Naturally, the European tendency to group Indians together despite differences in socio-economic status, caste and religion, also impacted the way Indian doctors were perceived. In general, even well-qualified Indians were not regarded by the Imperial government as suitable role models to pass on public health messages or to be appointed to prominent governmental medical roles. As Jeffery has argued with regards to the situation on the sub continent, Indians were held back from promotion despite impressive qualifications because they were routinely characterised as 'susceptible to insanitary practices themselves, and they lacked the conviction and status needed for producing change'.[72] This type of discourse was also evident within East Africa and was based upon a long history of routine discrimination. Especially since the hardening of attitudes towards Indians which occurred after the Indian mutiny of 1857, examples can be increasingly found of Europeans refusing to be treated by Indians, feeling that this was culturally inappropriate, especially in terms of European women being attended to by Indian men.[73] In the East African context this belief was compounded by the idea that only doctors in the most need of employment (and by implication those who were least competent) would ever really consider a career outside India. Indeed, in some sense this was borne out by reality as Indians too frequently complained about restrictions on the practice of medicine and law in East Africa that were felt to discourage 'high class Indians from settling'.[74]

This feeling of being somehow second class, despite qualifications, became explicitly evident from the legal rule that was put in place to regulate medical practice in Kenya in 1910: the Medical Practitioners and Dentists Ordinance. Under these regulations, two categories of doctors—'licensed' and 'registered'—were identified as legally authorised to practice in the Protectorate. These two types of doctor were not given the same rights, with Assistant Surgeons and Hospital Assistants (called Sub-Assistant Surgeons after 1910) employed by the Government Medical Department categorised in the lesser category of 'licensed' rather than 'registered' practitioners. In reality Indian doctors, even with GMC approved qualifications, were unable to become Medical Officers and many of the Indian doctors who joined the Medical Department as subordinate staff did not have GMC approved qualifications. Thus by practice and inference (and not by law) all Indian doctors in Government Medical Department were considered as inferior to Europeans.[75] Elite medicine was very much seen as the British preserve.[76]

As Clause 9(2) of the Ordinance explicitly stated, being 'licenced' meant that the right to practice was exclusively dependent upon being granted a licence by the PMO. Furthermore, the regulations stated that 'every licence issued under this Sub-Section may be any time be cancelled by the Principal Medical Officer and shall immediately expire' if the person ceased to be in Government service.[77] In short, Indians who fell under these terms were denied the opportunity to continue working in the Protectorate as private practitioners. Meanwhile 'registered' practitioners, Indians and Europeans, were entitled to be doctors in the colony, irrespective of whether they worked for the colonial medical department or not, because their qualifications were recognised by the GMC. Though the Ordinance did not explicitly discriminate against Indians the intention was clear as no licensed or registered Indian doctor was made a Medical Officer between 1910 and 1940.

Significantly the first draft of the Ordinance (which was submitted to a special committee of the Legislative Council in November 1909 after amendments were tabled in August 1909) contained no specific restrictions on the right of Assistant and Sub-Assistant Surgeons to practice after they had completed their government service. It was only at the insistence of European doctors and settler representatives, led by Delamere, that such specifically restrictive terms were included. Indeed, it is evident from the discussions undertaken by the special committee, which were published in the *East African Standard* in 1909 while the issue was debated, that PMO Milne held Indian doctors in high regard, feeling them to be an invaluable constituent of the Colonial Medical Service. In direct opposition of settler demands, Milne maintained that licenced Indian Hospital Assistants had to 'pass stiff examinations' and were often working independently in positions of large territorial responsibility. Besides which, to restrict the number of Indians in the Colonial Medical Service, Milne reminded people, would be folly as there were not enough European doctors to man all the Kenyan outstations.[78] Yet despite this line of argument the settlers got their way and it is notable that Jeevanjee, the only Indian representative of the Legco did not participate in any of the deliberations of the Council relating to the Ordinance to stop the proposal going through.[79] An insight into Jeevanjee's attitudes has been given by Zarina Patel. Patel claimed that Jeevanjee considered Legco attendance a hopeless task because he had to fight against 'tremendous odds'.[80] It was to await the passage of time for the issue to raise its head again in the heated debates over amendments to the Ordinance that became part of broader deliberations surrounding the rights and entitlements of Indians in Kenya during the 1919–1923 (see Chapter 6). Indian community representatives bitterly complained, through EAINAC, about these unfair stipulations.[81] These protests finally led to an amendment of the rules in 1921 to remove the restriction to practice privately post-colonial service employment as applied to Assistant Surgeons.[82]

The result of these generalised and specific attitudes towards Indians (as both doctors and patients) meant that many Europeans, by the First World War, increasingly regarded Indians in Kenya with suspicion as both 'pathological' (in the sense used by Warwick Anderson; see p.69) colonial subjects and culturally inappropriate doctors. Particularly when related to the growing class of educated professional Indians, this repeated trope became one of the key colonial dilemmas of empire. One of the key objectives of the Raj in India had been to introduce western standards and education. The problem was how to contain the ambitions of Indians once they had achieved this social and economic improvement? While educated Indian doctors, lawyers and teachers were useful for the running of empire, their presence (or worse a reliance on them) was also deeply erosive to embedded ideas of white racial superiority.[83]

The Priority of European Health

In order to understand these often-negative attitudes towards Indians, it needs to be understood that until the First World War, non-European health was simply not a priority in colonial Africa.[84] Tellingly, one of the first returns of the first Medical Officer of IBEAC in 1890 was 'glad to be able to report that the European community is in excellent health' though the Sudanese troops had suffered severely from fevers and pneumonia.[85] Similarly, Commissioner Eliot's Report of 1901 noted the death rate amongst the 'small population of Europeans in the last few years', omitting to make any reference to mortality or morbidity among the majority population.[86] Although African and Indians were labelled as the source of many diseases, they were nevertheless not considered important enough to direct colonial health policies towards their care.

It took some considerable time for the remit of health policy to widen and it was not until 1918 that PMO Dr Arthur Milne arranged for the appointment of the first European Medical Officer in a dispensary in an African reserve. It is noteworthy that the first systematic survey of the health of inhabitants of an African reserve (Digo) was not undertaken until 1928, thirty years after the establishment of the Medical Department.[87]

Yet, even after the initiation of the well publicized, and much celebrated, rural dispensary system, Annual Medical Reports until the late 1930s documented information which gave assurances about the good standard of health of the European community which was considered 'very much higher than that which prevails among either Asians or Africans.'[88] Furthermore, even as official policy extended to be more focused upon African health needs, Indians fell between two stools and became almost erased from official consciousness. The vague assessments presented in the 1934 Annual Medical Report were illustrative of the official attitude:

> It is difficult to say how Asians fared in Kenya in 1934. The majority of the Asian community are far from being well off, well housed or well acquainted even with all the more elementary laws of hygiene and the general impression which one gathers is that the average standard of health that prevails is poor.[89]

Given the evidently high levels of distain for Indian public health standards, it is hard to understand why so little official attention was focused upon improving their situation. John Langton Gilks, the PMO from 1921–1933, under whose tenure Indians medical staff was drastically reduced and barely mentioned (see Chapter 6), actively proclaimed his enthusiasm for re-orientating colonial medical strategy towards African health needs, and endorsed policies to recruit more African health personnel. That one non-white group should be suddenly favoured, while another became totally ignored needs explanation.

One factor might have been that the colonial administration was more worried about Indian power in the colony. Africans, although the majority, did not present such an articulate and comparatively organised community, at least not until the 1940s, and therefore (theoretically at least) presented more of a passive group on which to apply so-called colonial philanthropy. To help Indians might be to acknowledge them as a viable permanent community in Kenya, or worse, might empower them further to want to expand their economic and political power. Alternatively, winning the confidence of the African population, which was resentful of the imposition of hut tax and the coercion into joining the workforce, might have been judged to more strategically important in terms of successfully promoting the 'duty of trusteeship' of the British colonial regime. Most colonials did not regard East Africa to be the historic home of Indians, so—despite their economic and military contributions to British success—they came to be side-lined in policies concerning the future of the region. Furthermore, Indian traditions of self-help and of giving substantive philanthropic contributions to community projects, also served to take the pressure off the colonial government as they were thought to be a relatively self-sufficient group within society.

Whatever the motivation between the 1900s and the 1940s the government priorities of health (but also in other realms) decisively changed from an exclusive focus upon Europeans to one which also included consideration of the health of Africans.

Race, Indians and Medical Policy

These attitudes played out in practice through interconnected medical policy decisions in a two-pronged line of attack: namely the enforcement of racial segregation because of fears of cross-racial contamination and the monitoring, and sometimes even closure, of places where disease was regarded to be rife—which in the case of the Indian community of Kenya meant their bustling bazaars. In both cases, health decisions were underpinned by legislation and regulations that limited the activities and movements of populations regarded as being central to disease spread (Figure 4.1).

Since the earliest British colonial incursions segregation of European population from the natives 'had been an accepted axiom in the planning and laying out of all towns in the tropics'[90] In East Africa, it was a model of demographic organisation imported from India and one that figured prominently in town planning decisions.[91] Segregation touched upon the

Figure 4.1   Conditions behind River Road Indian bazaar, Nairobi, 1915
Credit: W.J. Simpson, *Report on the Sanitary Matters in the East Africa Protectorate, Uganda and Zanzibar*, London, Colonial Office, African No. 1025, 1915, The British Library Board, General Reference Collection B.S.7/35.

organisation of hospitals, residential areas, schools, sports clubs, cemeteries and jobs.[92] Similarly, parliamentary and municipal seats were all allocated on racial lines.[93] From the times of the earliest plans for Nairobi, segregation was taken for granted. Not only did this mean that the broadly different ethnic groups were physically separated from each other, but it meant that the best land, and best facilities were reserved for the Europeans.

An early description by Colonel T. Gracey who visited Nairobi in 1900 at the behest of the Whitehall-based Railways Committee described the houses of the superior European officers as excellent dwellings 'of masonry and tiled roofs on the brow of a hill', whereas the site for the subordinates was judged as 'suitable' though he witnessed the prevalence of rats, jiggers and fleas in much of the accommodation making it 'quite unfit for habitation'.[94] Upon seeing the inequity of conditions Gracey made recommendations for urgent action to raise the floors of the dwellings for 'natives of India', improve the drainage and conservancy of the settlements. Sanitary measures, he observed, 'cannot be organised too early' if the health of the residents was to be preserved. He additionally urged the appointment of a Sanitary Committee with a medical officer to oversee the necessary improvements.[95] Gracey was not alone; other medical reports between 1900 and 1907 directly criticised the undemocratic distribution of resources in the planning of Nairobi. The 'total unsuitability' of the site was condemned particularly for the unhealthy swampy nature of the site in which the railway subordinate staff were expected to reside.[96]

Such concerns over the planning of Nairobi filtered back to London and led the Secretary of State to commission an expert from London, G.B. Williams, to further investigate the matter. The resulting 60 page report pulled no punches and concluded that the decision to locate the railway workshops and non-European quarters on black soil plain rather than the hills would only increase health problems.[97] Williams' preferred solution was based on a major relocation scheme but he also detailed the option to move the Indian bazaar from the proximity of the swamp to a new location on the other side of the river where proper investment could be made into its drainage and sewerage systems. Although his recommendations would have helped the conditions of the bazaar the usual level of paternalism permeated his recommendations. Indians, he stated, should be helped because the 'ideas of the Asiatics in the Protectorate are so far removed from those usually held amongst civilised people'.[98] Yet, despite these protests, the Railways organisation refused to spend money to move non-Europeans to more suitable areas and the plan never came to fruition, an inaction that Jeevanjee later noted as an early failure of colonial public health policy towards the Indian community.[99]

Frustration over the lack of investment in improving non-European housing and amenities continued, with members of the early European medical department, such as Clare Aveling Wiggins (served Kenya 1901–1909), also recommending the removal of the Indian bazaar to a more suitable location with better facilities. Despite Wiggins doggedly pursuing this project throughout his term in East Africa, his proposals were never taken up because they were again claimed to be too costly.[100] The Medical Department additionally claimed that Indians opposed the proposal. Again the Indian characteristic of avarice was seen as being at the heart of this. Indian owners were said to prefer to 'stick closely to old feus [sic. mostly likely "fence"] probably in the hope of enforced removal and increased compensation thereby'.[101] This theme of Indians being responsible for their own fate was constantly reiterated. Indian landlords were accused of subdividing and subletting their limited floor space in order to get higher rents, while simultaneously sacrificing the health of their own community members. The informal subletting system did undoubtedly exacerbate conditions, although broader factors encouraged this, namely the fact that residential and ownership restrictions on land confined Indians to a small acreage of the municipality.[102]

One of the greatest disease fears closely associated with the Indian community, and their propensity for living in unhygienic and cramped conditions, was plague. The outbreaks of several plague epidemics at the turn of the century coincided with the establishment of British rule in East Africa and led to great alarm within the European community eliciting draconian responses from the colonial administration. As ever, the emphasis was on safeguarding the health of the European residents rather than on implementing measures to secure the welfare of the whole population.[103] A number of ordinances were hurriedly passed (see p.82) in order to control plague spread, principally through limiting movements, enforcing inoculations, sanctioning evictions and property destruction. Indians were felt to be instrumental in plague transmission not only because of their insanitary lifestyles, but also because they could not be trusted.

The medical department claimed that Indians concealed infected cases and would not allow preventative measures such as taking of spleen smears.[104] Furthermore, Indian trading activities were thought to encourage plague, as plague-infested rats were thought to have been imported on ships with goods from India, but the traders themselves were reluctant to acknowledge plague victims, as they feared the impact of such a diagnosis on their livelihood. On one occasion at a meeting in Nairobi in October 1916, Indian community representatives publicly supported the motion to restrict the movements of suspected Asian plague victims, but the Indian Association subsequently wrote to the Secretariat of the Protectorate to point out that the two Goans who were present at the meeting could not be seen as representatives of the Indian Association more generally.[105]

The notion that Indians carried plague into East Africa from India was not unanimously held in European circles. Even Arthur Milne, PMO from 1909–1921, noted that it was likely plague in the region pre-dated the arrival of the 'European or the arrival of the railway, with its close communication with plague-stricken India'.[106] Another Medical Officer, Dr Ronald Nelson Hunter (served Kenya, 1921–1934), who carried out a large study on the epidemiology of plague in Kenya in 1928 concluded that 'of actual fact and unprejudiced observation upon which to draw definite conclusions there is little.'[107]

Yet despite these assertions, the associations continued and action to tangibly improve conditions for Indian life were not forthcoming. This was despite increasingly urgent calls to sort out the awful health condition of the non-European sectors of Nairobi. An article in *The Leader* newspaper drew attention to the open drains in and near Government Road, describing them as having their origins in the Indian bazaar.[108] The lack of progress with drainage was also noted in the Annual Medical report of 1911 that stated that bazaar conditions were still as bad as reported by Bransby Williams in 1907.[109] So despite the continued appearance of reports about the shockingly bad location and poor health of the Indian bazaar, it remained as it was originally sited, with little tangible improvements to its condition.[110]

Bertham Walter Cherrett the Medical Officer (served Kenya, 1910–1918) was shaken by the state of affairs in Nairobi.[111]

> The whole place is in a shocking insanitary condition, in fact it is a huge evil smelling swamp, due to escape of liquid refuse from the house, drains and evil flowing slumps. The cause of the trouble is that there is no drain in River Road or sanitary lanes except an earth one that we have recently constructed. Houses have been built or are being built all over the estate and not the slightest provision for drainage has been attempted on the part of the authorities.

Although the focus of most criticisms was on Nairobi, Mombasa was also frequently spoken about in similar terms.[112] A Medical Department report of 1914 recommended that strong action should be taken to improve the living and working conditions of the Indian inhabitants of the coast, although quite squarely blaming them for their lamentable health predicament, declaring that '[t]he Indian population has to be taught the rudiments of municipal cleanliness and no matter how unpopular such a measure should prove it had to be carried out, opposition or no opposition'.[113] In all of this responsibility was directed at the Indians themselves, rather than at the colonial government's policy priorities. Indeed, in some extreme examples, it was felt that Indians deserved their fate. One Medical Officer went so far as to report in 1911 that it 'was satisfactory to note that not a penny of compensation was paid' to the Indian residents against whom harsh anti-contagion measures were enforced in the bazaars.[114]

By the time Henry Belfield took up his term as Governor in 1912 the state of inaction concerning health conditions of the non-white population in urban centres was clearly critical. After consulting the Secretary of State,

Belfield agreed to seek the help of an internationally renowned sanitation expert to advise the East Africa Protectorate on necessary sanitary measures. The person selected for this task in April 1913 was Professor W.J. Simpson. Simpson, a member of the Advisory Committee on Tropical Medicine, was the natural choice for the job. By the time of his appointment he had already been on missions to investigate sanitary conditions in Gold Coast, Sierra Leone and Nigeria on behalf of the British colonial government.[115] Most importantly, in terms of directly influencing his findings about the Kenyan Indian community, Simpson had also been appointed in his career previous as the health officer of Calcutta, a position in which he quite brusquely vocalised his antipathy to the attitudes of the Indian residents of the city towards sanitation and hygiene and was embroiled in no small amount of controversy for his opposition to the Calcutta Corporation.[116] With these experiences no doubt firmly implanted in his memory, Simpson visited Kenya in 1914 travelling extensively throughout the country for six months to make his health assessment. In compiling his decisive account he drew upon many interviews, although notably few with non-Europeans.[117]

Simpson's final report of 1915, illustrated with town plans and photographs, was unambiguous in its support of racial segregation and became much cited as official justification for colonial health policies thereafter. He was unswerving in his recommendations:

> Lack of control over buildings, streets and laws and over the general growth and development of towns and trade centres in East Africa and Uganda, combined with the intermingling, in the same quarters of towns and trade centres, of races with different customs and habits, accounts for many of insanitary conditions in them and for the extension of the disease from one race to another... Also that the diseases to which these different races are respectively liable are readily transferable to the European and vice versa, a result specially liable to occur when their dwellings are near each other ... it is absolutely essential that in every town and the trade centre town planning should provide well defined and separate quarters or wards for Europeans, Asiatics and Africans, as well as those divisions which are necessary in a town of one nationality and race, and that there should be a neutral belt of open unoccupied country of at least 300 yards in width between European residences and those of the Asiatic and African.[118]

The Simpson Report (Figure 4.2) was reviewed several times at the Nairobi Municipal Committee between the middle of 1914 and the end of 1915. Eight European officials as well as two representatives of the Indian community (Dr Rozendo Ribeiro (the Goan representative) and Mr P.K. Ghandi (the Indian representative)) attended these sessions and within this forum the policies that should be undertaken with regards to the bazaars were

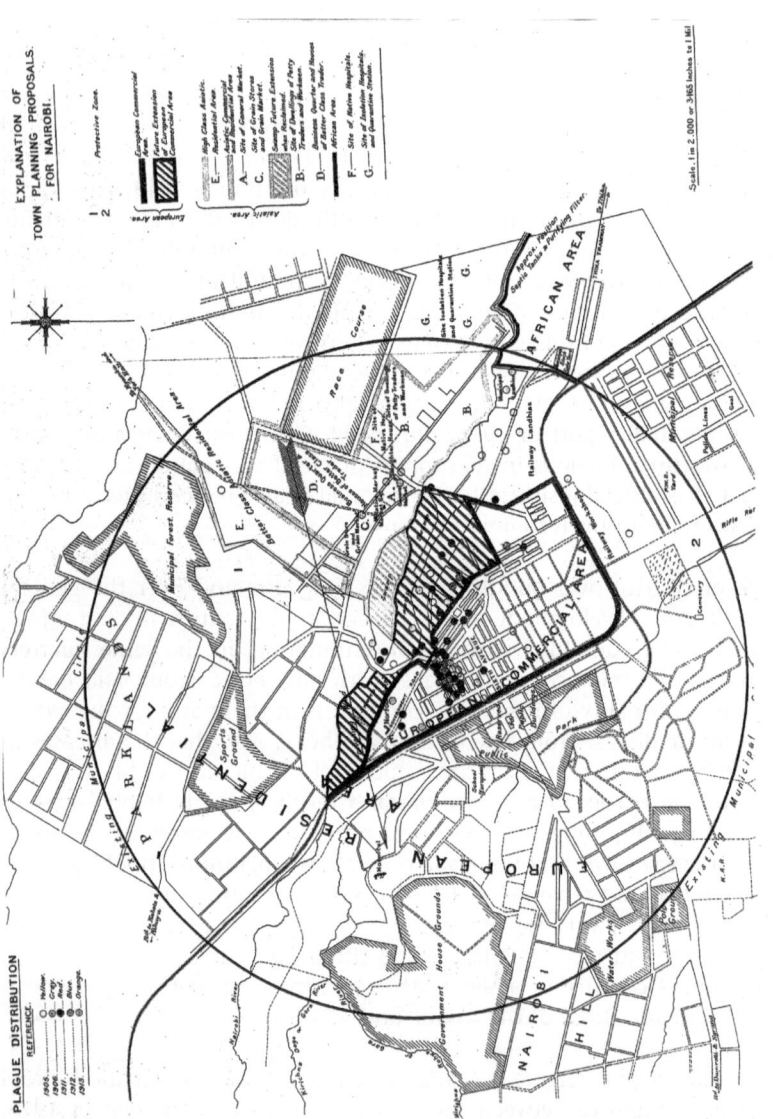

Figure 4.2 Map of segregated Nairobi, 1915
Credit: W.J. Simpson, *Report on the Sanitary Matters in the East Africa Protectorate, Uganda and Zanzibar*, London, Colonial Office, African No. 1025, 1915, The British Library Board, General Reference Collection B.S.7/35.

extensively debated.[119] As the meetings progressed and unfolded, Ribeiro and Ghandi could be seen to have held different views. By the meeting of 17 April 1915 Ghandi, anticipating his absence had tabled his unequivocal opposition to segregation between 'Asians and Europeans' in advance. At the meeting, however, the committee, including Ribeiro, endorsed the general principle of segregation enshrined in the Simpson Report. In their decision, they cited the medical evidence of Drs Milne, Burkitt, Haran, Radford and the opinion of Alidina Visram, a prominent Asian trader, as justification.[120] The Municipal Committee, whilst backing the general principle of segregation, stopped at directly supporting absolute segregation, declaring this full enactment of the ideal to be too expensive to implement in the commercial area.[121]

Despite Ribeiro's consent, the conclusions of the Municipal Committee in support of the Simpson Report were opposed by the vast majority of Indians. Segregation was felt to be irrational, unfair and impractical. Gandhi was so incensed by the decision taken in his absence that he resigned from the Municipal Committee as a direct result of their decision to support segregative measures.

The debate vigorously continued with a new Goan committee member, J.M. Campos, playing a prominent role between 1915 and 1918. Campos vocally opposed the policies dictated by the European majority of the Committee, adopting a radically different position on segregation than his predecessor:

> Asiatic interests have not received the attention they deserve. Add to this his various disabilities such as the lack of an Asiatic hospital, which necessitates that the Asiatic converts his bedroom into a hospital ward; the inevitable dumping of thousands of Africans on this area for want of a native location; defective drainage whereby owners are required to construct drains round their houses without at times an outlet into the public main drains, with the result that filth is allowed to pollute the surroundings or a public road like a portion of Race Course Road facing the Metropole Hotel where stagnant water is remaining to be washed off by rains.[122]

Campos was adamant the discriminatory regulations imposed on Indians were the main reason for their high rates of diseases and poor health. He also pinpointed the practical limitations imposed on Indian access to land, which was responsible for the shortage of housing, giving rise to overcrowding, high rents and poor sanitation. How could it be, he argued in 1920, that 'a plot in Major Grogan's estate in River Road measuring 50 by 75 feet costs about 10,000 pounds whereas a European can acquire an acre on the Hill or Parklands for 100 pounds. The difference is appalling.'[123] One European councillor, Mr J. Riddell, swayed by the evidence presented, even declared his support for Campos, agreeing that it was wrong that Indians should be confined to a 'miserable patch of a few acres', while Europeans had monopoly on the thousands of acres of ideal and fertile land of the Hill and Parklands.[124]

Five years after the Nairobi Municipality deliberations on the Simpson Report, the neglected condition of the bazaar was still a contentious political issue in Kenya, one so powerful that debates reverberated in London and Delhi right up until the 1930s, and continued to be debated within the Kenyan press.[125] But despite controversy, the Simpson Report became used as the authoritative medical reference to defend Kenyan policies of racial segregation. For example, in response to Indian complaints the Colonial Secretary, Viscount Milner, invoked Simpson's recommendations as a justification for the Public Health Ordinance of 1921 that officially endorsed segregation.[126] This, Milner said, was based on the opinion of 'the highest living authority on tropical medicine'.[127] Furthermore, it was a blessing to this colony, as 'Indians have been the chief carrier of plague and it is mainly due to them the disease has spread over the country'.[128] Finally and with no sense of irony, Milner went on to claim that the separation of races would result in the avoidance of 'social conflict' causing instead 'social convenience and social peace'.[129]

Public Health Ordinance of 1921

Between 1900 and 1913, a number of public health ordinances and amendments were passed, the most important being Plague (1902), Infectious Diseases (1903), Sleeping Sickness (1903), Plague and Cholera (1907), Vaccination (1912), Mosquito and Malaria in Townships (1912) and Immigration Restriction (1913).[130] Limits were placed on the movements of Indians and Africans while Europeans were not subject to the same restrictions and detentions.[131] Under the terms of the 1903 Ordinance, Non-Europeans could be subject to a prison term of up to one month, if they did not comply with its terms.[132] Similarly the Ordinance exempted the non-European staff of Europeans from travel restrictions, showing how flexible policy could when it was deemed inconvenient to Europeans.[133] It has been argued for some time by the medical department, that all these various ordinances should be amalgamated in order to simplify procedures. Although this idea was first mooted in 1911, it took a decade for the final Public Health Bill of 1921 to solidify.[134]

The first direct recommendation came in the form of a short 11 section Bill introduced in 1918 that lapsed. When John Langton Gilks came to the position of Acting PMO in 1920 he quickly resurrected the matter. Gilks looked to South Africa for the inspiration, letting his deputy, Dr Arthur Rutherford Paterson (served Kenya, 1920–1941), who had just returned from a trip to the Dominion, to model the Kenya Ordinance on the situation there. As a result of Paterson's positive experience 'the South African Ordinance was hurriedly adopted to local conditions' in 1921. Due to settler pressure and the Governor the Bill incorporated clauses supporting racial segregation in townships. The Health Ordinance was described by W.M. Ross,

Head of the Public Works Department as having 'no other object than to empower Government to give effect to segregation of races'. Moreover, it seemed to him to be a thinly veiled attempt to secure 'political end by adopting a moral attitude on the subject of public health and well-being.'[135] When Gilks came to present his account of the path of the Ordinance in his retirement he omitted to mention that that the racial segregation clauses, which the settler leaders insisted upon stirred up a huge controversy in Kenyan and British governmental circles.[136]

In Kenya the proposed ordinance was referred to a subcommittee of the Legislative Council, which was headed by Gilks, and contained only one council member nominated by the Indian community, V.V. Phadke.[137] After much deliberation, the subcommittee unanimously decided to delete the clause that related to segregation, Clause Fifteen. Gilks seemed at first to see no major problem with this and agreed with the subcommittee that public health issues in the non-European areas could be resolved by adequate sanitary regulations alone.[138] However, the settler community responded powerfully, with Lord Delamere leading a fierce attack on the proposal to remove the clause.[139] At first Gilks, supported by Public Works head, Ross, stood his ground, arguing that the 'stringent sanitation regulations in the Bill' already afforded 'ample safeguard against the professed fears'.[140] It did not take long, however, for Gilks to change his position in alignment with both the Governor (Northey) and the settler community. In the debates that ensued, one settler directly attacked Gilks for not being 'in the class of Simpson', upon which Gilks quickly capitulated with the words 'I cannot say anything as the Committee is against me'.[141] The motion to delete Clause Fifteen was defeated by 20 to 1.[142] Significantly the vote was taken in Phadke's absence, leaving only Ross to put up a spirited and robust defence of opposition to the Clause.[143]

As it turned out it was authorities in London that finally put a halt to the contentious clause. Winston Churchill, the Colonial Secretary, well versed in the politics of Kenya sensitive to the parliamentary sensitivity on the Indian Question and having seen the hazards of the South African model, made sure, in 1921 the controversial segregation clause was deleted before the Bill received Royal assent. The Ordinance was sent back to the Kenyan Governor for the removal of the relevant clauses.[144] As Governor Northey was not in the country at the time, Acting Governor Notley had no option than to pass the amended bill (without any segregation clauses) on 6 September 1921, against the wishes of the settlers. Intriguingly, Churchill ensured that Northey, who had displayed embarrassingly blatant pro settler policies on a number of occasions, was finally removed from Kenya in June 1922.[145]

Racial segregation of township areas in Kenya did not end with the removal of Clause Fifteen. Europeans found other ways of maintaining *de facto* segregation and the administration did little to intervene and ensure compliance. Instead subtle ways were found to conform to most resident

Europeans' preferences to live in a segregated society, for example by turning down planning applications by non-Europeans, and refusing to sell land to Indians. Officials in Whitehall were aware that the law was being circumvented, but were content to turn a blind eye. A handwritten comment on an internal Colonial Office memo is revealing: tacit support to these continued practices of segregation was provided as long as a way could be found 'of avoiding official correspondence' on the subject.[146]

***

It has been argued that had the Indians in Kenya been less heterogeneous, less insular, less conscious of caste and religion they would have been more successful in gaining political rights.[147] Indians, stuck in the middle, between white and black, found their rights were never championed in London with the same persistence and vehemence as African rights and the duties of the colonisers in terms of trusteeship. Indeed, once it had been reconfirmed between 1919 and 1922 not to offer land grants to Indians in the Kenya Highlands, and once their demand for rights and political representation equal to Europeans was officially curbed in the Devonshire Declaration of 1923, their plight in East Africa was not to receive serious political attention again until the end of the colonial period despite periodic efforts by the politically split community.

It is in this context that Indian doctors found themselves increasingly marginalised. Sentiments such as 'whenever one finds the Indian in Africa, he appears so dirty' were commonplace and profoundly affected the way Indians were regarded not only as patients, but also as private and government doctors.[148]

Compounding and extending these racialized discourses was the discriminatory nature of much colonial health legislation. Although the Colonial Office in London eventually tried to limit the explicit enforced segregation of non-whites from whites, the practice nevertheless continued in colonial Kenya. Even after policies started to be more inclusive to African needs, few financial resources were directed at improving the situation of the Indian community.[149] From the Colonial Office perspective, Indians were a bothersome sector of the community in Kenya and one that did not fit easily with new colonial emphasis on expanding African rights. The irony of this was not lost on the Indian leaders. Jeevanjee pointed out in 1915 the illogicality of identifying Indians as in need of health reform but simultaneously denying them the funds for improvement.[150] Moreover, Indian leaders were fully aware that labelling them as the harbourers of disease was a means of legitimising an increasingly anti-Indian colonial political agenda. By the time of 1923 and Devonshire their fate was already all but sealed. This broader social and political context makes the quiet, yet forcible, exclusion of Indians doctors from the Colonial Medical Service all the more explicable.

5
Indians in the Colonial Medical Service

> [T]he Government had no option but to employ assistant
> and sub assistant surgeons in certain parts of the colony,
> and that when such officials had rendered meritorious ser-
> vice, it would appear to be inequitable and indeed contrary
> to public interest in view of the small number of European
> doctors, that they should be debarred from continuing to
> practice.[1]

It is very curious that Indian doctors have been ignored in the colonial
medical history of Africa. Although scholars have examined the European
doctors of the various African Colonial Medical Services, non-white personnel have received comparatively little attention.[2] Some studies have looked
at the lower ranked African personnel, but the experiences of the Indian
doctors that worked contemporaneously in higher status positions have
received no attention.[3] Indeed, the studies of non-European doctors in
the Indian and African Empires have been so infrequent that one could be
forgiven for thinking that Indians and Africans had no access to medical
education and therefore were not employed in anything other than positions that did not require professional qualifications. Mark Harrison has
briefly touched upon the Indian staff cohort of the Indian Medical Service
(IMS) and Ryan Johnson has examined the progressive exclusion of the
small numbers of black doctors from the West African Medical Service from
the beginnings of the twentieth century.[4] But these studies are exceptions,
with the majority of the historiography focussing on either the white elites
or the black subordinates, with little or no acknowledgement of non-white
qualified practitioners. Indeed, even the broad histories of the East African
medical administration written by Anne Beck in the 1960s and 1970s did
not consider the work of Indians within the colonial health department.[5]

This exclusion of Indians was reflected in a number of contemporary
descriptions. Lord Hailey, describing the medical services of Africa in his *An
African Survey* of 1938 also described a health structure made up principally

of African subordinates, with the minority of Europeans in the leadership roles:

> The medical service ... can be envisaged as a pyramid, the base of which is formed by a large body of African Subordinate Staff, the apex by the fully qualified Medical Officers, and the central part by the African 'auxiliary' doctors or 'medical aids'.[6]

These persistent omissions, both in historical and modern retellings, failed to recognise that by 1915 (until 1922); Indian doctors were almost double in number to the European doctors working in East African Medical Service. They omit to describe the key roles Indians had in administering British East Africa and fail to depict the multifaceted administrative reality than captured in the stereotype of a 'country of blacks ruled by blues'.[7]

***

Set up of the Colonial Medical Department

Like many other legal and administrative frameworks adopted in East Africa, India provided the blueprint and the fledgling medical department of the East Africa Protectorate was directly modelled on the IMS. Just as in the IMS, it consisted of a head of the medical department, the Principal Medical Officer (PMO), European Medical Officers or Senior Medical Officers (MOs or SMOs—dependent upon rank) as the senior staff who supervised and provided policy guidance, while Indian Assistant Surgeons and Sub-Assistant Surgeons performed the bulk of the routine medical activities. In East Africa, this core team was further helped by African medical assistants, dressers and European and non-European Nursing sisters. Additional to the Indian Assistant Surgeons and Sub-Assistant Surgeons were a small but significant cohort of Indian support staff: chiefly consisting of administrative clerks and compounders (a formulator of drugs below that of a pharmacist in status and training).[8] In both of these non-qualified positions Goans had a dominant presence, in line with their reputation amongst the British for being efficient and trustworthy.

But, although strong organisational similarities existed between the two health services of the Indian and African British Empires, some crucial differences were present. For example doctors joined the IMS via a competitive examination, while the EAMS recruited its doctors on the basis of performance at interview. This led to the (by-in-large warranted) impression that the IMS was the more prestigious service, as only exceptionally gifted candidates were able to pass the difficult entrance before 1914.[9] Furthermore, in East Africa all healthcare operations were centralised upon the one government medical service, while in India the IMS was the most important of three different medical services, each targeted at different sectors of the population.[10] The IMS, catered principally for European patients and recruited mostly European doctors, the Central and Provincial Medical Services (CPMS) and the Subordinate Medical Service (SMS), dealt mainly with non-European patients and relied on locally recruited Indian physicians.[11] As the only medical service in East Africa, the EAMS arguably came under more focused pressure to provide comprehensive geographical healthcare coverage, as there were no alternative services supporting the majority indigenous poor.

In both the IMS and EAMS Indian doctors were an assumed component of the health personnel. Although Indian recruits were in the minority in the IMS they nevertheless were, since 1860, theoretically entitled to apply for inclusion in the government medical services on the same terms as their white peers.[12] An 1880 list of the 26 successful entry candidates to IMS, for example, included 8 Indian names.[13] This was quite an impressive proportion when the other circumstances that militated against Indians joining the IMS are considered. Above all, it was difficult for Indians to gain access to the financial resources to take a trip to London to participate in the IMS entrance exam.[14] Furthermore, even if they achieved a position, European cultural prejudices meant that most colonials preferred not to be treated by Indians, but rather by practitioners 'of their own race'.[15]

In East Africa, in contrast, Indian Assistant Surgeons and Sub-Assistant Surgeons (Hospital Assistants prior to 1910) were present in the medical department as a much more obvious force since its inception. As discussed in Chapter 3, Indians were part of the medical provision in the construction of the Uganda Railway and were immediately part of the colonial medical department, which was founded in 1895 and four IBEAC Medical Officers had their appointments transferred to the new Colonial Medical Service.[16] Some of the earliest Indian medical staff included Edward Oorloff, who had joined the medical department in 1897, E.W. Rodrigo and G.P. Vinod who commenced their employment in 1898 and Maula Buksh who joined the department in 1899.[17] It was apparent from very early on that, even if the healthcare focus was primarily on the needs of the European community, insufficient numbers of European doctors were available and the department needed supplementary staffing.[18] The shortage was to become even more apparent when military campaigns also intervened to take Medical Officers away from their official posts. Between 1897 and 1900 the military 'struggle in the North', for example, was said to have consumed all the available European medical personnel.[19] Furthermore, as Crozier has pointed out, the East African Colonial Medical Service was by no means a popular career and the service had its own recruitment difficulties. The job, although attractive to some, was nevertheless perceived as something of a last resort for the newly qualified British doctor. Pay, although stable, was low and working conditions were perceived as being both inadequate and uncongenial.[20] Given the relative unpopularity of a medical career in Africa the inclusion of

Indian doctors in the Colonial Medical Service was a natural and explicable way of compensating for staff shortfalls at a lower cost.

Correspondence between Kenya, London and India confirms a heavy dependence on Indian staff for the provision of medical services between 1891 and 1922. A letter from the PMO in January 1907 urged the Colonial Office and the India Office for stronger action to secure Indian medical staff.[21] The concerns voiced were quite commonplace for the time:

> Having regard to difficulty of procuring Hospital Subordinates for the Medical Department of East Africa and Uganda through Agent General in India I have the honour to suggest that the Colonial Office be requested to open negotiations with the Government of India to obtain loan of Assistant Surgeons and Hospital Assistants for service in East Africa and Uganda. There are at present two vacancies for Assistant Surgeons and two for Hospital Assistants, and after April next there will be vacancies for five additional Hospital Assistants who should arrive as soon after that date as possible.[22]

There were further telex exchanges spelling out the urgent need to negotiate a loan of subordinate medical staff from India, throughout 1907.[23] In general the India Office's reluctance to quickly comply was a cause of agitation amongst East African colonial administrators. Reminders were frequently sent to India emphasising the urgency of requests and stipulating that experienced, qualified and higher grade Indian Hospital Assistants were required.

The same themes can be discerned in earlier and subsequent years. The 1908 Annual Medical Report highlighted the shortage of Indian subordinate medical staff; bluntly stating 'as usual there was great difficulty in keeping staff to required numbers.'[24] In the same report, the PMO praised the work of the Assistant and Sub-Assistant Surgeons and postulated that the lack of Indian candidates offering their services in Kenya was because 'pay and prospects offered are insufficient'[25] Indeed until 1922 this leitmotiv continued to run throughout the correspondence and *Annual Medical Reports* of Medical Officers Mackinnon, Moffat, Will and Milne. In 1895 Mackinnon had made a special request for the extra recruitment of 'very capable men for the ordinary diseases of these countries' and Milne recollected that during this period, when the medical department was just getting on its feet, that many Indian Surgeons and Sub-Assistant Surgeons had devoted 'faithful service' to the cause.[26] In 1921, even John Langton Gilks, the PMO who was to actively squeeze Indians out of the Colonial Medical Service employment, had remarked that there were insufficient numbers of Indian medical staff in his first year at the headship of the medical department.[27] At one stage the shortage of subordinate staff from India was considered so serious that the Colonial Office contemplated starting a recruitment drive in other colonial dependencies, such as Egypt, although a lack of appropriately qualified doctors available for export meant that these efforts were unsuccessful.[28] Worries about adequate numbers were further compounded by concerns that the quality of candidates despatched for service in Kenya was not so high. Particularly, there was concern that India was reserving the best candidates for work in its own country, and only offering the lower quality staff for overseas service.[29]

The idea to recruit medical assistants from India was also stimulated by the fact that no equivalently trained medical workforce existed within East Africa at that time. Whereas the medical training of Indians in western medicine had been underway since the first third of the nineteenth century, this was not the case for East Africans where access to even a rudimentary medical training did not begin until 1917 (at Mengo Hospital, Uganda). In 1923 a six-year training course for Senior African Medical Assistants was initiated at Makerere training centre (later University), but it was to take until 1931 for African doctors to graduate from this institution. Even then, these doctors although licenced to work for the government medical department, were denied the right to practice privately, even if they were to stop working for the government medical department. Indeed, this privilege that was not granted to African doctors until 1950, and then only after they had first undertaken a satisfactory period of employment in government service.[30] In Kenya the training of African staff was regarded as much more difficult than in Uganda because of the lack of Kenyan recruits who had received a secondary school education.

Terms and conditions of employment

It is not known when the first formal regulations concerning the employment of Indian subordinate medical staff in East Africa Protectorate and Uganda were produced. Regrettably no copy of this document can be found. Instead, the earliest set of employment regulations that have been located derive from 1917, presumably tightening up the criteria and contractual obligations that were outlined in the original guide.[31] Even though we cannot have the benefit of seeing the original guidelines, the 1917 terms and conditions nevertheless provide an important overview of policies and staffing preferences with regards to Indian medical recruitment. It seems reasonable to suggest that the most of the views expressed within this document were ones that had been *de facto* adhered to since the beginnings of the government medical department:

2. *Qualifications:*
   (a) Good character, physique, qualified in English.
   (b) Religion, immaterial
   (c) Assistant Surgeons must possess a University Degree or Medical Diploma.

(d) Sub-Assistant Surgeons must possess Certificates of Competency from recognised Indian colleges and schools, list of which to be kept and supplied to the Agents in Bombay.

3. *Period of Engagement.*

Three years from the day on which the employee lands at Mombasa until the day of his departure therefrom on his return to India. This term is subject to extension owing to exigencies of the service.

4. *Pay.*

(1) Assistant Surgeons Rs. 250/- per month rising to Rs. 300/- per month by annual increments of Rs. 25/- per month.

(2) Sub-Assistant Surgeons

(a) Senior Sub-Assistant Surgeons Rs. 220/- per month rising to Rs. 240/- per month by annual increments of Rs. 10/- per month.

(b) Sub-Assistant Surgeons Rs. 200/- per month by annual increments of Rs. 10/- per month.

(3) Compounders Rs. 75/- per month by annual increments of Rs. 5/- per month....

5. *Quarters.*

Government quarters free of rent but not of rates and taxes or other similar outgoings when available but when no such quarters are available no allowance in lieu will be granted. When free quarters are enjoyed the value of the quarters is not reckoned as part of the pensionable emoluments of the post.[32]

Additional stipulations were made with regards to conditions for leave, passages and pensions. Leave was to be three months on half pay after 33 months of continuous service or five months on half pay after 43 months continuous service. Second class sea passage was paid between India and Mombasa for both Assistant Surgeons and Sub-Assistant Surgeons: although this allowance was not extended for any accompanying family. All posts of Assistant Surgeons and Sub-Assistant Surgeons were in principle pensionable. However Indian holders of the posts were considered to be on probation for three years and were only eligible for pension benefits thereafter if their work was passed as being considered satisfactory by the PMO. Finally, it was written into their contracts that all Indian medical personnel had to be available for military duty at the discretion of the PMO, despite their employment in the civil medical department (unsurprisingly, despite this clause, no provision for injury pensions were made in their contracts).[33]

As would be expected, these conditions of employment were much less favourable than those offered to European Medical Officers. Nonetheless that fact that they were formalised offers further contributory evidence that Indian staff were regarded as an integral part of the colonial medical department and, at least in 1917, were anticipated as remaining so. The vagueness of some of the recruitment stipulations, however, is nonetheless quite striking. The exact qualifications needed for entry into the service, for example, were only imprecisely addressed in the 1917 guide, with no mention made of the specific medical degrees that doctors were expected to have achieved. As the quotation above shows, in 1917 Assistant Surgeons were required to have achieved a 'university degree or medical diploma', while Sub-Assistant Surgeons needed to have 'a certificate of competency'. As this was at a period when several Indian medical colleges offered medical degrees recognised in the UK, it is possible that the criteria were kept deliberately vague as a means of giving the PMO some leeway and room for personal discretion when granting appointments. Certainly in the very early years of the colonial medical establishment occasionally individuals were taken on without having achieved medical qualifications. There were exceptional instances of a compounder being promoted to the rank of a Sub-Assistant surgeon (for example, the case of S.F. Da Costa).[34] Whatever the reason to not be too specific about entry requirements was, some reticence to make minimum qualifications too specific was in evidence. In fact, qualifications to be an Assistant or Sub-Assistant Surgeon were not actually formally specified in the Medical Practitioners Registration Ordinance of 1910 and left the decision in the hand of the PMO. Before that point, the declared recruitment standards of the colonial medical department were even more inexplicit; namely, to seek 'experienced and fully qualified' individuals from specific Indian schools that had high entry standards rejecting those of inferior education, such as 'class 4' subordinates.[35] In reality Assistant Surgeons were expected to possess a university degree or diploma and Sub-Assistant Surgeons a certificate from recognised Indian colleges or schools. The 1917 guidelines merely formalised the entrance qualifications that had existed. Under these new terms, the minimum qualification for an Assistant Surgeon was a medical diploma, or Licentiateship, which was achieved after a minimum of three years medical training in India

More significantly the racial climate of the time, indeed right up until independence, meant that even Indians doctors who had achieved higher than this stipulated minimum and had equivalent qualifications and experience to their European colleagues were unable to become full Medical Officers at equivalent salaries. The unfairness of this became particularly apparent as the twentieth century progressed and an increasing number of medical students in India took the MBBS degree, which was—theoretically at least—regarded on parity with the General Medical Council (GMC) approved MBBS that could be taken in the UK.[36] It is evident that some extremely experienced and well-regarded Indians practised medicine at the Assistant or Sub-Assistant Surgeon level despite having qualifications equal to their senior European counterparts. This was evidently not palatable to all. Dr A.C. L.de Sousa, for example (who subsequently became a well-known general practitioner and public figure) resigned from Government service specifically because of lack of career prospects for him to be promoted.[37]

Dr Nair, who worked in Kenya between c.1918 and c.1928, provides another example. Despite the fact that he travelled to the UK and obtained the qualifications of LM (Dublin) and DTM (1920), he found to his frustration that he could never be promoted beyond the grade of Assistant surgeon.[38] Similarly, Dr Bakhtawar Singh, who had qualified with the GMC recognised MBBS degree from Punjab University (Lahore) in 1929, worked for several years in the Kenya Medical Department as an Assistant Surgeon with no prospect for promotion despite the fact his degree was at the same level as that possessed by his European superiors.[39]

The issues of remuneration, inequality of status compared to Europeans and limited prospects for advancement were persistent bones of contention for the Indian doctors throughout the colonial period, but of all of their contractual terms, perhaps issues of pay were the most persistently debated. Although Indians working in East Africa governmental service received more than they would have working in the CPMS and SMS, the debates over the inadequate remuneration of Indian doctors in East Africa mirrored the long history of the pay issue for native doctors in India, where salaries were also felt to be problematically low.[40] Notably however, the Colonial Service was thought of as poorly paid by Europeans as well as by Indians, although of course European Medical Officers were in a much better financial position than their Indian counterparts. While both groups had their passage to East Africa paid (albeit in different classes), the salaries offered were far from equivalent. While European Medical Officers were paid £400 *per annum* (in 1939 this changed to £600 *per annum*), the salary of the Assistant Surgeon was approximately £200 and that of Sub-Assistant Surgeon was under £70.[41] Furthermore, unlike the European doctors, who were provided with government housing, subordinate Indian staff had neither accommodation allowance nor guaranteed government housing, no gratuity for long service, no passages paid for spouses and family, and no formal provision was made for their pension, unless they first passed a three-year probationary period.[42]

The lowness of the salary combined with the difficulty of the job in an unfamiliar climate, meant that there were persistently severe recruitment problems between 1897 and 1922. Consequently, the issue of revising the pay of Indian doctors in East Africa was the subject of debates lasting many years between Whitehall, the Protectorate and the Government of India. The first small improvement to one element of the pay package was finally made in October 1915 when the Secretary of State, Bonar Law, approved a recommendation to grant pension rights to all Subordinate Surgeons employed in East Africa subject to the Governor retaining a right of refusal. Although initially Bonar Law's recommendations only applied to staff on secondment from India, by 1921 all posts of the subordinate medical staff of the protectorate became eligible for a pension.[43]

More substantive and major revisions to pay were not made until 1924, by which time other polices had intervened in Kenya to discourage the employment of Indians. The fact that the change in policy took so long to emerge was partially to do with political nervousness over angering the powerful Kenyan settler society, who was concerned that if the job was too attractive that too many Indians would immigrate to East Africa. Indians had earlier expressed their frustration, with the *East African Chronicle* commenting in 1920: 'A fair minded observer would admit that if a non-European subordinate service is not up to the mark the root defect lies in the salary and conditions offered, and not in the personnel.'[44] Sir Robert Coryndon, Governor of Uganda had noticed the difficulty of 'fill-ing vacancies at lower rate of pay' and been lobbying London to increase the starting level of salary by 20–25% for Indian Assistant and Sub-Assistant Surgeons in the region since 1921, but his suggestions were met with resistance in Kenya.[45] It took until the ascension of the Labour MP, J.H. Thomas to the position of Secretary of State for the Colonies for the new pay structure to finally gain approval in 1924. This stipulated that Assistant Surgeons should receive an annual pay of £360–£420, £300–£360 for Senior Sub-Assistant and £200–£240 pounds for Sub-Assistant Surgeons. In con-trast, by 1939, the pay of the European doctor improved significantly.[46] The 1924 decision to modify the pay structure of the Indian doctor therefore should be seen as a relatively minor improvement, particularly in Kenya where the rank of *Senior* Sub-Assistant Surgeon was simultaneously removed. A marginal improvement, furthermore, which confirmed the political real-ity of an entrenched policy to maintain major salary differentials based on race. Stimulated by the fear of possible legal challenges from the League of Nations, the differences in salaries and terms of employment between Indian and European professionals were finally published as a set of rules in 1947. Although these rules carefully made no explicit reference to race determining salary, the wording of the document was such that it, in effect, it allowed for the continuation of a situation that has been in existence for several decades.[47]

Additional to the complaints about the level of the core salary were complaints about the extent to which Indian doctors were dependent on the personal discretion of the senior colonial administrators in Kenya. The PMO and the Governor retained overwhelming power in the inter-pretations of all the regulations regarding remuneration. For example, in 1918 the Governor's Executive Council reviewed and granted War bonuses to five Assistant Surgeons who supposedly 'look and behave as English Gentlemen'.[48] The Executive Council minutes of the 1916 reveal that the Council concerned itself over individual dismissals as well as on the alloca-tion of perks, such as additional allowances, including horses to travel or the right to train tickets.[49] Overall this system was felt to be too subjective and too dependent upon the personal good will of the PMO.

Amongst the additional benefits, it was restrictive access to the right to private practice which caused the most upset. Again, debates about this were not confined to the Indian subordinates and access to private practice was a notoriously contentious issue amongst European Medical Officers. In general, however, although government policy was never quite explicit about its formal policy on this, most European MOs were allowed to practice privately so long as it did not interfere with their government work.[50] With regards to Indian doctors the debates around private practice were similarly ambiguous. In Uganda, for example, it was decided that 'it would not be possible to deprive' the much lower paid Indian medical staff of privileges which Europeans were permitted to retain.[51] The situation in Kenya was different because the country contained more resident general practitioners, who did not want the competition. The engagement of Government doctors in private practice was fiercely resisted by European general practitioners and became a contentious issue in Kenya.[52] Resentments towards this policy were also evident among Indian members of the colonial medical department (see Chapter 7).

The Medical Department placed a very high premium on the obedience of the Indian doctor to his European superiors. There was a process for appeal and complaints—although because of the prevailing ideological commitments to ideas about race—in practice Indians had very little chance of instituting change. Individual petitions seeking readdress from alleged unfair treatment by members of the Medical Department could be made to the Governor or the Colonial Office in London. It seems that despite the low odds in gaining a fair hearing, Indian doctors had enough faith in the system to submit numerous petitions via this route. Unsurprisingly, perhaps because of the PMOs membership of the Governor's Executive Council, almost all the appeals and petitions submitted by Indians were unsuccessful. A few examples in East Africa provide an indication of grievances. For example, in 1926 Dewan Chand, a Sub-Assistant Surgeon, appealed against a decision to reduce his salary by 10% because of perceived disloyalty and insubordination. He was described as having performed his duties diligently but 'not with fidelity'.[53] Another example was the petition by Sant Ram in 1932 against a decision to dismiss him from service, when he claimed that his employment had been terminated because of an unjustified personal bias against him.[54] Similarly in 1935 Dr A.V.R. Rao pleaded a case for the supplementary discretionary payment for meritorious service for retired assistant and Sub-Assistant Surgeons. His petition was against the decision of the PMO not to grant the award and the appeal was rejected.[55]

It is clear from the evidence that remains that Indians were far from unquestioning about the restrictions placed upon them and that they were often indignant about the terms under which they were employed. In some senses, this was a normal response to working in any department of the British Colonial Service where pay and conditions were commonly felt to be inadequate. European Medical Officers, also regularly complained about their poor salaries, lack of access to decent facilities and their annoyance towards the extensive bureaucracy. A big difference, however, is that while the European MOs enjoyed the support of the local branch of the BMA and the PMO to lobby the Colonial Office for improvements, the Indian cause was not championed by this medical representative body in the same way. Indeed, quite the reverse was true.[56] The BMA's lack of support for pursuing similar reforms for the Indian doctors who possessed recognised qualifications was partly a result of the fact that Indians at this time were banned from membership of this professional body, but also offers a powerful example of how ingrained racial distinctions were when making assessments of professional equitability.

***

Despite the fact that Indians were regularly included on the medical staff, and were even praised for their contributions before 1922, nearly every positive remark made about Indians was made in a climate that concurrently also assumed that they were neither as able nor as desirable as European doctors. These profound presumptions—particularly in evidence in the highly racially charged environment of settler-dominated Kenya— hampered Indian claims for equal treatment in all realms of government service, whether in terms of receiving an equitable salary, or in terms of serving at an equivalent rank to Europeans. When, for example, in 1931, Mohamed Asan (a railway worker), appealed against the fact that his promotion within the Railway department was turned down, it was noted on the file that 'no qualification, merit or efficiency' would enable an Asian to rise from a subordinate to a higher position.[57] Going further Colonial Office exchanges on the same file of 1931 went on to state:

> In East Africa the conditions of service for persons of different races vary as regards leave, pensions, etc., largely on account of the varying effects of climate on persons of different races. Theoretically there is no reason why a suitable Indian or African should not rise to a high appointment, but in practice this would obviously give us rise to considerable difficulty... In practice appointments filled by Asians have so far been of subordinate nature...there is no racial discrimination in the selection of candidates for any appointment.[58]

Officially at least the justification for restricting the appointment or advancement of Indian medical staff typically centred on the lack of suitable applicants, but the evidence reveals more fundamental prejudices. For example, some Indians tried to enter the Colonial Medical service via the normal recruitment routes of selection via interview in London, hoping no doubt that this would be a means of being appointed to the superior rank of Medical Officer. The reaction to these applications by the Director of Colonial Service Recruitment, Ralph Furse is illustrative: '[I]n dealing with Indian applicants our method so far had been to do all we can to avoid telling them openly that they are ineligible on racial grounds.' Rather brutally this memo describes how the Colonial Office in London would allow Indians to fill in all the application forms for a position as a Medical Officer and then should find 'some excuse or other' to turn them down.[59] Consequently, when P.L. Gupta (MB, ChB, DPH, FRCS) from Gwalior State, India enquired via letter in 1923 whether there were any vacancies suitable for him either in the West or East African Colonial Medical Service, the Colonial officials decided that 'any such appointment would be undesirable'. However, it was recorded that Gupta should not be explicitly rejected on racial grounds. Clearly it was preferable that the Colonial Office should pursue a quiet policy of refusal, without making any open declarations over racial ineligibility.[60] A similar note in a colonial office reiterates the persistence of this informal policy. In 1931 Governor of Kenya defended the way this recruitment policy was subtly managed in the Colonial Administrative Service in a letter to a senior official in the Colonial Office:

> If all European posts were thrown open to Asians I should hesitate to contemplate the results. Let us imagine a situation where the Administrative Officer in charge was an Indian and the MO was a European. The difficulty would be insurmountable...the idea of throwing the Administrative service open is quite impracticable. It is not colour of a man's skin but whether he can command the respect of both his European and African subordinates. I cannot conceive any Indian being able to survive the test in Kenya, whatever his qualifications.[61]

Finally, another supplementary argument sometimes put forward for the lesser suitability of Indian doctors: namely, it was claimed that Indians were inherently unsuited to treating African patients because they simply could not command their respect in the same way as Europeans could. Such views were enduring. One memo of 1951 asserted that Indian doctors did not have 'the qualities for exercising control and discipline over the African staff.' The same document also went on to describe the way that 'Africans would undoubtedly resent the Indians' if they had to come under their medical care.[62]

This state of affairs continued in Kenya throughout the colonial period. It was felt that few Indians could never 'either clinically or administratively' prove equivalent to Europeans.[63] It was only as late as 1955, that the Medical Department in Kenya created a new grade entitled Assistant Medical Officer for Indians. This was meant as a salve to ease discrepancies, but in fact the creation of a separate job grade (and one moreover that was still subordinate to Europeans) only served to further reveal how deeply unwilling officials were to make Indian doctors of equal pay and status to their European colleagues.[64]

***

Distribution and numbers

Several sources of information can be utilised to make an assessment of the number of Indian doctors who were engaged in Government Medical Department. These include the Blue Book, the Medical Register in the *Official Gazette* (from 1910), The Annual Medical Reports (from 1911) and the *Medical Directory*. Unfortunately all of these sources have shortcomings because of their lack of consistent availability over the entire period that this study covers. They also, when crosschecked have been found not be consistent with each other and to variously contain differing definitions of medical staff over time. However a reasonable assessment of the numerical strength of the doctors can be made using the staff numbers listed in the Blue Book until 1910 and in the *Annual Medical Reports* from 1911–1937. As detailed in Appendix 2, Indians were an important part of the medical staff until the early 1920s, peaking in around 1919/1920 at 72 staff.

In Simpson's 1914 report he outlined the regional distribution of staff. His results showed the way that European doctors were, as frequently as possible, located in urban areas and how Indian doctors were sent to, the less favourably regarded, remote outposts such as Nyanza and Kenia. This decision to use Indian doctors to provide healthcare in 'more remote' areas was part of a very deliberate staffing policy whereby Europeans were preferably located in the 'centres of European populations' and 'provincial head quarters' such as Nakuru and Eldoret.[65]

With an average ratio of 1 European MO for 200,000 people, the typical rural African had little prospect of receiving treatment from a European doctor but would have been more likely treated by an Indian. However, even then, given the large indigenous African population the coverage provided by Indian doctors was skeletal and barely adequate. Some prominent Indians noticed this inadequacy of provision. A memorandum forwarded to London in 1922, by Nairobi-based politician Mr B.S. Varma, pointed out the extremely low number of doctors proportionate to the native population. He drew attention specifically to the unfair tax burden levied on African people in the light of the inadequacy of their health provision. A population of 80,000 Africans in one Province of Kenya was provided only two doctors, Mr Varma claimed—despite the fact that the hut and poll tax raised in the same region was £600,000.[66]

By 1937 the composition of the medical staff had altered radically. The employment of European MOs had substantially increased, while the Indian subordinate staff had more than halved from their peak of 72 in 1919. Basic rural healthcare—which had been expanding steadily since the end of the First World War—instead came to be conducted by African dressers who had been introduced by the Medical Department as a result of adoption of a new policy in 1924.[67] The number of African dressers went up to almost 700 in 1937 compared to a handful in 1925.[68] The policy which envisaged no explicit role for Indian doctors was summed up by the PMO as one of 'educating and encouraging African natives to play an ever increasing part in the work of the Department, that this policy has been successful because as teachers, and leaders, and guides, Government has been able ... an enthusiastic and highly trained European staff....'[69]

But although to all intents and purposes they disappeared from official colonial office records, and although the policies to actively stimulate the recruitment of Indian doctors died a dramatic death after 1923 (Chapter Six), some Indian members of the Colonial medical department remained.[70] As late as 1937, about a third of the medical facilities in the native reserves were still under the charge of an Indian doctor.[71] Although Indian names were no longer listed in the *Annual Medical Reports* (nor was any mention made of their roles), about two dozen Indians continued to be noiselessly and unobtrusively employed by the Government in the late 1930s.[72]

Remit and experiences

As so few records remain regarding the official employment of Indian Assistant and Sub-Assistant Surgeons (no individual employment contracts could be found for example), it is difficult to estimate the average length of service. Record show that a few individuals such as, Anant Ram in 1909 were so disappointed with the reality of life in East Africa that they returned to India almost immediately upon arrival, but others indicate that individuals remained in Kenya for many years.[73] In general, most Indian medical staff stayed for at least two years. Contracts were not issued for more than three years in length, however, again revealing a discrepancy between Indian and European Staff, longer service did occur, but was dependent upon the discretion of the PMO. Correspondingly the average length of service for an Indian doctor to work for the colonial medical department was lower than the average term in office experienced by European Medical Officers (eleven years).[74] There were instances however where individuals stayed in service for many years. Sayyid Wilayat Shah from the Punjab, for example, joined the Colonial Medical Service in 1913 and stayed in government service for 32 years before setting up his own private practice in Nairobi.[75]

The lack of source material makes it similarly difficult to gain a picture of the typical experiences of Indian doctors. Similar to European doctors, most Indian doctors posted outside of the main townships conducted very independent professional lives, able to make their own decisions and often responsible, at a comparatively early stage of their careers, for thousands of patients.[76] District medical reports from between 1915 and 1923 from Meru, between 1914 and 1922 in Malindi and 1921 in Kabarnet give an indication of the large levels of responsibility many Indian subordinate doctors had and contradict much of the European rhetoric in the press that Indians were insular and uncommitted to helping African communities. Disease prevalence is described in detail and the Indians in charge complained about the lack of resources and pitiful facilities.[77] For instance, an Indian Assistant Surgeon was responsible for the hospital at Fort Hall during 1919 and a Sub-Assistant Surgeon was in charge of the Machakos hospital in the Ukamba reserve for several years before 1922.[78] Although theoretically Indian subordinates were always under the supervision of the local MO, in reality those in remote locations were only infrequently visited, in some instances only once a year.[79]

The range of medical challenges that any government employed doctor faced in East Africa was formidable. Indians were regarded as having an advantage however, because of their knowledge of tropical diseases, gleaned from their first-hand experiences in India. As, one scholarly assessment of the white community of colonial Kenya pointed out: 'The doctors were reliant on Indian Sub-Assistant Surgeons, frequently very skilled and superior to qualified new recruits from Britain in their knowledge of tropical ailments.'[80] This predisposition was fortunate as no evidence can be found of any specific training being given to newly recruited Indian doctors in terms of adapting them to the East African disease environment. It is likely that Indians were employed and thrown into their duties at the deep end with no induction or training.[81]

Similar to what we know about the lives of European Medical Officers, the accounts that remain attest to the fact that the job was arduous. Not only was disease rife, but facilities were poor and patients too numerous to give individual care and attention. One common feature of the Indian colonial medical experience was that Indian subordinates were switched between stations as needs arose with little apparent concern for maintaining a continuity of personnel locally. There is evidence, for example, that Sayyid Wilayat Shah was in nine different stations throughout his career— Kisumu, Kericho, Fort Hall, Nyeri, Kitui, Nakuru, Malindi, Kisumu (again) and Kajiado.[82] Similarly, Sub-Assistant Surgeon A. N. Bowry and his family were moved around a number of distant dispensaries throughout his career during the 1930s. His inconvenience of having to work at Garissa, Lodwar, Marsabit, Wajir, Meru, Lamu, Nanyuki and Thika as well as Nairobi actually earned him the rare privilege of a salary bonus for the hardships he had to endure. These were adventures, he recounted, that once even literally saw him and his family fleeing from lions.[83]

Furthermore, the Indian Subordinate, however well qualified, was not encouraged to participate in research. This was a 'civilised' pedagogical pursuit considered only appropriate for Europeans. Restrictions were tangible. If an Indian subordinate wanted to publish his research findings in the most important local medical journal, the *Kenya Medical Journal* (after 1932, *East African Medical Journal*) the work had to be accepted by the editor, which in the racially exclusionist climate of the time, was not always easy.[84] Nevertheless, although state-employed Indian doctors seem to have been more inclined to secure private patients to supplement their income in their spare time rather than pursue research, it seems that a substantial number nevertheless were regularly research active. Between 1922 and 1940 fifteen different Indian medical department colleagues contributed to the *Kenya Medical Journal* reporting on topics as varied as pellagra, pneumonia, surgical methods and memory loss, although it should be noted that articles tended to be shorter than those contributed by Europeans, or came in the form of notes and comments.[85] Some individuals undertook large surveys of their local African communities and were committed to the improvement of standards of care and the expansion of knowledge about African diseases and their mitigating factors. As with most European research, most of this published research was related to the treatment of African patients. Although this is in itself not surprising, it nevertheless forcibly contradicts one of the dominant stereotypes propagated against Indian medical practitioners: namely that they were uninterested in African patients and therefore fundamentally unsuited to medical roles of responsibility in the African region.[86]

Other evidence also seems to contradict this idea, chiefly propagated by members of the settler community, that Indians were somehow disengaged with medical research concerning the local African community. For example the 1913 *Annual Medical Report* describes in significant details the anti-plague campaigns of three Indian doctors: A.N. Nyss, K.H. Bhatt and Murari Lal.[87] Another article from 1927 by Assistant Surgeon T.D. Nair, described his extensive yaws eradication campaign along the Tana River.[88] An additional medical report authored by Minoo Dastur reveals in vivid detail his substantial initiatives to improve public health provision in the Baringo district of Kenya.[89] Although ultimately the European management dictated what the public health or disease eradication programmes would be, the descriptions of the social conditions described in these reports (along with the accounts of the responsibilities routinely managed) confirmed that subordinate doctors in practice acted independently of their European seniors, carrying out large disease surveys and widespread public health campaigns. In the 1913 Plague survey, for example, it was acknowledged that Indian subordinates were responsible for attending to all reported cases of plague, performing inoculations to all those without a certificate of inoculation and also were responsible for making the decision to close individual dwellings if deemed necessary. Additionally subordinates were called upon to take trips out of the town to investigate rural outbreaks of plague and, where necessary, set up camps. Murari Lal was to conduct precisely this sort of rural investigation in 1913. His resulting report is particularly revealing of his personal pride over the success of the plague inoculation campaign. He was happy to declare that under his supervision 'Each and every person of village has been inoculated'.[90] Assistant Surgeon Nyss similarly described in detail his experiences of plague control at the village of Tsanganzani. His notes indicate that he was meticulous in gathering evidence on the differing levels of infection between peoples of different tribal ethnicity, particularly mapping differentials in incidence between the Wadigo and the Kavirondo. The way his report was structured provides a clear indication that Nyss hoped his findings would be useful as part of broader medical research over East African plague. His efforts were appreciated by the colonial medical department and faith in his abilities was confirmed when he was put in charge of the important Fort Hall Hospital in 1919.[91]

Although not a majority a number of Indian government doctors clearly saw themselves as full engaged, research-active medical practitioners, rather than general unscientific dogs' bodies who only helped out in subordinate medical roles. For example, a fascinating account of a hundred day long expedition along the Tana River in second half of the 1920s conducted by Assistant Surgeon Nair provides insights into the way a large yaws eradication campaign in what was regarded as the 'most unhealthy and cheerless part of the colony' was undertaken.[92] In the course of this campaign Nair reported he had personally treated over 3,000 patients with the help of just six African dressers. The level of detail in Nair's report also attests to his serious scientific interest in the subject. Nair provided extensive research data on the results he achieved from the microscopic examination of smears from typical ulcerations and stressed the importance of relating his local findings to findings in other parts of East Africa.[93]

The medical report for Baringo district, authored by Parsi Minoo Dastur (served between 1932 and 1941), provides a good example of the professional attitudes of an Indian doctor employed by the Government Medical Department in mid 1930s.[94] In a lucid, wide-ranging report Dastur described the geography, population, social conditions, education, sanitation and the medical services provided in the region. Drawing on his experiences in charge of the small 20-bed hospital at Kabarnet and four dispensaries located at Marigat, Emining, Nginyang and Mukutan Dastur wrote with considerable sensitivity and professionalism about the medical problems of the different peoples within his station. As an active proponent of improved public health education, he insightfully recommended that education programmes should be undertaken through regular informal meetings with the community (called barazas) as the 'real work lies in the Reserve proper'.[95]

Insights into the professional activities of Indian doctors can also be gained through reading the district level reports submitted (via the District Officer) to the central medical department each year for the consideration of the PMO for possible reporting in the *Annual Medical Report*. What is particularly indicative of attitudes during this period is that although these local reports by Indian doctors were submitted on time, they were rarely used in the final medical reports supervised by the PMO.[96]

Lastly, there is no need to conjure up too romantic a view. It is clear that not all Indian doctors employed by the medical department were upstanding and honest. Some individuals were found guilty of criminal activities, such as the doctor who as sentenced to a three-month prison sentence after he was found guilty of possession of cocaine and theft of government property.[97] Another was found to have committed perjury and an accessory to a murder in 1931.[98] Needless to say, in the racially prejudiced world of colonial Kenya, the European press took up such reports with barely-hidden glee.

Social standing

Before the 1920s individuals working in East Africa, India and the UK regularly praised the quality of Indian staff. E.B. Horne who was the District Officer in Meru, for example, was immensely impressed by the performance of Abdulla Khan who commenced work in Meru in 1915, describing his 'relations with the natives' as 'excellent'. Horne further commented that because of Khan's professional efforts and good personal relations there had been a substantive increase in patient consultations under his tenure.[99] Similarly, a report issued in 1921 concerning the remote Kabarnet station which was considered to have 'deplorable' facilities and to be 'notoriously unhealthy' is fully appreciative of the changes that occurred under the Indian doctor's charge.[100] Six months after Gokul Chand's appointment to the station, the District Commissioner was happy to report that 'his work has been eminently satisfactory; the sanitation of the station is looked after by him with great care'.[101]

In his published reminiscences of 1928 'The Rise of the Colonial Medical Service' former PMO, Arthur Milne also extoled the contribution of Indian doctors and Goan clerks in glowing terms, describing them as 'the two main-springs which have kept the wheels of the department turning'.[102] He singled out a number of individuals for their gallantry and dedication to the establishment of colonial medicine in the region, and included Assistant Surgeon de Cruz as one of 'those never- to- be forgotten comrades who laid down their lives in the building up of these colonies'.[103] Other European Medical Officers provided similar positive testaments of the Indian medical staff. R.S.F. Hennessey, made particular note of the vital role of the Indian doctors in running of the hospitals and their importance in undertaking much of the routine surgery.[104] Another senior European MO, Peter Clearkin, who worked for sometime at Kisumu Hospital, described some of the Assistant Surgeons he worked with as 'very good indeed', making specific reference to the outstanding efforts of one individual, Kartar Singh.[105] Norman Jewell was appreciative enough of the services of two Sub-Assistant Surgeons he worked with (Abdul Kadir and Zorawar Singh) to make specific reference to them in glowing terms within his private autobiography. [106] Indeed, the praise of Indian doctors in matters relating to tropical diseases was sometimes extended even to the point of controversy. For example, even as early as 1861, British member of the House of Lords, Lord Ellenborough reported after a visit to India, that he felt Indian doctors to be superior to Europeans for medical work in the tropics due to their first-hand knowledge of local diseases. This claim enraged some British doctors who responded enthusiastically in the pages of the *Lancet*.[107]

Despite positive, first hand descriptions, the one element that remained a consistently controversial point in the debates over Indian doctors concerned the quality of the medical training offered in the Indian medical schools and colleges.[108] Most British officials and academics generally considered not only the shorter 'subordinate' training but also the GMC approved courses offered by colleges to be inferior to that available in the West. So much so, that the poorer quality was frequently cited as sufficient explanation for appointing Indian doctors on a lower grade than their European counterparts. Leading these criticisms was the BMA, which argued that training within Indian medical schools varied drastically, was not subject to regular audit inspections and could not provide a reliable equivalent standard to British medical degrees. While the BMA conceded that some institutions provided a good medical education, it firmly maintained that others were of a 'deplorably bad standard'.[109] Overall the BMA lumped doctors with the GMC approved degree qualifications from Universities and those with the truncated training from Medical Schools in its judgement of Indian doctors: 'his standards are still far from being of his British brother'.[110] Such opinion had long been held by leading members of the IMS, such as Sir G Gifford, who in 1924 claimed that throughout India dispensaries were staffed by subordinate doctors who were 'inferior and imperfectly trained' having graduated from medical schools in India which were 'shamefully inefficient'.[111] In a display of barely concealed prejudice Gifford argued that improvements in colonial health could only be made by the 'instrumentality of European medical men.' He poured scorn on the recommendation of Lee Commission into Indian Medical Service of 1924, which would lead to Europeans serving as 'an equal with Indians who would certainly be in most cases not as well qualified'.[112] As Roger Jeffery has shown, it was to take some time and persuasion for Indian doctors (registered, not licenced) to have their degrees recognised as being the same level as those gained in Western based university locations.[113]

The Kenya branch of the BMA echoed many of these views, often contradicting the positive comments that emanated from individuals within the medical department. In one of its memoranda to London the Kenya Branch of the BMA's close alignment to the prevailing European sentiments of racial superiority was made explicit:

A specific question having been asked by the Commission as to the efficiency of Indian sub-assistant surgeons, the Branch wishes to express the opinion that though they may fulfil a useful function when working under the supervision of medical officers, yet, owing to their attitude towards the African they are as a rule unsatisfactory for independent medical work amongst natives. At the time of retrenchment the establishment of sub assistant surgeons was greatly reduced. This was a step in the right direction and the Branch considers that the eventual replacement of Asiatics is desirable. The replacement of an Asiatic sub-assistant surgeon in charge of an outstation by a medical officer entails additional expense yet the increase in value of the public health service rendered is out of all proportion to the increase in cost.[114]

This highly negative assessment of the suitability of Indian doctors fitted with the new mood of the mid 1920s, but nevertheless contradicted many earlier testimonials of Indian subordinate work and much of the flattering historical evidence found within regional and national medical reports.

Yet, as late as 1955 a memo by the Director of Medical Services, in a move to recruit GMC qualified Indian doctors to fill Medical Officer level vacancies decisively reported that, despite the economies that could be made, Indian doctors should not be 'selected for appointment as Medical Officers by the Secretary of State'.[115] In 40 years attitudes towards Indian doctors seemed to have remained fundamentally unchanged.

***

Although accounts left by Indian members of the medical service are few and evidence has had to be pulled together from disparate sources, it has nevertheless still been possible to build up a picture of the conditions and experiences of Indian doctors in the Colonial Medical Service. Indian doctors, although employed on less favourable contractual terms—and typically posted to the less popular remote stations—nevertheless were undoubtedly an extremely valuable part of the health service infrastructure, easily outnumbering European doctors. Despite operating in a climate that routinely discriminated against them, several of them showed themselves to be active researchers, ambitious in their professional scope and intentions. They often worked independently and in positions of considerable responsibility. Many led campaigns against specific diseases, conducted large-scale disease surveys or managed key hospitals and their staff. Despite the increasingly hardening racial attitudes within East Africa, to all intents and purposes, before 1922, the situation of relying on Indian doctors within the government medical service looked to be one that was unlikely to change.

6
Squeezing Indians Out of Government Medicine

> [The Indian] deprives the African of all incentives to
> ambition and opportunities of advancement.[1]

Things altered very much for the worse for Indians in the Kenyan colonial medical department during 1922 and 1923; simultaneously the early enthusiasm for the employment of Indians in all departments of the British Colonial Service in East Africa dramatically waned. Winston Churchill (hardly an advocate of Indian colonial rights) anticipated the dilemma this would throw up with uncanny accuracy as early as 1908:

> Is it possible for any Government with a scrap of respect for honest dealing between man and man, to embark upon a policy of deliberately squeezing out the native of India from regions in which he has established himself under the security of public faith?[2]

Also in the realms of the Colonial Medical Department, suddenly, and largely without any satisfactorily documented explanation, the work of Indians no longer figured in the *Annual Medical Reports*. Similarly, within the colonial archives, discussions over the Indian members of the colonial medical service—neither supporting nor bemoaning their presence—almost completely dried up after 1923. Although evidence indicates that about a third of Indians continued to work as government doctors in outlying regions until 1940, their presence was barely acknowledged.[3] Indians had been completely eliminated from official discourses about government medicine in Kenya.

The groundwork for this sea change had been gradually established since the beginning of the century, from which time (as discussed in Chapter Four), attitudes and policies in the East Africa Protectorate towards its Indian community had been progressively becoming more negative. As mutually reinforcing phenomena, legislation and social attitudes evolved in ways that were more overtly discriminatory against the Indian community as the twentieth century progressed. The settler community in Kenya, worried by discussions over the possibility of making land grants to Indians, particularly made it their concern to limit Indian influence in a region where they wanted to be the only elite landlords (namely in the white highlands, although Indians were allowed to buy land in areas not favoured for white settlement such as near the coast or by Lake Victoria). Ultimately the power the settlers had over the local press and political influence they held over key British parliamentarians meant that they were largely, although not entirely, successful. By the 1920s Indians were perceived as a problematic, potentially even volatile, sector of Kenyan society. Richer Indians were said to be exclusionist and snobbish while demanding too much politically, economically and socially, while the majority poorer sectors of the Indian community were frequently represented as a filthy, under-developed, health liability.

Although the general social and political climate was increasingly unsympathetic to Kenyan Indians, a few turning points can be recognised as notably influential in the dramatic shift in the constitution of colonial medical staffing. The first centred on the struggle, between 1919 and 1922, over amendments proposed to the 1910 Medical Practitioners and Dentists Ordinance. Although in this instance Indians gained a small victory in obtaining an improvement, the debates surrounding this issue nevertheless can be seen as indicative of a heightening pressure to limit Indian influence. A more wide-reaching and decisive change occurred as a result of the 1922 Bowring Committee which—after surveying the poor economic condition of the colony—specifically, proposed, amongst other budgetary measures, the reduction of Asiatics in government employment as a way of addressing economic concerns. Finally, the suggestions of Bowring fed into the 1923 Devonshire Declaration, which marked the decisive set back to Indian political ambitions in Kenya.

Although very few sources explicitly explained why Indians were squeezed out of the government medical service, three interconnected factors seem to have brought things to a head. First, this cull of Indian medical staff was fitting with the politico-economic climate of the time—one that had been progressively marginalising and socially denigrating Indian community members. Second, and as knock-on effect to the first point, the spirit of the times acted as a spur to changes in the declared priorities of the colonial government in Kenya. The new emphasis was to Africanise the lower branches of the colonial service, a reorientation that directed both attention and funding away from Indian medical personnel. Lastly, at a much more individualistic level, it strongly seems that the head of the colonial medical department, John Langton Gilks (1921–33) supported—perhaps even engineered, after an initial hesitation, (succumbing to pressure from the Governor and settler leaders)—policies which effectively curtailed the employment of Indians in government service.

***

Amendments to the 1910 Medical Practitioners and Dentists Ordinance

Restrictions on the practice of medicine, specified through the 1910 Medical Practitioners and Dentists Ordinance, were based on the way qualifications were formally recognised in the UK. The terms of this ordinance strongly implied the relatively higher worth of UK gained medical qualifications over those received in India, although this ranking was never stated in explicitly racist terms. Nevertheless, many members of the Indian community found the limits imposed on those qualified in India worryingly problematic. The ordinance effectively discriminated against Indian medical qualifications and also questioned the competency of Indian doctors working for the medical department by stipulating that licences for Indians to practice were dependent on the discretionary approval of the supervisory (necessarily European) Medical Officer. It additionally limited Indians' freedom to practise privately. This was also at the discretion of the PMO, and—particularly contentiously—any individual Indian's license to practice medicine whilst under Government employment would be immediately revoked if he left Government service. In short, this stipulation gave licenced Indian doctors a serious restriction to make a medical living in Kenya colony after a period in government employment.

In response to increasing disgruntlement over the terms of the Ordinance, Indian leaders including M.A. Desai, A.M. Jeevanjee, B.S. Varma, and I. Sham-su-deen mounted a spirited campaign to get the Ordinance amended after 1919.[4] The Indian Associations of Kenya presented a much-publicized seven-page long list of grievances to the Viceroy of India, which mentioned, among other issues, the need for changes in the treatment of Indian professionals, including doctors.[5] The regulations surrounding the licensing and registration of practitioners were declared to be 'a serious check and discouragement to the high class Indians settling in the country'.[6] Furthermore, the letter also pointed out that Indian patients in East Africa were irrationally compelled to go to European private practitioners, who did not understand their languages, because of the restrictions placed upon private practice for those trained in India.[7]

The matter was deemed to be so pressing that it was specifically referred to in the debates of the British Parliament on the wide-ranging issues termed the 'Indian Question'. In 1920 Viscount Milner, Secretary of State for the Colonies (1919–1921), provided a number of policy clarifications regarding the list of Indian grievances, but although he appeared to seriously consider the terms of the Ordinance he was ultimately unsupportive of amending it. A letter to the Governor of the East Africa Protectorate is revealing:

> The position in regards to the medical practice is determined by Medical Practitioners and Dentists Ordinance 1910, which was sanctioned

after consultation with the General Council of Medical Education and Registration and the British Medical Association. I understand that the greatest care is taken by the General Council in considering any medical diplomas brought before them and I see no reason to doubt that the holder of any diploma, which is a guarantee of fitness to practise, would be declared by the council to be eligible for registration in the UK. I am not therefore, disposed to suggest to your Government any amendment of the existing Ordinance, which already provides for the practice of therapeutics native, Indian and or other Asiatic method by persons recognised by the community to which they belong, to be duly trained in such practice.[8]

Milner's unwillingness to amend the Ordinance was a major setback for Indians. Desai, writing in the *East African Chronicle* during the same year outlined the unfairness of the policy:

What I can not understand is the anomaly that a raw Sub-Assistant while in Government service can have private practice but after a number of years experience the same gentleman if he happens to leave Government service can not practice. It is so simple and self evident that the unwillingness to remove it is an insult to common sense.[9]

Although the British parliamentary position seemed to have been clarified, the Dominions Committee of the BMA debated the issue during June 1922.[10] The local Branch declared itself unsupportive of any changes; reiterating the gulf that existed—and should exist—between the two grades of doctors that worked in government employment: Indian licenced doctors and European registered ones. It declared its position in unambiguously hostile language:

(i) that the proposition amounts to making an unqualified person a qualified one; (ii) that the medical Practitioner was registered not only because he was qualified but also because he was honourable and understood the ethics of the profession, whilst the Indian was not honourable, but seditious and had no affection for the Native; and (iii) that the Indian Assistant and sub-assistant may be all right under Government supervision but otherwise not.[11]

Members of the BMA failed to mention that even Indians who possessed registered qualifications, still had to work in capacities subordinate to Europeans in the hierarchy of the colonial medical department.

In general the BMA supported its Kenya Branch in all of its decisions, although it is notable from minutes taken at the time, that individual members of the committee held viewpoints that were more polarised. Some members of the Dominion Committee were clearly concerned by the intrinsic unfairness of its terms. A minute from one of the meetings of the Committee noted that other pressures might have forced the PMO and the Government of Kenya to resist changing the unfair stipulations of the Ordinance:

> It will be seen that there are three classes of persons ....This last class of practitioners usually an East Indian and it is presumed that he ranks pro-fessionally with the Sub assistant Surgeon class in the Indian Services.... The P.M.O. and the Government of Kenya consider it unjust that a man who has done good medical work often with minimum supervision, should be precluded from practice simply on the ground of retirement.... It may be that the Kenya Government is influenced by political pressure and the real motive underlying the opposition is fear of the economic competition of a class of practitioner who will naturally tend to ask lower fees, simply because he is not qualified and who will be enabled to do so because he is in receipt of a Government pension.[12]

The committee concluded that the main reason why the Kenya BMA and the Government of Kenya were opposed to changing the law was financial, most specifically that they feared Indian private doctors would charge less than their European equivalents. They decided that 'the obvious colour prejudice of the opposition should be discounted' however.[13]

Despite the recommendations from Milner, the BMA and the colonial Kenyan government to keep the Ordinance as it was, it was Winston Churchill, who replaced Milner as the Secretary of State in February 1921, who finally got the changes approved. He rejected the three BMA arguments point by point in a letter to them:

> (i) that in the financial circumstances of Kenya, the Government had no option but to employ assistant and sub-assistant surgeons in certain parts of the Colony, and that when such officials had rendered meritorious service, it would appear to be inequitable and contrary to public interest in view of the small number of Europeans doctors, that they should be debarred from continuing to practise; (ii) that licensing of Assistant and sub—assistant surgeons affords the benefit of services of persons trained in European medical work, to sections of the community who would not in any case call in a European practitioner; and who in the absence of these licensed practitioners, might have recourse to less desirable persons or to superstition or witch craft; (iii) that the persons to whom, under the new Ordinance, licenses may be granted, will not be placed on the medical register of the Colony, but merely be authorised to practise dur-ing such time as the Principal M.O. is satisfied that it is in the public interest that they should do so; and (iv) that under the circumstances Mr Churchill does not consider that the system is likely to give rise to

abuse, and he trusts that, on further consideration and in light of the foregoing remarks, the Association will concur in this opinion.[14]

The Medical Registration Ordinance Revision of 1922 appeared at Churchill's insistence in part fulfilment of his pledge in 1921 at the Imperial conference to confer equal treatment on all his majesty's subjects, irrespective of race or colour.[15] Under the new terms, the PMO was granted the power to license Sub-Assistant Surgeons to practise privately on retirement from Government Service after having discharged their duties satisfactorily. Although the PMO remained the final arbiter of each Indian's employee's fate, and the terms of the ordinance were far from putting Indian doctors at parity with their European counterparts, it nonetheless, represented a small and significant success for the Indian community.

An interesting footnote to this series of events, discussing the rights of Indian government doctors to practice privately, was that the memory of settlers' opposition to Indian doctors left a lasting impression in London. Parliamentary debates in 1927, led by Lord Arnold, confirmed that Lord Delamere had made specific recommendation to re-examine the issue five years earlier. His intention was clear: to put a halt to the hopes of Indian Sub-Assistant Surgeons (even those who had conducted 'meritorious service') to practice medicine privately in Kenya after retirement. Delamere, if nothing else, had been consistent in his opposition to Indian doctors having previously influenced the 1910 Registration Ordinance.[16]

The Bowring Committee, 1922

In the same year of this small victory, a much more damaging blow was inflicted on Indian colonial rights in Kenya. This came in the form of the Bowring Committee that was called to investigate the expenditure of the colony and to provide recommendations as to how to produce savings. The early 1920s marked the beginnings of a rapid deterioration in world prices, and presented the first stirrings of the worldwide depression that was to dominate the next decade. In this new climate, Churchill requested, after lobbying from Delamere, a Colonial Office study of the patterns of expenditure in the colonial administrations of several different African colonies: Kenya, Uganda, Nigeria and the Gold Coast.[17] The resulting figures indicated that higher medical departmental expenditure in Kenya was more than double that spent in Uganda and almost four times that of Nigeria taking into account the size of the populations.[18] Surprisingly given their low salaries compared to their expertise, Churchill attributed (amongst other factors), the high scale of costs in Kenya to the cost of employing Indian subordinate staff rather than local subordinates, such as was the practice in Nigeria:

In all branches of the administration, it is apparent that a large part of the Colony's expenditure compared to that of a West African colony, is due to the absence of any local subordinate service. Asiatic clerks, medical subordinates and artisans are employed at high salaries.[19]

This mood of austerity touched areas beyond medicine and was additionally seen by the settler leaders of Kenya as an extra opportunity for them to further secure their position. After the general survey commissioned by Churchill, a more comprehensive and formal survey of affairs was declared necessary. In 1922, a motion proposed by H.F. Ward, a settler representative on the Kenyan Legislative Council, to appoint a committee specifically to this end was debated.[20] Initially the proposal was met with some caution and suspicion, with several members of the colonial administration, including the PMO Gilks, expressing their concerns about such external interrogation.[21] Despite these misgivings, the Bowring Committee commenced work on 27 March 1922 with the active support of Governor Northey and Lord Delamere.[22] The terms of reference, agreed by the Colonial Office, were to investigate:

(a) The means by which production and exports may be fostered and increased. (b) The means whereby the cost to the community of imports may be decreased. (c) The present amount and incidence of taxation.[23]

The Committee was also charged with 'making such recommendation for the regulation of Government expenditures and department activities as may in their opinion be necessary to give effect to the conclusions arrived at by them on the subjects mentioned above'.[24] While three settler representatives sat on the Bowring Committee (Lord Delamere, E.S. Grogan and J.E. Coney), only one Indian was present, Ismail Shams-ud-deen, the Indian representative of the Legislative Council. Although Shams-ud-deen challenged many of the settler propositions, he realised that he constituted an easily outvoted minority. He appears to have ceased participating in the proceedings of the Committee after September 1922 out of frustration.[25]

To their unconcealed delight, Bowring readily met settler demands and secured their support. Grogan was so pleased with Bowring that he publically declared him 'possessing a large part of the brains of the Government'.[26] For reasons that are not entirely clear, attention was immediately focused upon medical expenditure as one of the key topics for discussion. In the early stages of the debate, quite rash—almost anti-European—employment suggestions were aired. An editorial in *The Leader* asserted that Kenya was unusual because it was one of the few British possessions that provided a free medical service for its officials and their relatives at the public expense. Kenyan Medical Officers were, it stated, paid 'large salaries with pensions at the back of them and frequent leave'.[27] The article went onto suggest that it would be much more cost effective to replace these government paid Medical Officers with private doctors who would be ready to settle with a 'little encouragement'.[28] Of course, this proved to be mere rhetoric—if jobs were to be sacrificed or economies were to be made, they were certainly not going to be made in anyway detrimental to European interests.

But it is noteworthy that a few European dissenting voices can be discerned from this period. W.M. Ross, the Director of the Public Works Department, for example, became widely disliked among the settlers for daring to raise his head above the parapet and complain about the racist insinuations contained within Economic Commission Report (1919), The Public Health Act (1921) and the Bowring recommendations.[29] The white settlers were unsparing in their scathing criticisms of Ross.[30] Despite the opposition of Indian leaders and Ross, the Bowring recommendations were accepted and soon became regarded as the new blueprint for the economic management of Kenya.[31]

The budget cuts proposed by Bowring directly inflicted the biggest hardships on Asiatics, who were deemed too expensive and not suitable to work in the administration.[32] It was declared to be more economical to replace Indians gradually with cheaper African labour. Furthermore, it was argued, only Europeans could act as Trustees for Africans, who were not yet considered to be in a position to represent their own interests. Indians were regarded as being fundamentally unsuitable for Colonial Service as they could not provide the necessary moral and social guidance to the majority African community.[33] Some Indians, such as Varma and Shams-ud-deen bitterly complained about the illogicality of these recommendations, not least as previously Indians had been relied upon so foundationally. But their voices were unheeded and the impact of the Bowring Committee became quickly far-reaching.[34] Other colonial departments such as Railways and Public Works also felt its effects, and the severity of the cutbacks were only fully revealed in the Pim Report published in 1936. Overall the Bowring recommendations were all blatantly pro settler and led to a curtailing of government expenditure, a reduction of income tax that favoured the European population, an increase of tariffs on imported goods (which was a form of regressive taxation penalising Africans) and fostered production of agricultural products for the export, rather than the domestic, market.[35]

The Devonshire Declaration, 1923

The most decisive watershed was to occur in the following year, however, in the form of the 1923 Devonshire White Paper.[36] This signified the formal pronouncement of the British Government's intentions for Kenya colony. Native rights were to be paramount in the long-term and, in order to prepare the indigenous population for eventual self-rule, the Governor's interim trusteeship was to be the short term focus. In this framework there was little space for Indian rights. Indians were not seen as the natural inhabitants of Kenya, neither through inheritance nor conquest, and their rights to political representation were severely curbed. They were denied land ownership in the Highlands as well electoral franchise on the common roll, they were allowed five elected seats on Legislative Council compared to 11 for Europeans, and the Governor maintained a right to curtail Indian immigration. One sole concession was granted to Indians however: Lord Devonshire called for the removal of segregation restrictions in the townships.[37]

But although Devonshire had appeared to resolve the 'Indian Question' in favour of European priority rights, it remained simmering under the surface of Kenyan politics for many years to come. Lord Francis Scott, a leader of the settler community, certainly still felt the need to state in 1935 that although the issue had '...remained dormant for some years. We must be on alert to see whatever head that pokes up and be ready to smack it down'.[38] Similarly, although they had been severely held back in their ambitions, some members of the Indian community continued to voice their discontent throughout the 1930s, 1940s and 1950s. In 1935 the protests of U.K. Oza, were indicative of the opinions of many other Indians. Indians, he said:

> ...will not forgo common roll, nor their legal rights in the Highlands, nor their right privilege of equal treatment of British commonwealth nor their right to their own culture, civilization and religion and social life nor their right to defend themselves as best as they can in case of attacks.[39]

Until independence, Indian rights in Kenya fell between two stools: those of the ruling whites and those of the Africans who were being allegedly prepared for self-rule. Some Indian nationalist campaigners and a few missionaries in Kenya as well as some liberal minded parliamentarians and humanitarian activists in the UK persisted in efforts to better the position of Indians, but little change actually materialised before 1940.[40] Their efforts, albeit lacking in national support and finance, did ameliorate some of the worst excesses of racial politics in Kenya however. The doctrine of 'native paramountcy' enunciated in the Devonshire declaration at least limited the potential for the creation of a white dominion in Kenya, along South African lines.[41] Nevertheless, for Kenyan Indians their social and political status in the colony remained secondary to those of Europeans.

***

Western Medicine in Kenya

In analysing the impacts of Bowring Commission and the Devonshire Declaration on the colonial medical department it is important to acknowledge what a closed shop western medicine in colonial East Africa was. In fact, a relatively small number of governmental and private European doctors constituted the whole of the medical establishment of colonial

Kenya between 1920 and 1940. The elite group included, at different times, several prominent individuals including: John Langton Gilks (Kenya, 1909–33), Murdoch MacKinnon, Ronald Wilks Burkitt, Christopher James Wilson (Kenya, 1911–28), James Harry Sequeira, Arthur Rutherford Paterson (Kenya, 1920–41), Fredrick John Johnston, Francis William Vint, Henry Laing Gordon, Arthur John Jex-Blake, Gerald Victor Anderson, James Bertram Clark, and John Sterry.

This relatively small group of men controlled all western medicine within the colony, with minimal interference from the outside world. The PMO, who automatically held a seat on the Legislative Council, might also be the editor of the *Kenya Medical Journal* (after 1932, *East African Medical Journal*) and could simultaneously be President, or Secretary of the Kenya branch of the BMA. There was no sense that the concentration of power in the hands of these few men might lead to any conflict of interest in carrying out their multifarious roles as clinicians, administrative policy makers and politicians.

The founder of the *Kenya Medical Journal* in 1923 was a doctor of candid and sometimes openly racist views, Christopher Wilson.[42] Wilson was also a senior Medical Officer in the Colonial Medical Service, having risen to the rank of Deputy Principal Medical Officer in 1923. He was secretary to the Kenya BMA in 1924 and he was appointed an unofficial member of the Legislative Council in 1938.[43] Similarly, John Gilks was the PMO from 1921 while simultaneously President of the Kenya BMA in 1921 and 1922. He was also the editor of the *Kenya Medical Journal* between 1929 and 1933.

The *East African Medical Journal* was very much the mouthpiece of a few prominent doctors, (Hicks mentions Uganda thought the Kenya journal was 'inclined to be a political organ' furthering 'the designs of the Kenya Medical Department')[44]. Similarly, the Kenya branch of the BMA represented the viewpoints and agenda of the small group of European doctors. Given the context of increasing racial discrimination against Indians it is hardly surprising that the views of this group were scarcely reflected in these official channels for discussion. Indeed, Indians were not even allowed to become members of the Kenya BMA until 1935. As one Indian doctor, H.T. Topiwala had taken up with the BMA in 1943; there was a glaring absence of Indian doctors on the Medical Planning Committee. Before 1940 none of the medical regulatory bodies in Kenya, in terms of professional registration, medical planning, drug registration or hearing cases of misconduct, had any Indian members, despite repeated requests from the Indian doctors.[45]

Departmental Level Changes

Changes to the staffing of the colonial medical department, stemming from the recommendations of Bowring, were almost immediately evident at a departmental level. Between 1920 and 1923 forty-four Indian doctors (out of a cohort of seventy two) lost their jobs as a direct result of the new emphases.[46] Little public mention was made of this one-sided redundancy policy. In reply to a parliamentary question regarding the staffing cutbacks within the Kenya medical department, the Under Secretary of State for the Colonies, merely admitted:

> ...that one senior medical officer, eight of medical officers, one of dental surgeon, and a number of other appointments have been framed so as to affect as little as possible the principle centres of populations, and they concern mainly the minor stations and those outstations where other facilities exist or may be made available for medical attention. They have been closely examined by the late and the present Governor, and, although the reductions have been effected with great reluctance, the Principal Medical Officer is satisfied that efficiency can be maintained.[47]

No overt mention was made of reductions in Indian medical personnel in the UK parliament. Indeed, the only reference to these changes published within governmental sources consisted of a few lines comment in the *Annual Medical Reports* between 1922–24 which had an extremely limited circulation and were published several months after the end of the year. Even when the cutback was mentioned in these reports, it was done in an unstated way, as if they were straightforward and uncontroversial, with Gilks stating merely it had been necessary to reduce staff and close down 11 sub stations which were in charge of Sub-Assistant Surgeons in order to reduce medical expenditure by £50,000.[48]

Two different arguments were put forward to justify the cutbacks. First and foremost, the staffing reductions were cited as a means of achieving the economic savings recommended by the Bowring Committee. The fact that it was cheaper to employ Indian rather than European medical staff was conveniently ignored within this pecuniary reasoning which did not make equivalent cut backs in its European cohort of staff. The second pretext for the unilateral action was that, under the new priorities of trusteeship, Indians were unsuitable role models for the African people. As PMO John Gilks asserted, Indian-African relationships were in general poor, and since Indian doctors (he felt) did not like treating African patients the new staffing directions would have beneficial rather than detrimental effects locally.[49]

This largely unpublicised shift in departmental policy did not escape the attention of the Indian community. Members of the East African Indian National Congress complained to the authorities in London in 1924 that 'government departments in Kenya and Uganda Railway have dispensed with large numbers of Indians' and in most cases replaced them not with Africans but by Europeans.[50] Although an internal Foreign Office note confirmed that the claim to be 'no doubt correct', it was evident that there was little that Indians could do to influence change given the political and social attitudes towards Indians that prevailed in colonial Kenya.[51] What is clear from the wording of their protests was that Indians felt that it was the threat of settler discontent, or even armed revolt, that implicitly underlay the government policy to remove Indians from positions of social standing in line with settler preferences.[52]

In concrete terms, staffing reductions, combined with the closing down of 11 medical substations (all of which had been in the charge of Indian doctors) in 1923, immediately reduced medical expenditure by the amount of £50,000 as recommended in the Bowring report.[53] Gilks had originally planned to reduce medical staff by 34 Assistant and Sub-Assistant Surgeons, and 8 European MOs. In the event only three European MOs were actually dismissed as three contracts came to scheduled end and two were reengaged. Furthermore, the number of European medical staff actually grew at an accelerated pace between 1924 and 1930.[54] The PMO reflected upon the economies, in characteristically phlegmatic terms, in an article reminiscing over his career written for the *East African Medical Journal* in 1935:

> A large reduction of personnel, in which all branches were affected, had to be made. No less than nine of the medical staff in addition to the Dental Surgeon, six nursing sisters, four sanitary and 29 sub assistant surgeons were retrenched or their services were dispensed with in other ways. Obviously activities in general had to be curtailed. This was effected without interfering more than was absolutely necessary with the recently established reserve centres, but 10 sub-stations at which sub-assistant surgeons had been in charge were closed down. In one or two instances these stations have since been re-established and at three of the centres native hospitals manned by medical officers and other staff have been erected.[55]

The Pim Report of 1936 outlined that European Medical Officer numbers rose from 24 in 1912 to 71 in 1928—nearly a 300% increase.[56] Indeed, contrary to any notion of economising, the increases of the salaries of MOs, PMO and Deputy PMOs were implemented in 1919, 1920, 1921 and 1927 as a means of improving the recruitment of European doctors to East Africa (and elsewhere).[57] The justification advanced by the PMO for increasing the European staff was unambiguous:

> the highest possible standard must be maintained because only in this way will the Africans, on whom ultimately any large extension [of the medical services] will depend, learn the necessary efficiency.[58]

There was no discussion about economies that might be gained through employing more Indian doctors. Over the same period the numbers of Indian personnel in subsidiary occupations, such as compounders and clerks, rose from 53 to 61.[59] It appears that the issue was not so much having Indians in the employ of the colonial medical department *per se*, but having them in positions comparable to Europeans. Ironically, at precisely the moment when the IMS in India was including more Indian doctors, the Colonial Medical Service in Kenya was becoming whiter, at least in its higher echelons.[60]

Despite the rhetoric of austerity, the reality was that there was almost a doubling in the total expenditure of the colonial medical department between 1925 and 1931, from £132,713 to £221,202.[61] This was to some extent the result of a significant rise in the number of European doctors, but was also partly due to a concerted effort to improve out-patient facilities in rural areas, which had been first enunciated in 1918. The number of dispensaries rose from a mere 20 in 1922 to 107 in 1932, when 646,000 African patients were said to have been treated in these new facilities.[62]

In contrast to the caution shown by the colonial medical department in publicising its policy to limit Indian medical staff, the local branch of the BMA was much more direct in its support of the policy.[63] In a memorandum to the London-based Dominions Committee the Kenya branch of the BMA declared that it considered 'the eventual replacement of Asiatics in the Medical Service by Europeans and Africans' as 'desirable'.[64]

There can be no doubt that the BMA supported and praised the strategy to remove Indians, no doubt reflecting the strong ties of many of its members to the settler community. Publically the new focus on employing Africans was lauded, but in actuality European recruitment continued apace. As the cost of employing a European Medical Officer was approximately two to four times more than that of employing an Indian doctor, racial politics clearly underscored these decisions.

The Africanization of the Colonial Medical Service

The desire to Africanise colonial medicine remained a central justification for changes in staffing policies, so it is no coincidence that the scheme to employ Africans in larger numbers started being formulated in earnest in 1922, the same year that Indian staffing was cut (although the scheme to recruit dressers did not effectively get off the ground until 1929).[65] If the right training facilities were available, Gilks had reflected earlier, he could see no reason why it would not be 'quite possible to train Africans to the necessary standard'.[66] Although it took until the late 1930s for the scheme to really bare fruit in the form of African doctors graduating from Makerere and other properly trained African medical assistants becoming available, there is little doubt that the aims and intentions of this programme were laudable.[67] The fact that it took so long to work satisfactorily was mostly to do with the deeply embedded inadequacies of Kenyan primary and secondary education that meant that the newly trained African dressers were for some time operating at a standard well below the competencies of the medical school or university educated Indian Assistant and Sub-Assistant Surgeons.

Nevertheless, the increase of African dressers within the colonial medical department was dramatic, numbers rose from a handful in 1920 to hundreds ten years later. By 1932 more than 1000 Africans were said to be working for the colonial medical department, over half of them categorised as dressers.[68] The 1936 *Annual Medical Report* stated that 153 European, 46 Asian and 1204 African staff had been employed (more than half as dressers) in the Medical Department in the previous year.[69]

Although *Annual Medical Reports* at the time were full of positive descriptions about the successful Africanization of the Colonial Medical Service, opinions about the success of this scheme were much more polarised than the official reports suggest.[70] By 1932 an internal Medical Department memorandum admitted there were serious problems with the quality of African medical staff, who were 'unquestionably far from satisfactory'.[71] Hugh Trowell (Kenya, 1929–35), who was put in charge of the African medical training programmes, wrote straightforwardly about the gap between rhetoric and reality in terms of the early history of African dressers. Trowell made no bones about his shock at their poor performance and was appalled by the brevity and inadequacy of their training. He found the quality of the average trainee so deplorable that he stated that no Medical Officer was prepared to have an African dresser on his ward. This, he admitted, was compounded by deeply held racial prejudices, as many Europeans believed that 'few Africans had the innate mental ability to become skilled in any complicated manner'.[72] The short training given to Africans with the barest primary education was in no way comparable with that undergone by doctors trained in medical schools in India, which was usually three years or more after secondary school. The 1933 *Annual Medical Report* supports this claim, stating that the majority of all cases of rurally experienced illness were treated 'not by doctors, but by African dressers whose standard of competency is not yet a high one'.[73]

The dual problems of there being a conspicuous lack of well-qualified Africans in Kenya, combined with the inadequacy of the medical training that could be provided—especially in technical matters and particularly in the early days of the scheme—was raised time and again during the 1920s and 1930s.[74] African education in Kenya consisted of only very basic primary education facilities until the mid 1930s, meaning that even if people were willing, it was hard to find suitable candidates before that time. Despite his obvious energy and commitment even the chief trainer on the dressers scheme, Hugh Trowell, was very vocal in his criticisms of policy directives surrounding its implementations. He felt instructions were vague and complained that 'medical headquarters had been unable to formulate a policy'. Without any clarity from his bosses, Trowell felt it was unclear whether he was to train 'hospital nursing orderlies, or sanitary inspectors or "bush doctors" to work at one of the numerous government district dispensaries'.[75] Furthermore, Trowell found Kenya far behind Uganda in this regard. Uganda (largely missionary efforts) had been actively developing a much more widespread secondary education system since the 1910s, and medical education of Africans came to be centred on that colony, at Makerere College (later university).[76] It was only from 1930 onwards that a concerted programme was initiated in Kenya to improve the training of Africans 'to a standard very much in advance of that of the previous dressers.'[77]

Gilks had described the African dresser as being inadequate, and commented in undisguisedly scathing terms that 'constant supervision will be necessary or the dresser will become slack and the work neglected.'[78] In reality, this need for supervision would have rarely been met, not least because Medical Officers, although supposed to visit outstations 'at least once a month', did so less frequently in practice.[79] As a result of their remote location, poor accommodation and the difficulty of direct supervision by European doctors, the rural dispensaries were still described as late as 1933 as being 'little more than dressing stations' where only extremely simple cases could be treated.[80] Gilks, the official proponent of the scheme to Africanise the Colonial Medical Service, reflected towards the end of his term as PMO, the comparative failures of the scheme:

> It has long been apparent that African staff is unquestionably far from satisfactory, in as much as it is with few exceptions partially trained, indifferently efficient and almost undisciplined.[81]

Unsurprisingly for the climate of the time, racial explanations were sometimes used to justify the perceived inefficiency of Africans. They were said to complain too much about their jobs and to terminate their contracts with little or no notice. Apparently, this innate (or cultural and social), lack of predisposition for the job was compounded by the absence of an African civil service to regulate the standards of the emergent profession. The result was that there was a lack of uniformity of pay and systems for promotions, leave and pensions. The report of 1932 concluded that these factors in combination 'made it impossible for the Department to attract a sufficiently good type of native from which an efficient and reliable service can be constituted'.[82]

As their training—and therefore duties—were not comparable, there was no immediate ambition for African dressers to be regarded as equivalent to Indian Sub-Assistant Surgeons.

> It is not considered necessary or desirable at the present moment to formulate any scheme that will qualify higher posts and those proposed for the Assistant Grade....Advancement by natives for posts commensurate with those now held by Indian Sub-Assistants is not contemplated.[83]

It is clear that many members and observers of the colonial medical department certainly did not share the optimistic claims made in official reports that
by 1932 Africans were rendering 'increasingly efficient service'.[84] Indeed,
although diligent and well-respected African individuals undoubtedly
existed, the services of the African dressers were generally looked down upon
by European contemporaries who concluded that they were unable to provide an effective subsidiary medical service. This had other impacts on the
composition of the medical department; for example, during this period
the role of European nurses was enhanced to meet shortfalls due to staffing
changes and to train African staff for native hospitals.[85] Whereas the medical department in 1919 employed only 15 European nurses, by 1936 the
number had increased to 57.[86]

A few Indian Assistant Surgeons and Sub-Assistant Surgeons continued to
operate in rural East African locations despite the official cull. In 1936, the
AMR indicated that almost a third of the hospitals in the remoter African
reserves were still under the charge of an Indian doctor, though their work
was barely mentioned in government reports.[87] A couple of dozen tenacious
Indian doctors, even if they were largely ignored and severely depleted in
numbers, remained as members of the colonial department (Chapter 5). The
ratio of European MOs to Indian subordinate doctors which was 1:2 in 1920,
therefore reversed to 2:1. This point was not lost on Indian members of the
Legislative Council, who poignantly pointed out in 1935 that favouring
European doctors over Indian ones was not only contra what was happen-
ing in Indian Medical Service and the Sudan Medical Service, but was also
vastly more expensive for the Kenya Medical Department.[88]

The Role of John Gilks as the PMO

The main impetus for these policy decisions was driven by the emergent
and racist socio-political dynamics of Kenya. At least so far as in that the
Devonshire Declaration and the findings of the Bowring Committee should
be seen as victories of the settler community in shaping Kenyan policies.
At the end of the day, however, the specific way that changes to medical
policy occurred at the departmental level relied upon the personal inclinations of the PMO. In this regard, the decisions of John Gilks and, to a lesser
extent, Arthur Paterson were decisive in sealing the collective fate of Indian
government doctors in colonial Kenya.

John Gilks was undoubtedly the chief implementer of the scheme to
remove most Indian doctors from government employment in the medical
service—although it cannot be surely known to what extent he made his
decisions independently and to what extent he succumbed to pressures
from the Governor and members of the powerful settler community in
Kenya. A graduate of London, St Thomas' and Edinburgh University, Gilks
had served since 1909 in various postings around Kenya, including Lamu,

Tanaland, Kisumu and Nairobi.[89] Gilks was a controversial, even somewhat eccentric figure (renowned for owning a pet leopard), but enjoyed high status among the elite of Kenya.[90]

By the end of his career, Gilks' unreserved allegiance to the settler community was plain, but during the first couple of years in which he was head of the medical department his position with regard to the 'Indian Question' was surprisingly ambiguous, even at times seeming pro-Indian in stance. Gilks was a strong promoter of the benefits of curative care, stating that although preventative medicine was important, curative care offered a more dramatic way to win the confidence of the local people. His initial plans to improve curative services, as outlined in his 1921 *Annual Medical Report* (penned in 1922) envisaged a role for both Indians and Europeans within his medical department, actually complaining about the lack of availability of the Indian medical staff and bemoaning that their numerical strength had become 'inconveniently low'.[91] In short, Gilks seemed at this point to be in favour of giving Indians a role in the colonial medical service particularly for treatment of Government employees though he did mention his reservations in the same report over the suitability of Indians for treating Africans given that he felt the 'inclinations of these Sub-Assistant Surgeons were not, as a rule, directed' towards Africans.[92]

It was during the discussions over the segregation clause contained in the Public Health Ordinance of 1921 that Gilks seems to have changed his position from one of moderation and support towards the Indian community, to one that was much harsher. When his views were sought at the meeting, Gilks at first (and in line with his other subcommittee members) clearly stated his opposition to segregation, declaring that public health measures were the superior means of combating disease, but he capitulated under pressure from the members of the settler society (and the Governor) however and adopted a position supportive of Indian segregation (see Chapter 4).

Similarly, during the Legislative Council debates over the amendments to the 1910 Medical Practitioners and Dentists Ordinance that occurred over the course of 1922, Gilks seemed to be still unsure as to where to position him. Remarkably, in the context of his actions later the same year, he opposed recommendations to limit Indian government doctors from private practice while they were employed in government service. Gilks instead defended their entitlement on the grounds that he felt that it 'would provide a reward for faithful service'.[93] Showing that, although he was never forceful in his support of Indian Sub-Assistant Surgeons, some of his early-recorded opinions reveal at least tolerance if, not sympathy towards Indian medical practitioners.

However, despite these instances of tacit support, an unmistakable anti-Indian position can also be seen running through Gilks' voting record on non-medical issues. This is an important consideration, as Gilks was the first PMO to sit on the Kenya Executive as well as the Legislative Councils, meaning he had achieved a political realm of influence beyond that of any PMO before him. Examining his voting record in these non-medical capacities, it is noteworthy that during the early 1920s he voted against the common franchise in favour of the communal franchise despised by the Indians.[94] On another occasion, when a controversial budget reduction was debated, he actively supported cuts in non-European salaries—a stance that provoked bitter criticism from the Indian council member present, V.V. Phadke.[95] In short, despite some initial wavering, by 1923 Gilks was certainly ideologically aligned to the settler-inspired goal to remove Indian medical staff from employment. He would never again soften his stance towards Indian members of the colonial medical service.

Gilks was extremely prudent when stating his intentions. In fact, the lack of mention of Indian doctors after 1923 is more striking than any direct evidence that he overtly criticised them. During the third year of his tenure, the work of Indians simply merited no mention what so ever and Gilks made no reference to any change of policy, although he had clearly adopted one. This was a notable contrast to the previous PMO, Arthur Milne, who on occasion singled out individual Indian doctors for praise in his medical reports.[96]

By way of contrast, the omission of Indians in the *Annual Medical Reports* issued by Gilks after 1923 is an indication of his attitudes towards the Indian employees within his department. Gilks simply made no reference to their work. Even at the point of his retirement from the Colonial Medical Service in 1933, Gilks, an adept politician, was happy to talk about his role in the expansion of the colonial medical service, or the achievements he had made in improving the salary scales of European MOs, but he made no reference to the political pressures surrounding the cull of Indian Assistant and Sub-Assistant Surgeons that coincided with his tenure.[97] The only sign of his attitude in his reminiscences can be found in the way he retrospectively categorised the staff that constituted the medical department when he arrived in Kenya in 1909. While European staff members were enumerated separately by rank, he lumped Indian storekeepers, clerks and compounders together with and Sub-Assistant Surgeons in one undifferentiated lower group.[98] Thus, departmental staff members were conceptualised on racial lines, rather than in terms of expertise; for example, Indian doctors held higher qualifications than European nurses.

Only rarely did Gilks speak publicly of the retrenchment of Indians in more explicit terms. In 1922 he opined in the Legislative Council that he felt some Indian medical degrees to be qualitatively worse than those held by Europeans: 'Certain degrees [of India] were not recognised. All sub assistants were not good doctors and some were not fit to practise without supervision'[99]

Posthumously Gilks has been written into the medical history of Kenya as having been the PMO who put African health at the centre of the stage. He should certainly be acknowledged for the extension of medical care in

Kenya to the district reserves through the Africanization of the Colonial Medical Service, however debateable the immediate results of this scheme were felt to be at the time. Ann Beck described him as being progressive in his 'positive outlook' towards African affairs and portrayed Gilks as vital in laying the groundwork for the post-war expansion and development of the medical services in Kenya.[100] Even a prominent Indian doctor, Dr Adalja, wrote highly of Gilks stating that he had been 'a man of vision, of understanding and of considerable ability'.[101] Other evidence, however, presents a different picture. Gilks was actively disliked by several of his European colleagues, some of whom negatively commented on his judgement and professional standards.[102] John Carman (Kenya, 1926–51), who knew Gilks well, questioned Gilks' administrative abilities despite conceding that he found him to be a competent clinician. In his memoir of his time in Kenya, Carman directly criticised Gilks for moving doctors from one location to another so that 'Africans never knew the doctor long enough to establish that vital relationship'.[103] Another colleague, Peter Clearkin (Kenya, 1916–25), pointed out Gilks' lack of substantive administrative and laboratory experience before becoming the PMO.[104] Clearkin made particularly damning remarks about Gilks in the context of describing a dispute they had had in 1921. According to Clearkin's account, when he was head of the laboratory in Nairobi he had successfully identified cases of diphtheria and typhoid. Gilks disputed these findings, stating 'there is none here' and 'you Irish are so impulsive and do exaggerate'.[105] Clearkin was incensed by Gilks' reaction, which undoubtedly coloured the way he subsequently portrayed him in his memoir, where he described him as 'one of those weak amiable people who wish to be popular'.[106] Gilks, he claimed, was too preoccupied with:

> trying to curry favour with settlers or their hangers on with something to sell who are to be found in a country with an expanding economy and money to be made overnight by methods best not mentioned. They resented loudly and indignantly any suggestion that Kenya was not a celestial paradise where disease was unknown.[107]

Another charge Clearkin made against Gilks was that, at a meeting in London in 1934, he falsely took credit for being the first to describe a case of typhus in Kenya.[108] Clearkin was furious and wrote to the *British Medical Journal*:

> Dr J.L. Gilks is reported as having said that the first description of typhus in East Africa was given by himself in 1920. This is correct, as far as it goes, but I am sure that Dr Gilks does not mean to be understood that he was the first to recognise...the disease, known for years in Nairobi as Dengue, was in reality typhus. The credit for the discovery belonged to Dr Anderson and the clinical work which established the diagnosis was

done by other members of the Government Medical Service and not Dr Gilks.[109]

Although these criticisms might simply come from disenchanted colleagues, it seems clear from this and other evidence that Gilks rubbed some of his staff up the wrong way. Needless to say, Indian leaders were also openly harsh in their verdict on his lack of concern for non-European communities.[110] Indeed, Indian leaders Varma and Sham-su-deen were so incensed that they actually tried to have a motion accepted in the Legislative Council to reduce Gilks' salary by £100 per annum because of the Medical Departments failings, under his directorship, to care for non-Europeans.[111]

After his career in Africa the British Government awarded John Gilks the CMG.[112] On his return from Kenya he continued to use his colonial expertise by becoming a council member of the BMA and the Chairman of the BMA's Dominions Committee.[113] Controversy continued to follow him in the UK; for example, he publicly defended the work of the eugenicist, Dr Henry Laing Gordon, who had formed many of his professional conclusions about the reduced mental capacity of Africans in Kenya.[114] An exasperated senior Colonial Office official, Fleet, was driven to make a damaging assessment of Gilks after a particularly fraught debate over his lobbying for Gordon's work in 1934: 'I have formed the impression that if there is any "ballyhoo" in Parliament he is at the bottom of it....it appears that he has always been an enthusiast of this particular line. I have now ceased to trust him'.[115]

Although, always careful in talking about his governmental work, once retired Gilks spoke more candidly about his views on racial differences, and innate European superiority. In a letter to *The Times* in 1933 Gilks, in support of Gordon and Vint's controversial work, talked of the need for more research into the mental capacities of 'backward races', exposing a deeply embedded irony underscoring his status as the chief engineer in the Africanization of the colonial medical department during the 1920s and 1930s.[116]

Many of the policies that Gilks started were continued in 1933 by his successor, Arthur Paterson, who had worked with him since the early 1920s. Although Paterson thought highly of Gilks and whole-heartedly supported his policies including the enthusiasm for the eugenics research, the annual reports which emanated from the successive Gilks and Paterson headships reveal their very different styles of operation.[117] Whereas Gilks had preferred to give minimum information and to be crisp and positive in his tone, Paterson, in contrast, was much more expansive and sometimes doubtful about progress made. Indeed his doubts were sufficiently articulate as to cause members of the Colonial Office to note their concerns that he talked too freely.[118] Paterson has been mainly remembered for his contributions to the promotion of public health for Africans within Kenya.[119] He was passionate about education, authored several works and produced an educational film to promote his public health ideas.[120] Although the emphasis was different, in certain core matters Gilks and Paterson appear to have been cut from the same cloth. Despite his focus on improving the health of Africans, Paterson's tone was, like Gilks', always paternalistic and—by modern day standards—tinted with racial considerations.[121] What is significant in its very absence is that, throughout Paterson's tenure, he never voluntarily made any mention of the Indian component of the colonial medical staff, despite the fact that about two-dozen Indian doctors were still working within the medical department during his headship. A rare tacit acknowledgement made by Paterson was in 1939 when he opposed granting Makerere University trained African doctors GMC recognised status. In his defence of this decision, Paterson claimed that this move would be unacceptable because it would imply that the newly trained African doctors were as good as European and Indian doctors. He also went on to say that such a change would unfairly increase the competition for Indian registered doctors.[122] Yet, if he did find the Indian contingent of his personnel useful and worthy of defence, he hardly mentioned them, despite being criticised within the Legislative Council for neglecting Indian health.[123] He also omitted to discuss their contribution in the official reports of the medical department. Thereby, rendering them invisible within official discourses.[124]

Given this stance it is no surprise that under Paterson nothing was done to re-admit Indians into the Colonial Medical Service. In fact, the unofficial bar on appointment of Indians as Medical Officers lasted until the late nineteen fifties.[125] Paterson, like many others in the Colonial Medical Service, expressed (privately) his concerns that African medical assistants were not comparable to Sub-Assistant or Assistant Surgeons. As was typical of prevailing colonial beliefs at the time, medicine to Paterson firmly remained the preserve of white men with Africans helping in the subordinate roles.[126] As he said in 1932: 'no matter how much the efficiency of African staff may improve ... that staff will not for many years to come to take over staff will not be able to take over the higher supervisory, directional and professional duties...'.[127] No mention was made of the valuable assistance Indians might be able to provide.

While Indian personnel had previously been the subject of urgent requests from the Nairobi or London to the Indian government for more recruits, abruptly in 1922 the policies of two decades were swept away and Indians subsequently became instead the targets of regular criticism for their alleged lack of suitability for medical work in Kenya. So powerful was this new focus on training up African healthcare workers instead that by the 1930s Indian contributions to government medicine were acknowledged even less frequently than before.

7

Indian Private Doctors in Kenya

> During this quarter of a century, I have travelled almost
> all over the East African Territories overland and watched
> with great admiration the work of our Indian colleagues ....
> I have seen them patiently going through their work of
> bringing civilisation to these bush lands of Africa facing all
> the horrors of tropical diseases, such as Malignant Malaria,
> Blackwater fever, Plague, Dysenteries, Sleeping-sickness
> and host of others.[1]

Very little has been written about the Indian general practitioners who were
part of the medical world of Kenya and yet they were an absolutely vital part
of the colonial (and for that matter, post-colonial) landscape of the country.[2]
Admittedly, in the early years of British rule the numbers who made up this
cohort were small—just as they were for privately practising European doctors in the colony—but after 1920 Indians became a relatively organized and
well-established part of the medical landscape. Furthermore, unlike most of
the European private doctors that worked in colonial Kenya, Indian doctors
commonly planted permanent roots in the country, remaining there with
their families long after the transition to national rule. A large number of the
doctors named in this chapter became the forefathers of a medical community
of professional Indians, some of whose descendants still live in Kenya today.

The scarcity of references to Indian private doctors within medical histories
of East Africa belies their contributions to healthcare in the colony. As one
prominent Indian general practitioner recollected in his memoirs, their role
was similarly important:

> There is a crying need in these backward countries not only for the highly
> qualified specialists to work in the larger centres but for soundly trained
> general practitioners able to tackle all but the most specialized tasks for
> the benefit of the vast untouched populations who are living beyond the
> reach of all but the most primitive medical care.[3]

Furthermore, the lack of mention of private doctors also fails to acknowledge that, over and above the responsibilities of their general practices, many Indian doctors were educated and respected community leaders; ones who frequently played a vital political role in defending Indian rights in the colonial context, and were key protagonists in establishing and funding important philanthropic initiatives. Historical source material is not always easy to find, but a few private doctors left vivid memoirs, articles or letters giving insights into their times in Kenya. The contributions of several individuals were also recorded in published obituaries or in unpublished family tributes.[4] A significant number of Indian practitioners wrote letters and comments regularly to the local Kenyan press.[5] Lastly, a sense of individual experiences and opinions has been woven together by searching for the names of these Indian doctors in places where they might not necessarily be expected to be found: such as within European memoirs, or within government and non-governmental minutes or publications.

The Earliest Pioneers

Dr Luis Lobo seems to have been the first Indian doctor to set up a private practice in British East Africa Protectorate after landing and settling in Mombasa in 1898 or shortly before.[6] Lobo was a Portuguese citizen of Goa and set up a private practice in Mombasa and subsequently also acted as the Portugal's Vice-Consul.[7] Rozendo Ayres Ribeiro was probably the next to arrive on the mainland, landing in May1899. Ribeiro was the first Indian doctor to travel inland, and although little is known about his early years, he set up a practice in Nairobi six months after landing in Mombasa with an assistant, called Mr C. Pinto, for whom no other information has been found apart from his name.[8] Sources indicate that L.A. da Gama was next to join the group in or around 1903; he, like Lobo, decided to stay in the city of his entry, Mombasa and is claimed to have also held a diplomatic position with the Portuguese. Taking the long view, da Gama unwittingly became the first Indian doctor in Kenya who would go on to be registered to practice in East Africa under the Medical Registration Ordinance, which came into effect in 1910, since he had a recognised GMC qualification (the licentiate of medicine and surgery from the University of Bombay). Lobo and Ribeiro were licensed, as opposed to registered, to practice medicine (Table 7.1).[9]

Notably, these three early medical pioneers were all Goan, meaning they were most probably Catholic and were likely to have been relatively Europeanised in their tastes and attitudes through their Portuguese-influenced upbringing. This fits in well with the profile of some of the early Indian Assistant Surgeons and Sub-Assistant Surgeons employed in the Colonial Medical Service. As noted in early chapters, it was well known that the British colonial administration respected Goans, and liked to work with them. Western medical education had been available in Portuguese

Figure 7.1   Dr Rozendo Ribeiro

Goa since 1842 at the Escola Médico-Cirúrgica de (Nova) Goa, but progressively many Goans with medical ambitions crossed to British India for their medical education. Bombay in particular seemed to promise more social and economic opportunities. After graduation from an Indian medical school like Bombay it would have been a relatively small and logical step to look for work in other parts of the British Empire; especially in nearby British territories, such as Zanzibar or the East African mainland.[10]

The best known personality of this early triumvirate of pioneers, was Dr Ribeiro, who went on to have a medical career in Kenya for more than half a century and was well known, not only for his involvement in many public causes, but for his flamboyant behaviour. Several reminiscences from the period recount his famous habit of riding a zebra through the streets of Nairobi for his daily rounds and growing grapes in his garden.[11] Less is known about Lobo, although his appearance as a licensed practitioner in the 1911 and 1920 *Official Government Gazettes*, and Vice—Consul in the 1915 Blue Book suggests that he forged a successful career in the colony for maybe as long as two decades.[12] A local newspaper report survives about Da Gama, however, and describes him as having been a popular, energetic and well-dressed young physician in charge of a successful and thriving practice in Mombasa.[13] Evidence may be scarce, but to all intents and purposes it seems that all three thrived in their new colonial home. In the early years of British rule before doctors were centrally registered, it is difficult to quantify the number of doctors who practised in Kenya with any certainty, but no evidence has been uncovered attesting to the presence of any private Indian doctors other than Ribeiro, Lobo and Da Gama before 1905.

Numbers

It is difficult to be sure about the numbers of Indian doctors arriving in the colony, as it was only in the early 1930s that reliable data became available. For government doctors, a central authority enumerated their numbers, qualifications and locations, but no such records were kept for private doctors, be they from India, Europe or elsewhere. Some information concerning private practitioners can be gleaned from the published lists of doctors that appeared in the *Official Gazettes* of the colony. But it was only after 1933 that the editors of this publication started indicating with an asterisk the names of doctors who had stopped practising, so that earlier lists were often seriously out of date.

A confident estimate however puts the number of Indian general practitioners at seven in 1920, meaning that growth of this group was slow in the early years of the twentieth century, with only four new doctors joining the three earliest pioneers. The *Official Gazette* of 1920 named: L.A. Gama, E. Dias A.C.L. de Sousa and M.M.de Sousa (Mrs) as registered practitioners and M.C.S.L Lobo, R.A.P. Ribeiro and L.A.D.P. Rodrigues as licensed practitioners (Table 7.1).

According to the Gazettes, the number of registered Indian doctors increased at a much greater pace after 1920. By 1932 the cohort of seven had swelled to 40.[14] This six-fold increase of registered Indian doctors setting up general practices in Kenya between 1920 and 1931 directly reflects three intersecting trends: first, the mounting numbers of doctors graduating from Indian medical schools available for work; second, the growing and relatively more prosperous India settler community in Kenya (which provided a medical market); and third, the considerably reduced opportunities that were available for medically qualified Indians within the Colonial Medical Department after 1922, which meant that private practice was the only viable career for an Indian doctor immigrant (Chapter 6). The

Table 7.1   Private practising Indian doctors, Kenya 1920

Name	Qualification	Location
A.C.L. de Sousa	Lic. Med. Surg. Bombay	Nairobi
M.M. de Sousa (Mrs)	Lic. Med. Surg. Bombay	Nairobi
L.A. da Gama	Lic. Med. Surg. Bombay	Mombasa
E. Dias	LRCS, LRCP. Edin. LFPS	Nairobi
M.C.S.L. Lobo	Licensed (qualifications not specified)	Mombasa
R.A.P. Ribeiro	Licensed (qualifications not specified)	Nairobi
L.A.D.P. Rodrigues	Licensed (qualifications not specified)	Mombasa

growth in the number of Indian doctors slackened somewhat due to the economic slump of the 1930s but, despite this small blip, the contingent continued to grow to the point where 50 individuals were listed in the 1940 *Official Gazette*.[15] This is as reliable an estimate as any from the period, although *Gazette* totals did not always tally with the personal estimates of participants. For example, Dr Adalja estimated in an article looking back over his career, that the total number of Indian practitioners in the colony in 1910 was 64, whereas the *Official Gazettes* of the same year indicate a number of only 4.[16] This huge discrepancy in numbers could not be reconciled even if all licensed government Indian doctors were also included in Adalja's estimate. Indeed, this example shows how subjective personal memories can sometimes be.

A notable core of these general practitioners—certainly well over half—evidently found enough prosperity to settle down. Thirty-one of the 50 identified as practising in 1940 can be shown to have spent a minimum of ten years in colonial Kenya (Appendix 3). Of course, many stayed on well beyond this.

Qualifications

Three of the seven physicians identified as working in the colony before 1920, achieved their medical qualification in Bombay and one was trained in Edinburgh. However, in the *Official Gazettes* of the period, qualifications were detailed only for registered practitioners, rather than for licensed ones. Thus it is not possible to ascertain from the Gazettes alone the academic background of Lobo, Ribeiro and Rodrigues or to confirm where they qualified and what form their qualifications took (Table 7.1).[17]

Finding data on the qualifications of doctors that arrived in Kenya after 1920 is easier. This is because immigrants by this period increasingly held GMC recognised qualifications, which meant that they were fully detailed in the *Official Gazette*. The first thing that is apparent upon looking at the list of qualifications is the dominance of Grant Medical College, Bombay as the main training ground. Twenty-five of the thirty one Indian doctors— just over 80%—who had worked in the colony for more than ten years (see Appendix 3) had obtained the qualification of M.B., B.S in Bombay, with only six displaying alternative training paths: three qualified from Lahore in the Punjab (M.D. Gautma, F.C. Sood and Bakhtawar Singh) and three in the UK (two from London and one from Edinburgh). Although in a minority, the three doctors who qualified in Britain (E. Sorabjee, M.A. Rana and B.S. Sandhu) were not the first: more than a decade earlier, Edward Dias who, after qualifying as a doctor from Edinburgh and serving for a short spell with the Army, became in 1917 the first UK-qualified privately practising Indian in Kenya.[18] There is also evidence that others, who had worked for shorter spells before 1940, had obtained their medical qualifications in the UK. For example, A.H. Ismail, who arrived in the mid 1930s held M.R.C.S, L.R.C.P and D.T.M from Britain. In summary, just as for the Indian members of the government medical department, the backgrounds of private practitioners were more diverse in 1940 than they had been in 1920. Additional to the core seedbeds of Bombay and the UK, the *Official Gazette* for 1940 listed Indian doctors who had qualified in Punjab,[19] Calcutta,[20] or even Canada[21] to give just a few concrete examples.

As in the case of European doctors, specialist qualifications were not typical until after 1940.[22] By 1940 however, a handful of Indian doctors, including Ahluwalia, Ismail and Mandalia possessed additional credentials indicating their professional commitment to certain areas of medical specialism (in this case public health, tropical medicine and ophthalmology). Some saw the advantage of supplementing their qualifications by travelling to the UK to gain additional diplomas, such as Karve and Najmudin.

Backgrounds

All seven Indian private practitioners identified as practising before 1920 in Kenya were Goans. By 1940, however, the composition of the Indian sector of the profession in Kenya had changed. Between 1920 and 1940 Goans still accounted for approximately a third of all Indian private doctors working in Kenya (for example, de Sousa, Figueiredo, Dias, Raymond and Campos), but they had become ever more frequently joined by doctors hailing from different parts of India. This trend unsurprisingly reflected the wider variety of Indians that were increasingly entering medical education.[23] By the 1930s and 1940s names of doctors born in other parts of India became much more commonplace in the official lists. For example, Sheth came from Bombay; Amin, Hamin, Adalja, Topiwala and Patwardhan were all from Gujarat, Oorloff was from Ceylon and Sood and Rana originated from the Punjab.

In accordance with this movement towards diversification of backgrounds, an increasingly broad variety of castes and religion came to populate the Indian colonial medical sphere. Ismail (who arrived in the mid 1930s) was one of the few Bohra doctors in East Africa, Dotiwala was a Parsi, Rana a Muslim, Adalja, a Hindu with a strong affinity with the Oshwal community, Karve was a Maharastrian, Singh a Sikh and Topiwala a Hindu from Surat. As will be discussed in Chapter 8, these affiliations predictably influenced the sort of patient groups each doctor attracted and the kinds of national causes they supported. Despite coming from different provinces and of being of different religious groups, the thing that united most Indian practitioners was that they represented a first generation of Indian immigrants. Indeed, all but one of the doctors appears to have been born on Indian soil. The exception was Elchi Sorabjee, who was born in East Africa to a relatively wealthy business family.[24]

Female Doctors

As in all areas of colonial medical work at this time, and reflective of the structural gender inequalities that concurrently existed in terms of access to medical education, most doctors (of all nationalities) were men.[25] There is no evidence of any Indian women being employed within the colonial medical department, other than as nurses, so Mary de Sousa can be identified as the first female Indian doctor in the country. de Sousa came to Nairobi in 1919 where she established a private practice together with her husband A.C.L. de Sousa (Figures 7.2 and 7.3).[26] As was typical for women doctors of this period, de Sousa held a special interest in midwifery, her proficiency in which was recognised through the prizes she won at Grant Medical College.[27] Her expertise was undoubtedly highly sought after in a community that would have been culturally uncomfortable with consultations with a male doctor, particularly regarding health problems specific to women. Within a few years another Indian female with a Goan background, Helen Figueiredo, joined Mary de Sousa in Kenya. Figueiredo arrived in 1925, and like de Sousa, set up a medical partnership with her husband. Although little is known of Figuereido's specific experiences, the fact that she and Ernst Figueiredo achieved a long mutual career in Mombasa attests to some level of success, and perhaps even satisfaction, with the job and lifestyle.

Distribution of practices

The two biggest urban centres, Nairobi and Mombasa, attracted most of the general practitioners. Of course, doctors gravitated to places where they knew they would attract the most business, so these preferences logically reflected the main areas in which Indians settled. By the mid 1930's Nairobi had become the largest centre of Indian population in East Africa with most practices located there, although Mombasa—also known for its vibrant Indian community—did not lag far behind in terms of the population density of its Indian community.[28]

Figure 7.2   Dr Mary de Sousa
Credit: Lacty de Souza.

Figure 7.3 Dr A.C.L. de Sousa
Credit: Lacty de Souza.

In both Nairobi and Mombasa it was usual for Indian medical practices to be clustered together, a bunching tendency also evident in the urban distribution of European professionals. In Nairobi, the Reata Road was something of a Harley Street for Indian medical practitioners. Several doctors such as Gautama, Patel and Amin as well as the partner practice of Adalja and Topiwala were located there.[29] This hub extended out to the nearby River Road, which formed a T Junction with Reata Road and was home to the surgeries and dispensaries of a number of other doctors, including Drs Dias and Sorabjee.[30] A similar cluster can be identified in Mombasa, where the old town appears to have been a point for the convergence of a number of the Indian medical practices. Both Karve and Dotiwala for example, practised in Ndia Kuu Street there—in Dotiwala's case, for almost 30 years until his retirement.[31] Indian private practitioners did occasionally work in other areas—particularly in Kisumu on the shores of Lake Victoria in Western Kenya. Sood, Singh, Raymond and Oorloff all worked in Kisumu, achieving no small degree of social success and status there.[32]

Motivation to work in Kenya

It is difficult to generalise over the motivations for embarking on a medical career in Kenya, but a few recurrent themes can be discerned. In common with other immigrant diasporas, the prime motive was economic: emigration embodied the search for opportunities abroad, preferably with advantages beyond those offered in the home context. Since pay at the lower echelons of the Colonial Medical Service was low, private practice was seen as a more lucrative alternative, combined with the professional advantages of having the status and independence of being one's own boss.[33] Most doctors who came to work in Kenya from India came from modest financial backgrounds, usually having achieved their medical training with the help of substantial scholarships. In the recollections and obituaries that survive, the pull of the country was often recounted in predominantly economic terms. Dr Adalja was not alone when he recounted that he arrived in Kenya in 1926 with high expectations of bright economic prospects. He stated his disappointment and surprise at the tough competitive conditions he encountered in private practice in Nairobi:

> Somewhat to my surprise, I, who had left India to escape excessive professional competition, found that in Nairobi, there were already established doctors.[34]

Similarly G.V. Juvekar, who became a long serving general practitioner in Mombasa, did so only after a difficult start. Orphaned at early age, Juvekar supported himself throughout his medical education through a series of clerical jobs during the holidays as a means to top up the scholarships and bursaries he received. Immediately after graduating in 1922 he found himself so burdened by debts that he took work as a ship's surgeon. He saw moving to Kenya in 1927 as an opportunity, at last, to pursue a stable and profitable career.[35] In circumstances strikingly parallel, Topiwala, who was also orphaned at six when his parents died in a plague epidemic, fits this pattern very well. An Aunt brought Topiwala up, encouraging him to pursue his medical studies and he managed to get through Grant Medical College with a combination of her financial help and scholarships. Although he had first obtained work in Bombay he admitted to his daughter that it was foremost the pull of a vision of superior professional prospects abroad that persuaded him to take his chance and aboard a ship to Kenya in 1929 together with his friend, Dr Patel.[36]

Even those from relatively privileged backgrounds still saw Kenya as destination to pragmatically make a decent livelihood. Karve, for example, despite his impeccable connections, due to his father's high public profile in India, was also very clear on his economic motivation to come to Kenya: 'I am a doctor and could have got on very well in India but I wanted butter on my bread, and wanted whiskey'.[37]

But although the lure of improved economic circumstances was evident, and stability, if not prosperity, would have been among every immigrant's hopes, it was not in the forefront of every decision to work in East Africa. The most striking counter example is that of Elchi Sorabjee, a Parsee doctor from an economically flourishing family, who entered Nairobi private practice from a situation of privilege and counted Karen Blixen as one of his patients.[38] His father's successful business ventures in Mombasa (which included playing a key role in the construction of Mombasa cathedral) resulted in Sorabjee being born there and gave him a sense of loyalty to the region. It was certainly this advantageous financial situation that allowed his medical education in the UK, making Elchi Sorabjee part of a very small professional elite who had been exposed to international educational opportunities before 1930.[39]

As in Sorabjee's case, personal connections—whether through family members or friends—played an important role in influencing newly qualified Indian medical practitioners to consider Kenya. One of the earliest doctors, Rozendo Ribeiro, was said to have come to Africa on the advice of an elder brother who had immigrated to Angola as a Medical Officer for the Portuguese army.[40] Dr Figueiredo, similarly claimed that his 1925 arrival in Mombasa was based on the advice of a friend in Zanzibar who had informed him of professional opportunities on the East African mainland.[41] Analogous recommendations informed the decisions of Juvekar and Singh. Juvekar was said to have been advised by his lawyer friend who had already settled in Mombasa[42] and Bakhtwar Singh who had seriously contemplated sailing to Fiji after graduation, allegedly changed his mind when a relative residing in Kisumu, recounted its advantages.[43] Topiwala likewise mentioned that one of the reasons why Nairobi seemed attractive was that he had heard very positive reports of the place (along with Rangoon, Burma!) from some of the patients in Bombay, where he worked immediately after graduation.[44]

The lack of opportunities within the government medical department after 1922 meant that Indian doctors would have had very restricted opportunities to work other than as a private practitioner in Kenya from the 1920s onwards. Indeed, three Indian government employees left their official duties in favour of pursuing a private medical career. Upon graduation from Grant Medical College, de Sousa took up a job as an Assistant Surgeon in 1914. Later in his career he was posted to the government hospital in Kisumu, at which juncture he started to supplement his income through treating private patients.[45] Apparently, realising the limited prospects of advancement in British government employment, compared to what he would be able to earn through setting up his own surgery, he resigned and set up a private practise in Nairobi with his wife Mary in 1919 after an interim spell in India.[46] de Sousa was quite explicit about his frustrated ambitions: two decades later in an impassioned speech to the Legislative Council, de Sousa recalled that while he had been in government service the PMO had told him that despite his fine professional record he could not be promoted to be a government MO solely on the grounds that he was a 'coloured man'.[47] Dr Oorloff also began his East African medical career as an Assistant Surgeon in the medical department. It appears that he remained in post between 1897 and 1922, clocking up an impressive 25 years in government service.[48] Indeed, Oorloff seems to have worked in the colonial medical department until the early 1920s when Gilks implemented his drastic cost-saving retrenchments. He appears to have left Government service and obtained a license to set up his private practice in Kisumu in 1923.[49] A suggestion exists that Edward Dias also worked for the East African Colonial Medical Service, despite the fact, that his name never formally appeared in the departmental staff lists, probably because he worked in this capacity for only a year before resigning and embarking on private practice in Nairobi.[50]

The only doctor to move in the opposite direction, from private practice to government service was Bakhtawar Singh. For several years Singh left his private practice in Nairobi and joined the government medical department, where he was sequentially posted to Eldoret, Kisii and Mombasa as an Assistant Surgeon. When his European colleagues allegedly attempted to demote him to the rank of a Sub-Assistant surgeon he resigned and recommenced his former career—although this time choosing to set up his surgery in Kisumu.[51]

Some doctors came to East Africa directly upon graduation, while others did so after first testing their fortunes in medical jobs in their native country, but by and large medical immigrants were enterprising young men who came in the early years of their professional lives. Some individuals first gained experience of specialist subjects as junior doctors in a variety of hospitals in India before venturing abroad (Topiwala for eye, Mary de Sousa for maternity). Karve came to private practice less typically, after spending four years in the Indian army.[52] Others moved to Kenya after periods of work in other parts of the British Empire. Gautama, for example, worked first in Burma and M.A. Rana first tried his hand at private practice in Zanzibar before transferring to Mombasa.[53]

Lastly, the relative geographical proximity of East Africa to the Indian sub-continent combined with the long Indian migratory history to the African East coast would have made the move seem less radical to those who initially left; many of who had intentions to eventually return back to the land of their birth. This was a new medical community that although adaptive to local political and social differences was ultimately 'never cut off from its homeland even as the political and economic reality of Kenya framed the immediate context of its political imaginary'.[54]

Social Status

Crozier has described the increased social status the profession of colonial doctor gave British medical graduates and, although the dynamics were subtly different for Indian medical practitioners, the sense of being metaphorically a big fish in a small pond undoubtedly still played a part in marking out the attractiveness of the career.[55] Deprived of access to non-subordinate government positions, no Indian doctor could hope to enjoy the 'life of a country gentleman' that Colonial Office bureaucrat Charles Jeffries promised European recruits to the Colonial Service, but private practice nevertheless provided Indians access to certain social advantages concomitant with being part (even if only on the margins) of a professional elite within colonial society.[56] The very possession of a university education conferred an enhanced status on individuals who were part of an immigrant population largely characterised by their poor education. The Indian doctor stood out because he was educated, generally Europeanised in dress and demeanour, spoke English, and was frequently among the privileged few in the community who (from the 1930s) owned a car.[57] Recollections from the period recall how the doctor often had strong personal relationships with the families in the catchment area of his practice and was regarded as a respected figure of Indian community life, rather like the British country doctor of the time in the UK.[58] On home visits, the doctor was said to be formally greeted by the family elder at the entrance of the house and was addressed with the affectionate, but also respectful designation: 'doctorsahib'.[59]

Competition

The relatively high status accorded to Indian private doctors made them somewhat socially separate from the poorer majority of the Indian community that inhabited the bazaars colonial Nairobi and Mombasa. But, in the period before 1940, less educated Indians often feared western biomedical intervention and, even if they did not, many could not afford the comparatively high fees that an Indian general practitioner would charge. Moreover the Indian sick had other traditional practitioners available to them in the local medical marketplace. Indeed, the presence of *hakims* and *vaids* (traditional practitioners of Unani and Ayurvedic medicine) were protected by the British colonial authorities who legally permitted them to offer their services so long as they confined their activities to their own communities.[60] Although indigenous medical practitioners would, in theory, have targeted a slightly different sector of society than university trained Indian doctors, there is evidence that registered Indian doctors disliked this element of competition, so much so that they made some attempts at lobbying to have the services of *hakims* and *vaids* curtailed.[61]

Of more concern to Indian doctors was competition from practitioners who were perceived to hold an equivalent training, particularly from Indian Assistant Surgeons or Sub-Assistant Surgeons keen to augment their incomes by taking on private work. European private doctors also expressed their annoyance at the competition that emanated from within government service, although logically enough Europeans were more worried about other European Medical Officers encroaching on their market, rather than Indian Assistant and Sub-Assistant Surgeons. European and Indian private doctors both expressed their antipathy to the competition they felt the government quarter unfairly allowed. Several arguments were advanced, including an appeal to the colonial government's ever-present concern to save money. These moonlighting doctors, it was claimed, used government accommodation, medical supplies and time to conduct their private consultations.[62] Moreover, it was argued that private practice constituted an unnecessary perk of the job for European Medical Officers, as they were already in the fortunate position of being in receipt of a regular and respectable salary. Anything they received privately unfairly took business away from bona fide private practitioners, who had no natural constituency requiring their services and more of a struggle to attract regular clients. In outlying areas where work was inevitably more meagre, it became impossible for a private practitioner to make ends meet if he was being undercut by the extracurricular activities of a local government doctor.[63] Indian private doctors could also legitimately claim that most of them held higher qualifications than Assistant and Sub-Assistant Surgeons who made extra money in this way, since most Indians in the employ of the colonial medical department were licensed, rather than registered. The reasoning that many of the Indian doctors employed by the Government did not fulfil the more exacting GMC standards in their delivery of medicine, and thereby potentially exposing paying patients to inferior care had been a contentious issue in India itself.[64]

The issue of private practice by Government doctors was considered less serious in Kenya (compared to Uganda) but the Medical Departments in both the countries were reluctant to take action because the BMA demanded the payment of compensatory allowances to European doctors if their private work was curtailed. In Kenya both the PMO and the BMA had supported private practice as an inducement to attract European candidates to the Medical Department.[65] The option to deprive Indian staff of the privilege (while Europeans were permitted to retain the right) was floated but considered not possible. These debates, concerning the situation in Kenya, Uganda and Tanganyika, continued throughout the period and remained unresolved, but they had a high enough public profile in 1931 to be brought to the attention of the Secretary of State to the Colonies by the Governor of Uganda.[66]

Philanthropy

Many doctors assumed leadership roles in community activities outside of, but sometimes related to, their profession. Numerous examples exist of Indian general practitioners serving on school boards, local social committees or donating their time to community based organisations such as the Goan Institute, Arya Samaj or the Social Services League.[67] Dr de Sousa played a leading role in the foundation of the Desai Memorial Library. Indeed, an interest in education was particularly prominent and a number of doctors throughout the period were closely involved in the charitable promotion of educational initiatives. Ribeiro, for example, spent much of his free time sponsoring improvements in the standard of education on offer to young Goans in Kenya. In 1931 he donated the large sum of 30,000 Shillings towards the cost of building a high school in Nairobi. In his honour, it was named the *Dr Ribeiro Goan High School*, which still exists in Kenya today (albeit under a new name).[68]

Ribeiro was not alone. Involvement in the establishment of schools played a part in the lives of other Indian doctors, including Dr Figueiredo, who devoted time, effort and funds to the construction of the School for Goan immigrants in Mombasa.[69] Dr Sheth was similarly pivotal in the formation of a private secondary school also in Mombasa; its focus was to help less privileged youngsters who had not passed the restrictive Kenya Preliminary Examination (KPE) required for admission to a government sponsored secondary school.[70] Dr Karve was heavily involved in the foundation of the Indian Girls' school in Mombasa following the footsteps of his father, Dr D.D. Karve, who had been something of a pioneer for women's education in India.[71]

Others used their social influence to work for improvements via political channels. The frequently politically active Mary de Sousa and Adalja subsequently served (although at different times) on the Government appointed

Advisory Council on Asian Education. de Sousa's activities in this forum show that she was interested in advancing the opportunities that were available for the schooling of Indian girls and for the appointment of Indian female head teachers.[72] Adalja used his position on this Council to campaign for educational reforms.[73] Topiwala was another active community leader deeply committed to the education of women, who regularly attended the school board of Adarsha Vidyalayai, a private Indian school.[74] These contributions fit in with one of the most enduring stereotypes of Indian diaspora communities: namely the strong cultural belief in the transformational effects of education.

Aside from an interest in the sponsorship of educational advancement in the Indian community, Indian private doctors were also key players in the delivery of medical philanthropic initiatives in the country. Their involvement referred back to a tradition of medical philanthropy on the Indian sub-continent, which had been transplanted to Zanzibar in the second half of the nineteenth century.[75] Almost mythological in their own lifetimes, two prominent Indian traders Siwa Haji Paroo and Tharia Topan played a crucial role in the foundation of early medical facilities in the East African region.[76] Tharia Topan famously commissioned the 'old dispensary' at the centre of Zanzibar Stone Town and also sponsored the founding of the *Jubilee Hospital* on the island, inaugurated in 1885. The Zanzibari Indian Paroo similarly founded the *Sewa Haji Hospital*, on the mainland in Dar es Salaam, in 1895.[77]

This habit continued in colonial Kenya. In his *Annual Medical Report* of 1941 Principal Medical Officer, Arthur Paterson, acknowledged that the financial provision for the building of a permanent hospital in Thika (including 40 beds for Africans and 6 for Asians) was only made possible 'by the generosity of an Indian merchant living in the town.'[78] Reported in the same year, was the addition of an Asian ward at Kitale hospital that was also attributed directly to financial aid from the Indian community.[79] This quiet reliance of the British government upon Indian endowments for medical developments is substantiated by other accounts: Salvadori, for example, detailed how an Indian community group, the Rattansi Trust, heavily subsidized the missionary hospital at Tumu Tumu.[80]

The organisation that repeatedly appears in recollections and accounts detailing involvement in community healthcare is the Social Service League (SSL). Little is known about the early history of this organisation, but it appears to have been one of the first charitable groups in Kenya specifically aimed at the Indian poor. Founded in Mombasa in 1921 and in Nairobi in 1933, the SSL concerned itself with educational and cultural improvements as well as the provision of free basic medical care and public health information for 'poor members of the non African communities.'[81] Under its guidance a dispensary was set up in Nairobi in 1934. The costs were financed from the proceeds of local Indian fundraising and a matching grant of 50 pounds from Nairobi Municipal Council.[82] An Indian doctor was recruited to staff the dispensary, which distributed medicines free of charge, and offered injections and home visits at a rate below that of private practitioners.[83] Many Nairobi doctors were known to have been closely involved with the activities of the Nairobi SSL, including Adalja and Topiwala.[84] Despite its intent to help Indian community members, however, some Indian practitioners argued that the dispensary formed a competitive threat to their own private practices because it sometimes treated patients who were not genuinely impoverished. As Patel stated: 'It was against all canons of social service that 'rich and poor should benefit from donations which are solely meant for the poor'.[85] Notwithstanding these hiccups the Nairobi SSL and the Mombasa SSL were sufficiently successful to warrant the opening of regional branches. Both Dotiwala and Hamin had volunteered their services to the dispensary of the SSL in Mombasa.[86] Sood was also said to have been President of the Nyanza SSL, although precisely when is less certain.[87]

The most striking medical contribution that occurred as a direct result of Indian philanthropy was the foundation of the first maternity hospital for Indian women, the Lady Grigg Maternity Home in Nairobi.[88] This initiative began when Lady Grigg, the wife of the Governor, called for improvement of the situation of all women of the colony and to this end founded the Lady Grigg Welfare League in 1926.[89] Immediately a committee of 20 people which included the European doctors, G.V. Anderson, Jex-Blake, Arthur Paterson, Christopher Wilson and three Indians, including Mary de Sousa as a doctor and Y.M. Ganiji and V.V. Phadke, spearheaded development of separate schemes for Indians, Africans and Europeans and started fundraising for their individual causes. It was decided that a maternity home was the most pressing need for Indian women of the community. Mary de Sousa and her husband played a key role in all aspects of the scheme, but particularly in its fundraising initiatives.[90] A glance at the figures from an early fundraising fête held in the grounds of Government House sheds light on how much the project touched the hearts and minds of the Indian community in Kenya. Sh11,668 was received as the Indian share of the fete, the Aga Khan (the leader of the Khoja community) donated Sh 30,000, the Indian Women's Committee collected a sum of Sh15,283 and there were additional sums from other donations of Sh26,000 (Figure 7.4).[91]

Despite evidence pointing to various internecine squabbles over the precise constitution of the board of the hospital, the development was hailed as one of the high points of the late 1920s.[92] When the hospital finally opened its doors in 1928, reports in the local press reflected the enormous enthusiasm generated, even though the initial capacity was modest at only six beds.

Beyond the practicalities of contributing to better maternity facilities for Indian women in Kenya, the progress of the project illustrated the

Figure 7.4 The Indian Maternity Hospital, Nairobi, 1933

Standing: Nurse Khatijabai, Nurse Jenabai, Mr S.T.Thakore, Dr K.V. Adalja, Mr Lahori Ram, Dr H.T. Topiwala, Mr D. Genovar: *Seated:* Mr U.K. Oza, Mr Salehmohamed, Mrs C.Diveton—Smith, Mrs Sorabjee, Mr B.S. Varma, Dr G.L. Gilks, Mr A.A. Legat, Mrs Maxwell, Dr M.M. Shaw. Credit: Harshad Topiwala private collection.

possibility for close cooperation between Indian private doctors and prominent Europeans. As the photograph of the trustees, Adalja and Topiwala with Gilks shows (see above), the PMO was evidently not adverse to working with Indian members of the medical community, despite the restrictive policies that he had tacitly engineered against Indian participation in the colonial medical service in the 1920s.

Professional Experiences

The account Angela Ribeiro gave of her childhood memories of her father, Rozendo Ribeiro, paints a picture of a rather leisurely and gentle lifestyle, certainly not one that seemed particularly pressurized.

> He would be at his surgery by 9 am, returning home for lunch at midday. After lunch he would relax with the newspaper before returning to the surgery, where he worked until 4 pm. Then before dinner he would he would go for a walk or take us children for a drive in his horse drawn carriage.[93]

Whatever the truth of Angela Ribeiro's description of a 'typical' day in her father's working life the work of an Indian private practitioner was surely not particularly stressful compared to the daily toils of the majority of their fellow nationals in Kenya. There can be no doubt about the diverse requirements of the job however. As Crozier described for European government doctors in the same period, the successful Indian doctor needed to be a jack of all trades in a context characterised by a meagre number of qualified doctors serving a large population and a paucity of facilities.[94] As one European wryly recalled, doctors in the colonial period had to simultaneously be 'nurse, dietician, health educator, dispenser and even ambulance driver'.[95] Adalja made a similar point when he noted that the successful general practitioner had to double as an expert in 'preventative medicine and as a family doctor as well'.[96]

In the light of this demand for professional versatility, the doctor's surgery came to serve many functions beyond consulting rooms, often additionally acting as an operating theatre, laboratory and pharmacy. Topiwala's daughter, Hansa, herself to become a physician, remembered helping her father with some of his routine surgical work:

> Many of his patients were Sikh craftsmen who handled machinery and got metal and wood splinters into their eyes or crushed their fingers in machinery. On weekends when he had no assistants and when I was 10 he would take me to the dispensary and I would help him hold a retractor while he worked. He always explained what he did—how to invert the upper lid of the eye and put local anaesthetic drops and then with

a magnifying lens he would delicately use tweezers like instruments to remove the splinter. He would wash the injured finger with some anti-septic solution and put local injections on the two sides where the nerves ran and then he would open the wound at the base of the finger, very carefully to take out the small crushed pieces of bone which would otherwise be recognised as foreign objects by the body and the wound would not heal until they all worked themselves out.[97]

Hansa's involvements in her father's medical practice went so far as to include helping him to mend simple fractures. She remembered 'helping him holding limbs whilst he applied plaster of Paris in the dispensary'.[98] Unsurprisingly doctors became locally well known for their skills in certain areas of medicine. Despite not having a specialist qualification in ophthalmology, Topiwala became known in Nairobi as an expert in treating eyes, another of his daughters, Mina recounted her father replacing cataracts and treating trachoma among his regular duties.[99] Juvekar in Mombasa became particularly recognized for his work with children.[100] Bakhtawar Singh was well known for, among other things, acting as the local optician and dentist.[101]

Partly because there were scant hospital facilities available to Indians, not only did some doctor's surgeries approximate to hospitals, they occasionally contained rudimentary laboratory facilities. Adalja and Topiwala were said to have been 'extremely proud' of the modest laboratory facilities that existed in their surgery in Reata Road, Nairobi. Topiwala spent 'considerable time to train an African technician who was then able to perform a complete blood count, urine and stool examination'.[102] Along similar lines, Dotiwala's Mombasa surgery was remembered for its interior 'lined with shelves which held large bottles of various colours as well as jars, boxes, tins, vials, and various pharmaceutical paraphernalia'.[103] Because few pharmacies existed in colonial Kenya, it was common for general practitioners to employ compounders on their premises to make up medicines on the spot. Indeed, Ribeiro's own formulations of malaria tablets were said to have become much in demand by both Indians and Europeans.[104]

Although most public health responsibilities fell under the remit of the Government medical department, there is evidence that some private doctors also played a role in this realm. Perhaps most famously was the contribution of Rozendo Ribeiro who was said to have identified the 1902 plague epidemic in Nairobi, and received a plot of 16 acres of land from the British government in recognition of his services.[105] Figueiredo, Karve, Juvekar and Sheth were remembered too for their dedication to public health, giving free lectures on the topic to the employees of the Uganda Railway.[106] Likewise, Sheth demonstrated his public health interests through his involvement as a founding member of the Family Planning Association of Mombasa and Adalja and Topiwala contributed articles on preventative medicine, nutrition and hygiene in popular Gujarati magazines.[107] Adalja and Topiwala were instrumental in the organisation of special training courses for Indian nurses for the first maternity hospital.[108] In an interview many years later Sir James Simpson recalled the names of three prominent Indian doctors in Kenya: Ribeiro, Adalja and Topiwala. He praised 'the Gujarati H.T. Topiwala' and claimed that he was

>...remembered for his probity in business as well as medical knowledge. Against all social norms of the day, the socially—conscious physician based his charges on his patients earnings rather than on fixed fees. 'What do you earn' he would ask, 'I have to know what to charge'.[109]

It is perhaps a consequence of the rich opportunities offered by the diverse nature of local medical practice that led several doctors to feel able to publish their medical findings despite having no specialised qualifications. Mary de Sousa became something of an expert in midwifery, publishing her research in the *Kenya Medical Journal* in 1924.[110] Karve too successfully submitted articles to the medical press in Kenya and the UK on various topics, including articles on midwifery and malaria.[111] Adalja contributed several letters and notes to international medical journals or BMA meetings on a range of subjects including Ayurvedic medicine, malaria treatments and haemophilia[112] Other examples saw Indian private doctors participating in the published professional debates over specific diseases[113] or the use of drugs such as M and B 693.[114]

Since most Indian doctors resided in the main townships, where their largest patient bases were to be found, they needed to travel regularly to rural areas to serve patients as part of their normal course of duties. In the early years of his practice in Nairobi, Dr de Sousa used to visit his patients on a motorcycle, before he—like a number of private medical practitioners—acquired a car in the 1930s. Accounts of his time in Nairobi confirm that Sorabjee paid regular visits to the Indian residential areas of River Road, Old Whitehouse Road, Ngara, Pangani and Eastleigh.[115] Topiwala, also based in Nairobi, was remembered by his daughter as often driving 'to outlying places in the countryside in case of emergencies to see patients where Asians had dukas'.[116] Adalja recollected that '[h]ours were irregular and... [n]ight calls common as were visits to outlying places'. Adalja and others regularly visited places such as Thika, Machakos, Limuru, Kiambu, Naivasha and occasionally even as far as Nyeri, which was no small undertaking in the first decades of the twentieth century as it was about a hundred miles away.[117] In some ways the need for long distance travel was even more pronounced for doctors not working in the capital. Sood working in Kisumu regularly used to travel up to 100 miles from his surgery to treat patients in outlying small communities in Nyanza.[118] Furthermore, Indian general practitioners would visit the typical residences for the majority Indian poor, where a family of six or more might share a single bedroom 'using tiered beds like bunks in a ship' in extremely basic living quarters.[119] Ribeiro's (son of Rozendo) reminiscences of helping at a home labour conjure up the unsophisticated environment in which he often had to work. He recollected he 'had to rotate and apply forceps with the aid of two small electric torches and an open flame light while my wife gave chloroform.'[120]

Unsurprisingly, Indian doctors mostly looked after Indian patients, although Europeans and Africans sometimes sought their services, as discussed in Chapter 8. Yet even though the doctors and their Indian patients would have shared the same broad cultural ethos, which is far from saying that they were entirely uncritical about the health problems (and attitudes to them) of the Indian populace. Indeed, through their training in biomedicine Indian doctors often found themselves somewhat socially and culturally distanced from the communities that they served. As Karve remembered, the bulk of the Indian patients visited him only when they were in 'serious difficulties', mostly because they were nervous of the potentially high fees, but also because they were used to relying on home medicines. He felt that Indian women were particularly difficult to convince as they 'are very much averse to having their confinements in nursing homes.....'[121] Karve described how he was often presented with illnesses only when they had reached an advanced stage, making them more difficult to cure.[122] Adalja analogously listed among his frustrations as a young Indian doctor in Kenya, the ignorance of many of the local Indians about the basics of elementary hygiene and nutrition. He bemoaned how many of his patients had set views on diet, the taking of medication and how they would sometimes avoid having blood tests for fear of the social effects of a positive diagnosis.[123] As supporters of western medical principles, many Indian doctors actively tried to counter what they perceived to be the superstition and prejudice that existed against biomedicine in much of the Indian community. Karve, for example, was recorded as saying that he was 'disgusted' by the unclean maternity methods he had witnessed among Indian communities and wholeheartedly supported the rejection of these practices explaining that he thought that the 'dayas [traditional Indian birthing attendants] have been rightly condemned by the health authorities in this country and India'.[124]

Yet despite some cultural gulfs, the ethnic provenance of the doctor gave them a natural relative sympathy for certain medical practices within Indian society. Karve, despite vehemently protesting against the dayas, also conceded that he felt the Indian midwifery method of 'posturing' the delivering mother was very beneficial.[125] Adalja, despite being a doctor convinced of the efficacy of western medicine, declared his support of a selection of Ayurvedic treatments within a public lecture to the BMA.[126]

A tacit cultural identification with their community meant that Indian doctors became expert in the health requirements of each sub-community, each of which held idiosyncratic attitudes towards western medicine and a relative propensity to certain diseases. This phenomenon of differentiating among community members was already familiar in Indian society, and also to Indian doctors in Kenya, who realised that Indians should not be grouped together for all medical purposes. Hindus, Sikhs, Goans, Parsees, Bhoras, Khojas, orthodox Muslims all had subtly different attitudes to doctors, hospitals, treatments and vaccinations.[127] Although there were exceptions to every generalization, by and large immigrant Parsees were characterized by their better access to education and comparative westernization. They were more likely to comply with biomedical demands. Similarly, Parsee women were more amenable to receiving treatment from male doctors, unlike most other Indian female community members who would have found this taboo. Khojas and Goans were also regarded as less conservative and more amenable to the modern methods (Khoja ladies were some of the first Indian women to put themselves forward for maternity nurse training). Among Muslims, Khojas were comfortable in accepting inoculations and were prepared to stay in isolation hospitals, while the Orthodox Muslims were not.[128]

Distinctive life styles among the diverse Indian sub communities sometimes led to them being differentially susceptible to certain diseases, often because of the different diets that their cultures demanded. Most Gujaratis, for example, were strict vegetarians and had religious taboos against eating meats and eggs. Patients from this group would refuse liver injections and any form of meat or fish extract. By way of contrast, Sikhs traditionally did not eat beef, were said to drink a lot of milk and also had a reputation for the consumption of excess brandy. In the same vein of customary generalisations, Muslims, eschewed pork, were fond of greasy food and their womenfolk who followed the custom of purdah suffered from lack of exercise. Goan diet had a tendency to be oily and highly spiced. Some Indian doctors became interested in studying these differences noting that high incidences of avitaminosis and anaemia were the result of diets that were too rich in oils and carbohydrates and lacked essential amino acids.[129]

Social Life and Recreation

Social conservatism meant that it was regarded as preferable for the Indian private doctor to be married. [130] In the early years of colonial Kenya this was not always easy and suitable wives were few and far between. Intermarriage was almost unheard of, and even marriage between different India castes was extremely uncommon. Some doctors, such as de Sousa initially left their intended partners behind in India, giving themselves time to establish themselves in Kenya. Securing suitable family accommodation was regarded as particularly difficult because of the shortage of well-constructed houses with adequate amenities before the 1930s. If unmarried, it was not unusual for an Indian doctor to return to India after a few years to seek a bride before resuming his career in Kenya; such as was the case of de Sousa and Topiwala.[131]

For the doctor's wife, life in the new continent posed formidable challenges. Prospects of employment were few and many Indian wives would have found themselves leading relatively isolated lives, often in climates much cooler than they would have experienced in their Indian homeland. One doctor's wife, Kalavati Topiwala, a university graduate, who had been raised in hot Surat, particularly mentioned the climate in her recollections of her early years in Kenya, adamantly attributing the cold as reason for her long-term asthma condition.[132]

The life of the Indian doctor bore no resemblance to the one enjoyed by his European counterpart. Indians were barred from all of the best hotels, clubs (which in Nairobi meant the Nairobi and Muthaiga clubs) and the elite enclosures at Nairobi racecourse. If, as Crozier has argued, European Medical Officers understood that a successful career depended upon conforming to the social structures of colonial society, it can be seen that Indians had a rather different, subordinate, niche within this context.[133] On one hand, Indians would have been barred from most of the smart social venues, but on the other, they did enjoy an elevated social status relative to the rest of the Indian community and many were active supporters of the British Empire, receiving honours and decorations for their services from the British government (Chapter 8). Indian doctors participated as much as they could in the local clubs and societies. A.L. Ribeiro was a founding member of a Nairobi Freemason Lodge, a group with which Adalja and Ismail held long associations.[134] S.D. Karve subsequently became one of the first non-white to become a Rotarian.[135] M.A. Rana was a founding member of the Mombasa Rotary Club.[136] V.V. Patwardhan was very closely involved with the Nairobi Scouts and the Desai Memorial Library.

Some of the doctors were enthusiastic fans of cricket and tennis but the facilities available to Indians in Nairobi were minimal until the late 1920s when the colonial authorities granted land to Indian organisations in Ngara and Parklands to enable the setting up of sports facilities (Figure 7.5). Except for the SVIG (Sulemen Virjee Indian Gymkhana, Nairobi), which was established in 1927 for all members of the Indian community, Indian clubs tended to be organised along local community lines. The most prominent clubs were the Patel Club, Ismaili Sports Club, Sir Yusuf Ali Sports Club, the Sikh Union, and the Goan Gymkhana.[137] Several doctors, such as Sorabjee, Adalja and Topiwala were active members of the SVIG.[138]

The eccentricities of several members of this group are evident. Most famous, is the image of Ribeiro riding his zebra, but others also had colourful personalities (Figure 7.1). Sorabjee particularly was remembered for his colourful manner, obtaining his pilot's licence at 31: an event that saw him featured in the local newspaper.[139] Oorloff too made a big impact on the youth of Kisumu sharing his hobbies, including kite making and flying, and collecting pictures for albums with locals. He was one of the first people to own a radio in Kisumu and installed a loudspeaker outside his front door for the benefit of people who were keen to listen to the news.[140]

150

Figure 7.5 Doctors versus Advocates, Nairobi, 1936

Standing: Dr Karve (Umpire), Dr M.D. Gautma, Dr Bakhtawar Singh, Mr. M.L. Anand, Dr M. Ribeiro, Mr C.M. Patel, Mr U.S. Kalsi, Dr V.V. Patwardhan, Mr S.G. Amin, Dr K.V. Adalja, Mr J.A.C. Burke, Mr D.D. Puri (Umpire). Sitting: Dr R.P. Patel, Dr A.H. Ismail, Mr B.S. Varma, Dr E. Sorabjee (Capt.), Mr N.S. Mangat (Capt.), Mr V.V. Phadke, Dr P.M. Amin, Mr S.R. Kapila. Ground: Mr H.A. Rahman, Mr H.D. Trivedi, Mr H.E. Nathoo, Dr H.T. Topiwala, Mr J.M.Nazareth, Mr B.T. Modi. Credit: Harshad Topiwala private collection.

A glance at the copious warm obituaries and tributes that were paid to a number of Indian doctors, confirms the regard in which they were held within their communities, and also, on occasion, by European colleagues (Chapter 8).[141] Perhaps the most striking example of the local esteem in which many of these doctors were held was a conference held in 1949 by the Indian Medical and Dental Association of Kenya, specifically to commemorate half a century of public service by Rozendo Ribeiro.[142] But numerous other individual instances are to be found within the tributes. Da Gama's 'energy and marked affability' with his patients was remembered; de Sousa for his 'fiery spirit' and 'leadership'; Karve for 'rendering great humanitarian services'; Juvekar was praised for his untiring hard work and diligence; Sheth for his 'outstanding personality and services'; Adalja for his 'kindness and consideration'; Sood for his 'outstanding public service'; Gautama for his 'sterling qualities'; Dotiwala,—known affectionately throughout Mombasa as 'yule dakatari Parisi' or 'that Parsi doctor'—was well known for his generosity.[143]

Many Indian doctors gave their entire careers to Kenya. To mention just a few: Ribeiro served for 51 years[144]; Hamin for 33 years;[145] VV Patwardhan for 40 years;[146] Dotiwala for 30 years[147]; de Sousa for 43 years[148]; Sood for 41 years;[149] Karve for 43 years; and Gautma for over 30 years.[150] Topiwala and Adalja were in partnership for 35 years and Topiwala treated patients in all for 47 years.

Many of these doctors were the first generation representatives of medical families the descendants of which still practise medicine today. One of Ribeiro's sons, Ayres Ribeiro, trained as doctor and developed a medical career in Kenya, initially taking over his father's practice and later becoming a well-known police surgeon in own right.[151] Adalja's elder son established a practice in Nairobi in the same premises on Reata Road that his father had occupied with Dr Topiwala.[152] Three of Topiwala's daughters were among the first Kenya-born Indian women to obtain medical degrees. One of Sood's sons became a doctor and had a distinguished career in Canada.[153] A.C.L. de Sousa, S.D. Karve, Bakhtawar Singh, and E. Sorabjee all had children with medical or dentistry careers.[154] Family dynasties of doctors are relatively unremarkable in the European context, but such career choices by the offspring of pioneering doctors in Kenya give them added significance. Most of the Indian doctors that came to Kenya between 1900 and 1940 had come from relatively humble financial circumstances, often forging their way through scholarships and grants. In the course of their careers many of them had created sufficient financial stability to be able to send their children abroad to medical schools. This represented no small expense before the 1960s when Nairobi Medical School was founded (1967) and the only option was to pay for education abroad. The number of these doctors that unwittingly founded medical families is lasting testament to their tenacity and success.

8
Private Doctors: Practising Medicine in a Segregated World

> That the standard of health which prevails among Europeans in Kenya is very much higher than that which prevails among either Asians or Africans, there can be no doubt.[1]

Indian private doctors experienced long and varied careers, they were held in respect and esteem in the Kenyan Indian community and they regularly acted as local philanthropists and community spokespeople. Yet, Indians—even if professionally qualified and ostensibly middle class in their tastes and demeanour—were always to constitute a second class to the white ruling elite.[2] As demonstrated in Chapter 6, Indians were ultimately, if unofficially, excluded from the Colonial Medical Service, but this was not the only site in which they experienced discrimination. Ample evidence also exists of the racial struggles of Indian private practitioners. Indian doctors were far from able to enjoy the social, political and professional benefits of their European counterparts with most finding that, even if they supported British governance, the colonial politics of race continually touched the remit and nature of both their public and private lives.

Race and Hospital Referrals

For an Indian doctor in colonial Kenya the most manifest proof that Asians were not regarded as equal to Europeans was demonstrated in the standards of medical facilities that were available to them. Although Indians could achieve access to facilities and treatments above those offered to the African members of the colonial population, Indians still felt indignant about the quality of medical resources available to them compared to those available to Europeans. When defending their medical rights Indians drew on a number of arguments. They declared that their lack of access to hygienic biomedical facilities was disproportionate to the social, economic and political contributions Indians had made to Kenya. Utilising similar arguments as those that were deployed in the struggles leading up to and around the

Devonshire Declaration, pointed reference was made to their unfair share of the tax burden relative to the number of medical facilities available to them.[3] There was a lingering belief that Indians should be respected and rewarded because of their active involvement in British military campaigns, and that recognition should be given for the way that they had been central in the infrastructural developments of the colony: such as building the Uganda Railway and staffing the colonial bureaucracy.

The complaints from doctors trained to support western standards and hygiene are unsurprising. Before 1940 the hospital and diagnostic facilities available to Indians were sparse and inadequate. In 1908, the Principal Medical Officer reported a situation in Nairobi where out of 'common humanity' some of the Indian clerical staff were exceptionally taken into the European Hospital for treatment. Notably, he felt this drastic measure was nevertheless somewhat 'objectionable as it introduces the colour question'.[4] The real picture in Nairobi, if anything, was even worse. To take a specific detailed example: before 1940, the only place that Indians could receive institutionally based treatment was at the Native Civil Hospital (established 1901 and later called the KAR Hospital). Although this hospital mostly catered for African patients, a section of it was designated for Indian use. Facilities, as reported by the Indian representatives of the legislative council, were described as so poor, however, that Indian patients were reluctant to use them. M.A. Desai said of Nairobi hospital that it was 'so bad that none of us would like to be there', Sham-su-deen's similarly pronounced conditions as 'no better place than would be used for treatment of animals'.[5] In 1913, the Governor himself had declared the hospital to be 'the filthiest sight he had ever seen'.[6] As late as 1929, the observation by Medical Officer Hugh Trowell that 'the hospital had no running water (except from two stand pipes), no electricity, no microscope, no test tubes, and a few drugs' seemed to suggest little improvement in the Native Hospital's environment.[7] For the Indian patients their own bedrooms or the doctor's surgery became the *de facto* hospitals.[8]

In the early 1930s hopes were raised when a scheme was aired for a proposal to build a new hospital complex for Nairobi, The Group Hospital. This was to include separate wards for Indians, Africans or Europeans, but would combine laboratory and personnel resources. Eagerly supported by Indian health community representatives, such as R.N. Shah and Dr de Sousa, the plan was debated in both the local branch of the BMA and the 1935 Legislative Council but stalled, and then entirely collapsed, because of the lack of funds.[9] Indians additionally bitterly protested via the letters pages of *The East African Standard* over the demise of the Group hospital.[10] Shah claimed that for more than ten years the colonial administration had deflected demands to provide suitable hospital facilities for Indians and the abandonment of this project was yet another example of the low priority accorded to public investments for non-Europeans. A suspicion was also aired that one of the main reasons why the project flopped was underpinned by racial anxieties. It was claimed that Europeans feared the concept of a combined hospital—even one with separate wards for each racial group—allegedly worrying that the shared blood banks and laboratory facilities might facilitate unwanted racial contacts.[11]

By the time of the Second World War, hospital provision for non-Europeans in Nairobi had widened with more beds in the Native Hospital and the establishment of the maternity homes, but was not drastically improved from the inadequate state of affairs described in the first decade of the 20th century. In 1937 four major hospitals existed in Nairobi: the European hospital of 31 beds, the native hospital of 256 beds, and the infectious diseases hospital of 149 beds and a mental hospital of 256 beds.[12] Because the European hospital was exclusive to whites, there was only one possible destination for sick Indians who were not forcibly quarantined or deemed mentally ill. The native hospital that had 20 beds reserved specifically for Indians had an Indian occupancy rate so low that it could only be surmised that a visit there was regarded as an unpopular one way ticket. The hospital buildings in Nairobi and Nakuru were described, in an official report, as 'unsuitable iron structures, exceedingly uncomfortable and cramped'.[13] Furthermore, unlike patients in the European hospital who could call in doctors of their choice, Indians were denied access to their western trained Indian family doctors (as well as nurses and food) in the native hospital, thereby depriving them of familiar sources of comfort and care.[14]

In effect, the poor facilities and the policy of only allowing certain ethnic groups in certain hospitals—with the best provision being allocated to the Europeans—meant that when faced with serious cases, Indian private doctors could neither check their patients nor even their sick family members into the hospital that best suited their needs. This led to all sorts of grievances. R.A. Ribeiro, for example, despite his place as a prominent member of Nairobi society met with resistance in getting his family access to the European hospital in the capital.[15] Edward Dias, notwithstanding being married to a French woman from the Seychelles, and having received his medical training in the UK, was denied urgently needed treatment in the European hospital when his daughter fell desperately ill. Her subsequent death in the 'dirty, dingy little' native hospital caused long lasting resentment within his family.[16] Furthermore, this discrimination was backed up in law and went on until independence. The Kenya Hospital Association the organisation, which managed the European hospitals between 1950 and 1961, had formalised the pre-existing tacit policy of admissions in a clause which stated: 'No person who is not of pure European decent and no person who is a lunatic or idiot shall be admitted to the membership of the Association'.[17] Complaints about access to facilities came mostly from within the non-European community who felt the direct negative impacts of this policy, but sometimes European doctors too found the racially-exclusionist nature of colonial medical provision unreasonable. Examples from the 1950s indicate that if the exclusionary practices were going on then, they most certainly were in the period before 1940. For example, Dr Hicks described his regret over an episode concerning an Indian patient who was refused the use of the superior operating theatre in Nairobi European hospital.[18]

As the *Annual Medical Reports* for the period 1910–1940 attest, the Medical Department was under no illusions as to the desperate need for improved hospital facilities for Indians.[19] Yet little tangible improvement was made. In an attempt to advance things, in 1931 Indians asked for the formation of a committee to advise the government on the Indian community's health needs.[20] The request was denied. The Petitioner for this cause complained in strongest terms, about the lack of interest, even going so far as to frame the lack of official blessing given to the idea as an example of the colonial government's criminal negligence.[21] It is not surprising that Indians relied, when they could, upon their own resources, although ironically, it was the community's strengths as well as their weaknesses that created circumstances that worked against them.[22] Deprived of access to the higher echelons of government that could reform medical policy, Indians struggled hard to make their complaints heard, but also, perversely, the fact that they seemed to cope (thrive even when compared to the majority African poor) and that Indians were associated with strong cultural traditions of self-reliance and community philanthropy served to take the pressure off the colonial administration to take action. Consequently, time and again, and despite repeated requests to the contrary, the colonial medical department put off plans to provide a decent hospital for the Indians stating that it did not have the money to provide the facilities for Indians.[23] Indian politicians had complained in 1925 that the budget allocation for Indian hospital provision was very low compared to that provided for Europeans.[24] Access to modern hospitals in Nairobi only became a reality in the late 1950s when Indians themselves established the Aga Khan Hospital and the M.P. Shah Hospital (previously the Parklands Nursing Home). In Mombasa improvements also took some time to come into being. It was not until 1947, in the form of the philanthropically funded Pandya Memorial Clinic (later Hospital), that general hospital facilities specifically for Indians were established.

Race and Private Practice

Racial politics also pervaded the surgeries of private practitioners. One thing that was regarded as particularly irksome by Indians surveying their professional competition was the fact that European doctors, irrespective of their qualifications were often accorded superior, consultant status not only by some Indian patients but also even by other Indian physicians who sometimes deferred to Europeans on matters of diagnosis and treatment.[25] Adalja and Topiwala, drawing on the experiences of their joint practice, complained that some of the wealthiest Indian patients seemed to place more confidence in the hands of European doctors, even going so far as to demand unnecessary referrals to them.[26] Adalja also ruefully noted the tendency of Indian patients to refer to 'any white practitioner whether good, bad or indifferent as "big doctor"'.[27] The racial politics of the time meant that these assumptions were more than empty prejudice. It was true that, compared to Europeans, the Indian general practitioner was thwarted in his work through his limited access to the local hospitals.[28] If an Indian could afford it the unfortunate reality was that the European doctor was more likely to have access to better diagnostic facilities and certainly would have had more local influence. The systemic bias against Indian private doctors was pragmatically managed by most as part of colonial life, but it clearly grated some. Karve rather poignantly described de Sousa as having reached the conclusion that his skin colour would never allow him to reach the top of the Kenyan medical profession. As an ambitious individual he looked around for other activities—ones 'where he would be acknowledged as a leader'.[29]

The demand for the purportedly superior European services was reciprocated in that many Europeans opened their surgeries to Indian paying patients, although usually diplomatically seeing them in a different consulting room, or scheduling Indian appointments at times different to those to which the European surgery operated. Ever with an eye on making a good medical living, even the notoriously racist Dr Burkitt—who had on occasion publicly expressed aggressively antipathetic feeling towards Indian health practices—was happy to have Indian patients referred to him by his Indian colleagues.[30] Occasional, albeit rare, evidence also exists that, when European influence was exerted, a degree of flexibility on the application of the colour bar in the European sections of hospitals was also permitted. For instance, Dr Sorabjee performed minor surgery on Karen Blixen's eyes, an operation that most likely would have been undertaken in the European hospital.[31]

Yet although Indian doctors complained about their lack of access to European standard facilities, and although they complained about the competition they felt from European doctors who, it seemed, only by dint of their skin colour, commanded the most social respect (and therefore the cream of the highest-paying patients), Indian doctors before 1940 in Kenya, were not always integrationist in their medical practices either. Indeed, Indians repeatedly demonstrated a strong preference to maintain the boundaries of their own sub communities with patients seeking out doctors who were of the same caste and religion as themselves. Often doctors actively engineered their practices around the needs of one sub community—feeling that the association enhanced, rather than damaged, their reputation.[32] Dotiwala, for example, was the doctor of first choice for the Mombasa Parsi community who were said to have called him 'Apra dokter Dotiwalla' or 'Our Doctor Dotiwalla' as a term of affection.[33] Towards the end of his career, Adalja expressed regrets that he had rather limited the community focus of his work:

> This did not make one's work any easier but I decided to aspire to become a good family doctor, largely confining my efforts to one community, a course which I now realise was ill-advised. It resulted in a parochial outlook on my part and it antagonized members of other communities.[34]

Yet, even if this was a persistent trend, not infrequent exceptions existed. Some Indian doctors developed wider communal, even multi-racial, practices. In the days before 1940 when few doctors with specialist qualifications lived in Kenya, a few examples exist of European patients specifically seeking out Indian doctors who had a high reputation in particular medical areas. Topiwala, known as an eye expert, was in sufficient demand to set up separate consultation rooms on Government Road for European patients.[35] He was not alone, R.A. Ribeiro, de Sousa, Sorabjee, B. Singh, A. Ribeiro, all listed Europeans amongst the visitors to their practices.[36]

If not sought out because of their expertise, another advantage of seeing an Indian doctor was his cost. Indian doctors were typically a little cheaper in their fees than European doctors and this saving, especially during the depression years of the 1930s, was given as a reason why a European might prefer an Indian doctor.[37] This could be done with the minimum of awkwardness. For the benefit of social etiquette, European patients were either seen by the Indian doctor in separate surgeries, or were told to wait in a separate waiting room.[38] As offensive as this may have been on one hand, on the other this habit held some tangible benefits for the striving Indian doctor. Separating patients based on the colour of their skin, ensured not only the sparring of European sensibilities, but also endeared the Indian doctors to more conservative members of their community who would not have felt comfortable with overt intermingling. It also held the very practical benefit of allowing the Indian general practitioner to unobtrusively operate a fee differential between the two ethnic groups. Naturally, this fee structure charged rather more to Europeans than to Indian visitors.[39] A major barrier to treatment of European patients by Indian doctors was that European doctors were only permitted to admit European patients to the European Hospital. However, as the colour bar restrictions eased in the 1950s, Indian patients would initially get a European doctor to admit them and then privately call in an Indian doctor for their advice, just 'in consultation' (as they were not officially allowed to work there).[40]

Although few would have been able to afford the fees, and would have relied mostly upon government dispensaries and clinics for their medical care, evidence also exists that Africans occasionally also came to see Indian general practitioners.[41] R.A. Ribeiro was said to have famously treated the wife of an African chief, who was presented to him with her lips hanging in ribbons after a fight with a rival female suitor. In despair at being confronted with such a messy wound, Ribeiro used only one enormous stitch to pull the two bits of her lip together. The chief was apparently 'very pleased with her appearance after Ribeiro's cosmetic surgery'.[42] Although this seems to have represented a particularly dramatic and likely embellished incident, de Sousa, A.L. Ribeiro and Topiwala all attended to Africans on a routine basis.[43] It was said by some commentators, that African patients presented a rather different set of challenges to either Indian or European patients. As, Ayres Ribeiro recollected: the appearance of an African patient in the private surgery was often when they were at the most advanced stage of their disease and after other avenues of healing had proved unsuccessful.[44]

Inter-racial communication on professional matters also ran between doctors, although a flick through the European medical memoirs of the period might lead a reader to think differently. Not one such account by a European doctor assesses the work of the Indian doctors that worked contemporaneously to them.[45] This omission is more remarkable given the referrals previously described between doctors working during the same period. Furthermore, as described below, many of European doctors served on Government institutions, commissions and committees that brought them in close contact with the Indian medical representatives on these bodies. Even Trowell, whose unpublished memoir confirmed his liberal views, did not mention contacts with Indian doctors that he had encountered. A couple of letters from Trowell to Topiwala show that Trowell consulted him for medical advice and regarded him as a colleague whose professional opinion he respected.[46] A similar reminiscence by Dr Sudha Young, distinctly recalls the memory of her father having received letters from Dr Burkitt asking for professional advice about an eye case that which had 'completely baffled' Burkitt.[47] Also illustrative, a formal group photograph of staff at the Lady Grigg Maternity Home, which includes Gilks, Topiwala and Adalja, shows that they clearly knew each other (see photo p.248).

Race and Politics

As was also common in India, many private Indian doctors openly played active political roles, although not always with a uniform attitude.[48] Some were fierce defenders of Indian rights and supported the creation of a united Indian front, while others focused their political energies on the public backing of their own sub communities and religions. As in India, others preferred to retain the *status quo* and were supportive of the British regime. Added to this multiplicity of standpoints there were many ways that political activism could be demonstrated. It could be informally acknowledged through social relations with radical Indian leaders, it could be formally conferred through the membership of government bodies, or it could be expressed more colloquially, though by no means less strongly, through letters against pervasive racial inequalities to the national press, such as *The East African Standard.*

From the early 1900s Indian political rights in Kenya had been campaigned for by a few ambitious reformists such as: Abdulrasul Allidina Visram (1851–1916), Alibhai Mulla Jeevanjee (1856–1936), Manilal A. Desai (in Kenya from 1915–1926), B.S. Varma (in Kenya c.1919–1935) and I. Shamsu-deen (in Kenya from 1900–1952). Each of these individuals represented slightly different views in terms of their political agenda, but in general, these activists were united in their ambition to secure fairer social and political recognition of all Kenyan Indians, irrespective or caste or religion.[49] The united Indian front broke down after Desai's death in 1926 and the colonial government encouraged and exploited the dissension between the Indian factions.[50] Although very little is known about one of the earliest Indian private doctors, da Gama, a photo of him with the leaders of the Mombasa Indian community including Allidina Visram and T.M. Jeevanjee (the more moderate brother of A.M. Jeevanjee), indicated that he was sympathetic at least—perhaps even actively supportive—of their campaigns for Indian rights.[51] In a similar vein, throughout the period certain Indian doctors kept up a close involvement with nationalist groups such as the East African Indian National Congress (founded 1914) and the Indian Association of Nairobi (founded 1919). Both Karve and Patwardhan held at various times the secretary-ship of EAINC or the Nairobi Indian Association. de Sousa, particularly in the first decades of his career, was a member of the Executive Committee and an ardent campaigner for EAINC.[52] Patwardhan was President of the Nairobi Indian Association. A 1934 photograph of the leaders of the Indian community in Nairobi included individuals with a wide spectrum of political views such as: Dr E. Sorabjee, Dr M.D. Gautama and the government employed Dr Minoo Dastur.[53] Unsurprisingly, not all of the Indian medical pioneers shared these political sympathies, but it was noteworthy that a significant minority identified themselves, at least nominally, with wider campaigns for Indian rights. By way of contrast, Rozendo Ribeiro provides a good example of the more conservative end of the spectrum and seemed, throughout his career, very much in favour of retaining equanimous relations with the white governing elite.

In terms of formal participation within the official bureaucracy of British colonial governance, the role call of Indian doctors who can be shown to have played an active political role in colonial Kenya—while never rivalling European medical participation—is nevertheless substantive proportionate to their numbers. Da Gama, Ribeiro, de Sousa, Adalja, Karve, Sheth, Rana, Sood, Patwardhan, Ismail and Gautama all left their historical imprint upon the colonial political landscape through their participation in colonial governing bodies. Since 1910 when A.M. Jevanjee (who became slowly but progressively disillusioned with the British government) became the first Indian to be selected to the Legislative Council, nominated Indian representatives were regularly forwarded to sit on various civic and national boards. The most important was the central Legislative Council in Nairobi, followed by the various regionally situated Municipal Committees. The exposure of many Indian doctors to the possibility of working on a government body closely mirrored the experiences of European doctors, many of whom held positions in both civic and national political institutions that they might never have dreamed of holding in the less upwardly socially mobile context of their home country.[54] The professional prominence of the doctor in society meant the doctor became something of a staple figure in British colonial institutions, and practitioners regularly found themselves participating in debates of a general matter as well as advising specifically on matters pertaining to public health. The strategic importance of Indian participation in government was understood by the British, who realised that the Indian community were more likely to accept official directives if they were seen to have been passed under Indian approval. Yet although lip service was paid to Indian participation, in reality, the minutes of the Legislative Council and Municipal Councils show that, all too frequently, the Indian minority opinion was outvoted, or simply over-ruled.[55] Consequently, Indians who were favoured for government positions were those who were perceived by the British administration as being handmaidens to their agenda, or at least those who were thought to be constructive in progressing the agenda of the colonial administration. This point is emphatically illustrated by the way the Governor ignored the election results of 1917 to the Nairobi Municipal Committee for the Goan and Indian candidates (the two were differentiated between in the official bureaucracy) and instead directly nominated two men he had personally selected (Mr Phadke and Dr Ribeiro) without going through the normal procedural channels.[56] Clearly Governor Belfield was more concerned with the qualities and attitude of the person selected than with following the democratic procedures. Repercussions for speaking too boldly could be far reaching. Although no direct proof could be found, Indian Legco members considered that the insinuation of Dr Sheth in a legal case of medical misconduct in the 1930s was directly stimulated by antagonism to his political views and suspect evidence of a rival politician Dr Rana.[57] Furthermore, doctors who accepted government positions faced ambiguous reactions from their own community. Participation in government, even with the intention to reform, nevertheless could seem to some to look like collusion with the white colonial overlords, even a betrayal of one's own people.

The list of formal political involvement was long and extends to involvement in many cases beyond the Second World War. Ribeiro was one of the first active Indian community members and served on the Nairobi Municipal Committee from 1905–17. Karve served on the Nairobi Municipal Committee (1925–27),[58] was then on the Mombasa Municipal Council (1930–45), and was nominated to be on the Executive Council in 1937.[59] He also became a nominated member of the Legislative Council (1952–56) and nominated member of Council of State (1952–56).[60] Sheth was also for more a short period also a member of the Kenya Legislative Council as well as being a member of the Mombasa Municipal Board for over a decade.[61] He was still listed, along with Rana as one of the elected members of the Mombasa Municipal Council in 1948.[62] Rana was also active in the Legislative Council during the 1940s.[63] de Sousa served on the Legislative Council (1934–38) and Nairobi Municipal Council (1925). Adalja was a nominated member of Nairobi City Council (1931–35) and the Legislative Council (1959).[64] Sood was a member of the Municipal Board of Kisumu (c.1940s).[65] The British Administration appointed Dr Ismail a nominated member of the Legislative Council in 1959, but he had already served on the Nairobi Municipal Council earlier in his career.[66] Gautama was elected to serve on the Nairobi City Council (1937).[67] Dr Patwardhan was a well-known Nairobi resident who served on the City Council, was General Secretary of EAINC and committee member of Nairobi Scouts Association.[68]

Many examples can be found that show the differing of Indian private doctors towards their political work. Rozendo Ribeiro's was one of the earliest Indian members of the Nairobi Municipal Committee, joining in 1905 (shortly after its constitution in April 1900), and serving there for 12 years.[69] As discussed in Chapter 4, Ribeiro was still a member of this committee in 1915 when the final vote supported the controversial recommendation of the Simpson Report for enforcing racial segregation in Nairobi.[70] Ribeiro, the only non-European present at the Council meeting, did not vote against or register any opposition to the scheme in the minutes. The Governor as an endorsement of segregation interpreted this stance to apply to 'all the races' in a communication to the Colonial Office.[71] This was despite the fact that the segregative policies were bitterly opposed by the Indian community, including prominent leaders such as A.M. Jeevanjee.[72] Ribeiro's passive acceptance of the measures was also contrary to the recorded wishes of the other (not present) Indian committee member, Mr Ghandi. In the absence of Ribeiro's own published account of the events one can only speculate about his reasons for the stance, the example is an illustrative one that not all doctors were willing to put their head above the parapet and speak out for Indian rights.

A.C.L. de Sousa, certainly during the early days of his career, could be said to represent a doctor more aligned to the more radical end of the political spectrum. From the beginning of his career in Kenya he supported many of the prominent Indian political associations. de Sousa was an active member of the East African Indian National Congress (EAINC) and was publicly involved with many prominent fundraising campaigns on their behalf during the 1920s.[73] Additionally, both de Sousa and his wife became well known for their hospitality to visiting political leaders from India, such as Sarojini Naidu, Pandit Kuzuru and Shrinivasan Sastri who were all high profile leaders of

Indian National Congress (INC). Despite being a Goan and associations with reformist movements, de Sousa got elected against the odds to the Legislative Council in 1934. Dr Patwardhan, himself a former President of the Nairobi Indian Association, observed that de Sousa was always slightly wounded that had never been given the chance to head the EAINC to which he had devoted so many of his years of political activism. The fact he was never nominated was said to have 'ultimately disheartened [de Sousa] and pained his inner sincere feelings'.[74]

He was said to be an eloquent, outspoken, defendant of Indian rights, unafraid to take a public stand against the settler representatives such as Scott, Grogan and Schwartze.[75] In a tribute another prominent Indian politician, N.S. Thakur, described de Sousa 'as a man of many facets, a courageous fighter and a gifted leader'.[76] Amongst his medically orientated campaigns were the challenges he mobilised against the Principal Medical Officer, Arthur Patterson in the 1930s concerning the below standard medical services available for non-Europeans, the registration ordinances for doctors and dentists and the lack of public health measures to eradicate malaria.[77] N.S. Mangat, a fellow elected member of the Legislative Council described de Sousa as 'fearless and unrelenting' in his criticisms of the hospitals and health services because he had more intimate knowledge of them than 'those responsible' for them.[78] de Sousa particularly pointed out that he felt the government medical department was negligent of the needs of Indian patients, employed expensive European doctors, located too many senior doctors in the head office and discriminated against Indian doctors. He accused Dr Paterson the PMO of pursuing race based policies 'such as hardly benefits a Government official in the position he is'.[79]

Such volatility and activism were unwelcome in the Legislative Council but interestingly, not just by the Europeans. de Sousa in fact found himself defeated in 1938 because of resistance to what was perceived as his partiality towards the Goan community of which he was a member. This was compounded by the fact that Goans who supported him were not able to demonstrate their backing due to severe limitations on their voting entitlement within colonial Kenya.[80] Disappointed in the abrupt end to his political career, after 1939 de Sousa moved his attentions away from the national defence of Indian rights to the more local community concerns of the Kenyan Goan community, editing one of their most important community newspapers *The Goan Voice*.[81] His role in building up support in Goan politics by 'playing up caste' in the community has been criticised by another prominent Goan politician.[82]

This move from relative radicalism to conservatism in politics can also be seen in the career of Karve. Karve, like de Sousa, had initially been a keen supporter of EAINC, even acting as the group's Secretary between 1926 and 1933.[83] On arrival in Nairobi in 1922 he had closely aligned himself with the campaigns for reform and enfranchisement led by Desai.[84] It did not take long, however, before a rift between the hardliner fraction led by Desai and the fraction led by Pandya and Karve developed with the latter advocating much less radical measures of protest in contrast to Desai's less compromising call for the total boycott by Indians of colonial institutions. In 1925, only a few years after joining the EAINC, Karve and others supported a motion calling for an investigation into the conduct of several EAINC officers including Desai on accusations of misappropriation of funds.[85] The meeting broke up in confusion triggering a fracture in the Indian leadership, which was to be the start of Karve's change of heart in favour of more moderate policies that advocated reform coming from within the institutions of colonial governance, rather than in direct external opposition to it.[86] Things came to a head again in 1928 at another public meeting when Karve, together with A.B. Patel and J.B. Pandya openly aligned themselves with moderate political policies and clashed bitterly with the hardliners Dass and Varma who, in this instance, were pushing for all Indians to abstain from joining the Municipal Committee.[87] The moderate Mombasa group of which Karve was a leading member took the view that participation on government councils and committees, however unfair, did nevertheless give a platform to Indian complaints. Non-cooperation, they felt, would only lead to worse outcomes with Europeans simply filling the empty Indian seats.[88] As a result of his criticisms and alignment with the more conservative group, Karve unsurprisingly antagonised the more radical Indian leaders who viewed him as a sympathiser of the unjust colonial system. This was reinforced by Karve's own large repertoire of roles within colonial government institutions, including the Nairobi Municipal Committee (1925–27), Mombasa Municipal Council (1930–45), the Executive Council (1937), the Legislative Council (1952–56) and Council of State (1952–56), as well as by the privilege accorded to him symbolised through the colonial government's personal invitation to him to serve on the 1939 Willan Commission looking into labour unrest.[89]

But although always willing to work within the framework of colonial governance, this is far from saying that Karve did not actively campaign for Indian rights. As the EAINC papers during the 1920s attest, Karve regularly made fiery representations in support of Indian rights.[90] He also used his positions of political representation to protest about the impact on the Indian community of the direction of colonial suburban development programmes, the lack of action being taken over the problem of community overcrowding and endemic infrastructural neglect he felt was accorded to Indian residential areas.[91] During the late 1920s Karve was one of the most active petitioners to the Kenya Local Government Commission, arguing for an increase in Indian representation proportionate to the size of their resident population and the amount of tax they paid.[92]

Yet other examples exist of Indian doctors using their position on the local municipal boards to campaign for health improvements for the Indian community, albeit with little tangible success. Drs Karve, Sheth and Rana were all on the Health Committee of the Mombasa Council and used their positions to campaign for the establishment of a non-racial Hospital or for improved housing schemes for non-European residents.[93]

Although consensus, or at least a majority, could be achieved on some common causes, it is difficult to generalise about the political stance of individuals. Ribeiro was loyal to the British administration, while de Sousa can be seen as having been much more critical, especially in the early years of his career. Some, perhaps most, like Karve, Sheth and Adalja took a moderate middle road, fighting for Indian rights, but ever careful not to align themselves too closely with the Indian hardliners, realising the disadvantages of offending the British elites and biting the hands that fed them. Whether broadly supportive or broadly critical of the colonial regime in many instances the way the Kenyan Indian diasporic consciousness developed would have also drawn on Indian based political movements.[94]

Doctors, Politics and the Press

While some doctors participated in the formal institutions of colonial governance, other preferred to air their political and social views through the letters pages of the local press or in public meetings. Topiwala was vocal in this regard and, from the 1930s onwards, his name was to be seen within *The East African Standard* and the *Colonial Times* passionately championing Indian rights.[95] During his long career in Kenya, Topiwala's writings and speeches covered many topics, from the white highlands settlement policy and the state medical service to public health and to the evils of alcohol and gambling. He was always hostile to policies based on racial grounds.[96] His espousal of certain social and medical policies were criticised by an observer as somewhat utopian and idealistic.[97]

Enthusiasm for letter writing evidently ran through medical practices, with Topiwala's partner, Adalja being another active correspondent to *The East African Standard* in defence of Indian public health. Perhaps his most heated campaign occurred in the mid 1930s when Adalja launched an attack on the practices of the Nairobi Health Committee, alleging that the town councillors (fearing bad publicity) had negligently failed to report incidences of malaria emanating from the central swamp.[98] Indignant at the lack of action on behalf of either the Nairobi Municipal Council or the government, he offered to personally escort councillors 'who may be in doubt' to view the conditions in which many Indian families resided in this swamp area.[99] The issue had been, earlier, raised by Dr de Sousa in the Legislative Council and his question over 'whether the Government considered the swamp as a source of malarial infections?' received prominence in *The East African Standard*.[100] Adalja had rounded off his attack by directly accusing the government of being preoccupied with new methods of tax collection at the expense of the public health.[101] The government's evasive response provoked a furious row in the columns of *The East African Standard.* With, one European, R. Ward, even alleging that Nairobi was analogous to a 'house with imposing façade and some fine rooms which has a backyard so dirty and squalid that the owner must hope that no stranger will see it.'[102] Ward went on to accuse the government of dereliction of duty in letting these dangerous breeding grounds for malaria remain. The campaign had some success, and as a result of this agitation by Adalja and others an emergency meeting of the Nairobi Town Council was hurried called in May 1935.[103] At this meeting, the government's Medical Officer of Health, Harold Tilling admitted the swamp was the key menace, with 70% of malarial cases emanating from the vicinity.[104] Settler leader, Ewart Grogan, who owned a considerable amount of land adjoining the Nairobi River, responded in his typical robust style claiming there was no swamp but conceding that an inquiry into malaria would be useful.[105] The acrimonious debate did result in the sanctioning of modest funds for emergency spraying but confusion persisted about whether it was the responsibility of the Municipal Council or the central Government to provide the substantial funds for the much needed drainage and new canals.

In a slightly different way, beyond letter writing, de Sousa also used the press to advertise his own political leanings. A keen journalist, de Sousa must have seemed politically dangerous to the British authorities when his access to the press was considered alongside his early sympathies for the EAINC. Over his career, de Sousa was closely associated with a number of provocative political mouthpieces, including regular contributions to the Asian journal *Democrat,* and the editorship of the *Colonial Times.* He also founded two community newspapers of his own: *Fairplay* and the *Goan Voice.*[106] In all cases, it was clear that he saw these newspapers as vital fora for the discussion of political issues pertinent to Indians.

Politics and Indian Sub-Communities

Political activism was not restricted to Indian political causes generally conceived and many doctors played a prominent role in furthering the causes of the sub-communities to which they belonged. Ribeiro, for example, despite his lack of forcible action in the Municipal Council on the issue of segregation, nevertheless expressed his support for his sub community through his participation as one of the founding members of the Goan Institute.[107] As mentioned earlier, de Sousa became during the course of his career increasingly focused upon Goan rights. The list of his tangible contributions to the Goan community is formidable and de Sousa was associated with the foundation of the Goan Overseas Association, the Goan Housing Estate, the Desai Memorial Library and the Goan Secondary School in Nairobi, as well as a newspaper for Goans (see above).[108]

Though the number of Indian general practitioners until the mid 30s was relatively small, plenty of them, a majority in fact, were involved in community organisations catering for the myriad of sub communities that existed in colonial Kenya. Dotiwala and his wife became respected figure-heads for the Mombasa Parsi community: representing the group at many official city and government functions.[109] Doctor Rana was the President of the Muslim Association and a fierce defender of Muslim rights.[110] Patwardhan, as well as being involved with general Indian politics through EAINC was also the President of the Nairobi Hindu Union.[111] Topiwala was the founder member of the Surat District Association whose membership was mainly the Hindu artisan community from Surat. He helped raise funds for a permanent social hall for the Association in Nairobi and laid the foundation stone for the building.[112]

Unusually, Dr Adalja, who was known for his strong affinity with the Gujarati speaking Oshwal community, nevertheless (although characteristic of his moderate beliefs) directed his energies in the later part of the colonial period in the promotion of inter-relations between the different races. One of Adalja's notable contributions in this regard was his close involvement with the founding of the United Kenya Club: a quasi-political organisation intended to bring the disparate racial groups together in the decade before independence.[113]

Government Honours

In recognition of meritorious public services rendered, several Indian doctors were awarded British government honours, the acceptance of which—although usually a matter of pride at the time—could have been tinged later on with some feelings of guilt because of the collaborative connotations such an honour espoused. The ambiguity of these feelings were likely to have become more pronounced as the end of the British Empire became more inevitable and the return of honours and titles—as by M.K.Gandhi the much revered Indian national leader—became a powerful symbol of untethering the shackles of European colonialism.[114] Many Indian doctors received government honours. Ribeiro received an OBE from the British government in 1932 and the award of Commander of the Order of Benemerencia from the Portuguese Government in 1947.[115] Sheth was awarded an OBE (1946) and the Queen's coronation medal in 1953.[116] Rana received MBE (1944) and OBE (1952)[117]; Adalja an MBE (1946)[118], Karve an OBE (1948).[119] These numerous acknowledgements of their service make their omission from the history books all the more remarkable.

Race, Politics and the BMA

The most important professional representation body for doctors was the BMA. Originally founded as the Provincial Medical and Surgical Association in 1832, the membership and importance of the BMA (which it became known from 1856) grew throughout the nineteenth and twentieth centuries. Correspondingly, as the British Empire advanced, the BMA opened regional branches throughout Empire with the dual stated aims of nurturing feelings of professional 'kinship' between colonial doctors and those working in the UK, and also with the intention of fostering common cause between doctors posted widely throughout the British Empire.[120]

As part of a wave of BMA branch openings throughout the British imperial possessions, the Kenya branch was founded in 1920 although its roots were in an earlier regional association called the British Medical Association of East Africa, founded in 1913.[121] Before the East Africa (Kenya) Branch was formally established a combined Uganda and Kenya BMA branch had been initiated in Uganda from 1913, led by key individuals such as the famous Missionary doctor Dr Albert Ruskin Cook (1870–1951), and government doctors Dr Clare Aveling Wiggins (served Kenya, 1901–09; Uganda 1909–23) and Dr George Cecil Strathairn (served Uganda, 1903–20). Open to all European members of the medical profession, be they missionaries, government doctors or private doctors, the role of the BMA branches was pretty much the same in remit and intent as that of the London based mother ship. The BMA branch associations acted as a regional professional trade union for doctors, defending doctors' employment rights, offering their opinion on medical projects and discussing scientific subjects pertinent to their profession. The branch association was also the place where more discrete local concerns were aired in terms of policies and research apposite to the particular region of Empire where the branch was based.[122]

The BMA branch associations were the most important bodies via which colonial doctors could make their voice heard internationally and many examples can be found of the Colonial Office reviewing its policies as a direct result of protestations from the East African Branches of the BMA.[123] Yet, although the BMA London was rhetorically proud of the way it promoted professional unity and identified with the common causes of all western trained doctors, it was fully aware that non-Caucasian members were excluded from many colonial branches. In Kenya, with its thriving cohort of Indian doctors, this remained the state of affairs until 1935.[124]

The basis of this exclusionist policy had been a point of contentious debate for some time within the policy-making boardrooms of BMA House. The London based Dominions Committee, who was the body responsible for overseeing the work of the various branch associations, had discussed on several occasions race-based restrictions on membership that *de facto* existed in many of the Branch associations. The subject became ever more pressing as western medical education expanded throughout the British Empire, dramatically swelling the numbers of non-whites who were graduating as doctors each year.

Things came to a head between 1913 and 1914 when the Dominion Committee passionately debated the topic, taking contributions from Hong

Kong, India, China and South Africa. The basis of the discussion came in the form of a four-page communication from Dr Campbell Watt (Natal Branch and the Honorary Secretary of the South African Committee) regarding a proposal to end racial discrimination in the membership policy of all the regional branches of the BMA.[125] The debates that ensued revealed a large amount of resistance to the proposal. The representative from Hong Kong voiced his concerns that the social standing of the BMA in the colonies would diminish if non-Europeans were admitted.[126] It was also declared that if 'Orientals' were admitted then the floodgates might open and the 'medical world in the East would be ruled by them from mere force of numbers'.[127]

But other evidence from the minutes indicate that the BMA knew that times were changing and were concerned about the negative fallout of pursuing such an overtly racially exclusionist policy. They particularly feared that exclusion from the BMA would lead to the formation of rival organisations, which would make it difficult for the BMA to centrally impose its professional standards or control patient charges.[128] For the BMA professional competition based on the charge of lower fees by non-European doctors was already a troublesome consideration especially in countries such as Kenya with multi-racial professional medical communities.[129] Watt also considered the merits of a suggestion that the organisational branches could be split into ethnically orientated sub-branches: according to this proposal there could thus be (for example) a Hong Kong (European) Branch and a Hong Kong (Asiatic) Branch, which could be justified as following the precedent already in existence in South Africa for separately organising churches into European, Half-Caste, Indian and Kaffir churches.[130] In the end no resolution was reached and the debates were shelved for some years, raising their heads only again with some force in the 1930s when 'coloured' African doctors complained about the operation of a 'colour bar' in the West African Medical Department. In effect the Dominion Committee sidestepped the proposal to end discrimination (for many decades) by taking the position that political decisions were a matter for the Colonial Office and reminding members that the BMA articles allowed individual Branches the freedom to govern themselves.[131] Unlike its position on many other issues the Committee refused to make any representations to the Colonial Office on this particular matter.

In Kenya, the admission of non-Europeans to the BMA Branch was specifically discussed in 1934 at the Nairobi Meeting which was chaired by Dr J Sterry and those present included J.A. Carman (Treasurer) and H.C. Trowell (Honorary Secretary) The following resolution was moved from the Chair:

Membership of Kenya Branch of BMA shall be open to any medical practitioner who is eligible for election by British Medical Association, that is he is registered or possesses a qualification, entitling him to be to be registered under the Medical acts of Great Britain and Ireland.[132]

Despite the fact that nearly 20 years had passed since the idea was first raised in the Dominion Committee it was evident that many European Kenyan doctors opposed, or were at least extremely cautious, about the proposal and after an intensive discussion no definite conclusion was reached and more information was requested on the powers of the Branch Councils.[133]

Change came just a few months later at the January 1935 Annual General Meeting of the Kenya Branch when the Kenya BMA branch finally opened its door to all medical practitioners who met qualifications of registration: which in Kenya of course meant its significant Indian medical population.[134] The most likely reasons for the lifting of the ban were to enable the European members to widen the principle of minimum fees to the Indian doctors.[135] The BMA Branch also needed more membership fees because of the need to fund the EAMJ.[136] Initially, at least, this move towards inclusiveness did not run very deep however and for the next 20 years the leadership and proceedings of the Kenya BMA were still dominated by its European members. The first Indian BMA Council member, Dr Adalja did not arrive until 5 years later, in 1940.[137] This started a precedent and by the 1950s a couple of the Indian doctors mentioned in this study (probably Ismail and Gautma) also became council members.[138] Adalja became the first non-European President in 1958, almost 40 years after the branch was founded and 23 years after Indians were first admitted to its membership.[139] Notably, however, it is evident from some private correspondence by Topiwala that Indians were not necessarily eager to join the newly racially inclusionist Kenya branch of the BMA, fearing that this inclusiveness was not necessarily going to improve their position and seeing benefit instead of retaining their allegiance with a purely Indian professional association, such as the Indian Dental and Medical Association.[140]

Out of Nairobi, in the coastal region of Kenya, racial intermixing was slightly less of an issue and the population of this coastal strip were much less enclavist in their attitudes, partly because of the area's long historic exposure to immigrant peoples and traders of other nations and partly due to the relatively small European population. When the Mombasa Division of the Kenya Branch was founded in 1929 Indian doctors played an active role in this institution as the colour bar issue did not feature in Mombasa to the extent it did in Nairobi where the large European population and the more numerous European doctors were known to be in sympathy with the race based segregation policies.[141] Sheth and Figueiredo were both founder members of Coast division, with Sheth going on to play an active role in its activities for almost 30 years in positions of Treasurer, Secretary and President.[142]

The contribution of Indian doctors to BMA activities did not go unnoticed. Sheth's service was formally acknowledged by the BMA in London by being made the first Kenya Fellow of the Association. Before the end of their careers Karve and Adalja, were also placed alongside the Europeans, Andrew Hicks and John Carman on the BMA Roll of Fellows in London, in a rare formal acknowledgement of the meritorious service of Indian doctors to the development of the Kenya branch of the BMA.[143]

Indian Medical and Dental Association

Deprived of access to the BMA until the mid 1930s, Indian doctors formed their own professional body to support their needs, establishing The Indian Dental and Medical Association (IMDA), probably in the late 1920s sometime.[144] As no archive material has to date been found, the history of this Association remains somewhat shady, although some evidence seems to suggest that it developed out of an earlier Indian professional representative body called the Indian Medical Union or in parallel with it[145] It is clear however that many Indian doctors became closely associated with the IMDA over the course of the colonial period. Although precise dates are unavailable, it appears that a number of prominent Indian doctors served as its presidents including: Ribeiro, Adalja, Sorabjee, Karve, de Sousa, Topiwala, Gautama and Ismail. The IMDA debated the wider medical issues of Kenya colony as well as issues close to the hearts of Indians specifically; it held several councils meetings a year as well as additional half-day workshops devoted to a wide range of professional interests including vaccination policy, infant mortality, disease control, protocols for referrals, treatments, hospital facilities, social problems, and the racial injustices to which Indian doctors were exposed in the colonial setting. Expectedly, once the BMA opened its doors to non Europeans in 1935, the power and purpose of the IMDA diminished and eventually a merger in 1962 subsumed the Indian professional body under the auspices of the Medical Association of Kenya.[146]

***

Indian doctors in colonial Kenya could not help but be embroiled in the politics of race that touched so many parts of their lives. Although circumstances, common to their colour, affected all of them—particularly in their restricted access to patients and facilities and also in terms of achieving equitable professional recognition—they exhibited a variety of political standpoints in their attitudes to their colonial overlords. Some doctors pragmatically chose to work quietly within the colonial infrastructure; others were more vocal in their campaigns for fairer treatment. Simultaneously acculturated by their national Indian identity, their particularly regional and caste identity as well as the specifics of their Kenyan situation, these Indian community leaders had a variety of political persona available to them. The colonial situation gave them work and opportunities, often far beyond those they could expect in their home context, and most were in the ambiguous position of being able to see the benefits of life as a Kenyan Indian doctor while also being heedful of its limitations and drawbacks. What can be sure is that these doctors constituted a relative elite sector of the Indian Kenyan community and, if they so wished, they could (and did) use their comparatively privileged access to education and social institutions to forward their political viewpoints, although not always with success.

9
Conclusion

The period between 1895 and 1940 was a profoundly important one in the history of the Indian professional diaspora to East Africa. Indian medical schools were bearing more fruit and, increasingly from 1900, their graduates were seeking employment opportunities further afield than within the borders of their subcontinent. Indian doctors, already familiar with the character and expectations of British rule, could conceive of a move within the British Empire as a logical step, even if it was simultaneously also a rather intrepid one. Whether coming of their own initiative, or at the behest of the British government, Kenya seemed to present certain prospects—its mild climate, relative proximity to India, comparatively fertile hills and already multi-cultural coastline leant it an appeal to Indians hoping for a better and more prosperous life.

As the preceding pages have shown, however, life in this racially segregated society was by no means plain sailing for most doctor immigrants. Despite having formal medical qualifications, Indian doctors were always fundamentally seen as second rank to the European ones and were never really awarded (even superficially) equitable treatment by a colonial state that paradoxically relied upon them. Furthermore the Indian population of Kenya as a whole was by no means a unified group and the lack of agreement and consensus between Indians of different sub communities in Kenya meant that they were rarely able to articulate any sustained demands.

Yet, despite these hurdles, the role of this community in the medical history of Kenya was absolutely fundamental. Both Indian government doctors and Indian private practitioners played important roles in the medical landscape of the time. Before the early 1920s the first route of government service was more common, but after that point—when the possibility of an Indian pursuing a colonial service careers became severely reduced—private practice became the main viable option. The total number of Indian doctors who entered Kenya in these two capacities between 1895 and 1940 was about 200 and while this was not a staggeringly large number, especially compared to the size of the rest of the Kenyan Indian population, it was still a substantial proportion (perhaps constituting more than a third of the total immigrants who entered Kenya with the aim of practising medicine in the period).[1] Given that about 400 European doctors worked in Kenya in the same time, Indian doctors decisively made up an essential part of the western curative services available in the country.[2]

***

The second main theme that this book has elucidated is the way that the racial politics of colonial Kenya impacted all areas of Indian medical work. Within the Colonial Medical Service, Indian doctors found themselves forced to work on lower pay in roles subservient to their European contemporaries. They were habitually excluded for appointment in, or promotion to, management positions. As *The Times* correspondent put it, in part this was to do with attitudes endemic throughout the British Empire, with the 'average Englishman' objecting 'to the possibility of being a junior officer under the Indian'.[3] In private practice too, however, although Indian doctors had relatively more autonomy, they found themselves routinely restricted both in terms of the patients they could attract and the medical facilities to which they had access. In some ways, the character of the situation resembled the experiences of Indian immigrants to South Africa or Uganda. But in Kenya racial antagonism was uniquely expressed through the voices coming from a politically powerful European settler community. Additionally, the country was home to the most powerful eugenics movement in Africa and, since the 1920s, had been at the heart of heated debates that occurred in London about the debt the British African Empire had to its immigrant diaspora community. All of these things created an environment inimical to non-white migrants, additional to the racism that was routinely experienced by the majority indigenous population.

Indian doctors accompanied the Indian regiments that assisted the British conquest of East Africa and Indian doctors provided medical services to the thousands of Indians who built the Uganda Railway. As the British bureaucracy began to formally establish and organise itself, Assistant Surgeons and Sub-Assistant Surgeons became, for almost three decades between 1895 and 1923, vital cogs in the machinery of the Colonial Medical Department. After Gilks dismissal of the majority of Indian medical staff in 1923, the significant role of the Indians in the operation of the colonial medical services greatly diminished.

With this avenue for employment effectively closed immigrant Indian doctors set up private practices in Kenya, increasingly frequently by the 1930s. Although their professional lives were not always easy, evidence suggests that many Indian private practitioners were respected community figures, often closely involved with local philanthropy and politics. Indians were side-lined from participating in many areas of colonial medical life, but themselves often identified primarily with the specific sub community to which they belonged. Attitudes are impossible to pigeonhole, the Indian general practitioner was as likely to be frustrated by the conservatism of Indian society as he was likely to be defensive of its rights. By the same token, not all white community members endorsed racially exclusionist policies. A number of government-employed European doctors, such as Robert Moffat, Clare Aveling Wiggins, Norman Leys and Peter Clearkin, possessed progressive and liberal views and on occasion fought hard, and publically to change racially discriminatory practices and attitudes, at considerable risk to their own careers.[4] On the other side Indian doctors can be seen to have worked as frequently within the colonial system as much as (sometimes more than) they fought against it.

Yet, despite their various views, education and allegiances, Indians were grouped together as one problematic group in most official decisions by the state. Their second-rate status was most powerfully demonstrated by the fact they were excluded until 1935 from membership of the most important medical professional body, the BMA. Ultimately, Indian doctors were not permitted any substantive involvement in setting colonial medical policies and were denied any positions of authority. They should still be acknowledged, however, for having provided care to hundreds of thousands of patients in towns and reserves.[5] Before 1923, Assistant Surgeons and Sub-Assistant Surgeons were as important as European medical officers in tackling large-scale epidemics, providing curative care and implementing public health initiatives.

As well as doctors being denied status within their own profession, Indians in general were also denigrated by the way the colonial state consistently failed to fund public health initiatives that would improve their conditions. Ironically the allegedly most disease ridden and disease prone sectors of society were also somehow not deemed worthy of being the recipient of health improvement schemes. Instead Indians were time and again characterised as money grabbing and unscrupulous, insanitary and unsavoury. *Annual Medical Reports* throughout the period routinely pointed the finger at Indian communities, particularly those residing within the bazaars of Nairobi, Mombasa and Kisumu, as the sites of filth and disease. Before the 1940s, suggestions were rarely made with any force or conviction to improve Indian living areas. Instead attention was diverted away from the medical department's responsibilities, through complaints instead about Indian habits, implying somehow that this community was innately culpable and responsible for its own ill health.[6] A regular criticism was made against profiteering Indian landlords for the state of overcrowding in the bazaars. Indian avarice was alleged to be so great that it was deployed at the expense of public health, with proprietors squeezing as many tenants as possible into their accommodation to reap the maximum profits.[7] Whatever the reasons, there is no doubt that conditions were very bad between 1900 and 1935, so much so that the Indian housing in Bazaar and River Road was declared unfit for human habitation. In 1928 the PMO caustically remarked 'even the rats die of plague'.[8]

The paucity of direct action directed towards Indian public health was partly because it fell between the two most obvious ethno-demographic foci in colonial Africa: European Whites and African Blacks. European health was always at the forefront of imperial thinking, but even after the 1920s—with the introduction of more progressive long-term ideas of extending medical care to the rural areas—the stated aim of the British government was to increase healthcare provision to the majority African population with little concern voiced over the provision for Asian society members. Somehow Indians (commonly characterised as successful and avaricious) were forgotten in the new rhetoric of expansion and development. In 1930, despite the numerous calls for greater attention towards Indian health that were made within the chambers of both the Legislative Council and the Nairobi and Mombasa Municipal Committees, medical infrastructure, sanitary conditions and health for Indians remained very poor.[9] In terms of healthcare resource allocation, Indians, Kenya's middleman, were largely ignored by the colonial state. Indians, averse to go to the native hospitals, relied largely on self-help for their medical care within their communities. It was in the minority of situations, only if their economic circumstances allowed it, that the average Indian could routinely afford to be seen by one of the general practitioners mentioned in this study.

Similarly action did not fully follow rhetoric in terms of moves towards inclusiveness within the medical departmental recruitment policy. As discussed in Chapter 6, from 1924, the Africanisation of the colonial medical department became one of the most publically lauded intentions for the colonial medical Department. But simultaneous to the opening of jobs for Africans Dressers in the government curative and preventative medical services, came the dramatic reduction of opportunities for Indian Assistant Surgeons and Sub-Assistant Surgeons. With the luxury of hindsight, this cull of Indian medical personnel might be criticised for being short term in its outlook. Most surviving evidence suggests that Indian government doctors were both well received and well respected in the African reserves to which they were posted, and furthermore African dressers could not be trained adequately enough, nor quickly enough, to equivalently fill the gap created by their removal. Mostly government medical reports remained quiet over any concerns they might have had over the early failures in the Africanisation policy, although the person responsible for the training of the African dressers, Dr H.C. Trowell had serious misgivings about their shortcomings, including what he felt was to be the very real and present danger of incorrect diagnosis and treatment of patients in the African reserves where there was inadequate supervision by doctors.[10] Similarly, in 1932, an internal memo by the PMO admitted the inadequate quality of these employees.[11]

Furthermore, money was still being vigorously channelled into European medical recruitment. In 1935, ten years after the policy of Africanisation was started, expenditure on European medical staff still absorbed the bulk of the department's budget.[12] The number of European Medical Officers more than doubled between 1924 and 1930 and European nurses increased by two and half times between 1923 and 1937.[13] The rationale was simply stated in terms reminiscent of earlier discourses of the beneficent effects of the civilising mission: 'only in this way will the Africans, on whom any large extension will depend, learn to appreciate the necessity of efficiency'.[14]

The picture that emerges is of a racially segregated society, in which Indians operated in a space on the margins independent of the most pressing colonial concerns. Despite the fact that Indian doctors in private practice held registered medical degrees and good social standing in their own communities, their suggestions for improvements were often disregarded within policy-making fora. On a daily basis too, Indian and European doctors largely functioned separately working only in their own racially compartmentalised spheres. Although some prominent examples of political collaboration can be seen, majority European, Indian and African communities lived in different residential areas and the three groups had little social contact with each other. The key window between 1895–1940 when the issues of Indians were actively, and seriously debated occurred between the end of the First World War and the early 1920s in the form of the 'Indian question'. Once their hopes were defeated, Indians were to remain tolerated (sometimes even individually appreciated by Europeans) but largely ignored, within colonial political and social circles until the 1940s.

***

After 1940 things became slowly, but tangibly, better for the Indian residents of colonial Kenya. Visitors in the late 1950s would have commented positively on the prosperity and access to modern medical facilities enjoyed by the bulk of the European population and a substantial number of Asians.[15] Although the position of most indigenous Africans had not greatly improved, the increased prosperity of Indian community members, along with a small, but nevertheless discernible, relaxing of racial boundaries, meant that many Indians (like their European counterparts) were taking advantage of the new private hospitals within Kenya (many of which were built post 1950 however through Indian philanthropic donations).[16] In terms of public health too, in 1939 Dr Tilling was able to express his cautious satisfaction to the Municipal Council of the overall improvements made for the health of Nairobi residents. Not only had the death rate had more than halved in the previous 15 years, but there was a noticeable 'gradual, steady and consistent' improvement in health overall, despite this headway being 'exasperatingly slow' in its advancement.[17]

Fraught with difficulties and fighting against the deeply ingrained racial prejudices of the time, this history is however something of a success story. Between 1895 and 1940 Kenyan Indians grew in both numbers and prosperity and while the Indian population of Nairobi was less than a few hundred at the turn of the twentieth century, it was estimated that there were 18,000 Indian residents in Nairobi by 1940.[18] Correspondingly, the areas they lived were becoming increasingly dispersed and reflected their rising economic fortunes. During the first quarter of twentieth century most of the Indian residencies in Nairobi were concentrated around the two areas of River Road and Bazaar Street. From 1925, Indians started to become upwardly mobile and bought up properties in the Parklands suburb, which had formerly been the preserve of Europeans.[19] The congestion in central areas was also eased by initiatives undertaken by the government to build new residential homes. Racially segregationist thinking still existed, and generalisations over cultural proclivities remained. For example, a concern that even in 'the new districts being developed by Asiatics' there was a unfavourable cultural tendency towards overcrowding' in each individual residency, could still be heard. Similar were the old complaints about systemic racial inequalities. The disparity in Municipal investment between European and non-European areas was to remain a persistent feature of the Nairobi political debates.[20] Indians complained bitterly about the poor state of affairs in their residential areas for almost 50 years after the publication of the Williams Report that looked into sanitation and housing conditions. A memorandum to a commission of inquiry into the affairs of the Nairobi Council in 1956 is telling. Despite improvements, an Indian resident described the sanitation and sewage conditions in Eastleigh as 'awful' with irregular latrine removal, unclean drains and overfilled sewage tanks. This he bemoaned (in a refrain many times reiterated in the history of Indian rights in Kenya) was despite 'Asian areas contributing most to the municipal taxes'.[21]

But despite evident difficulties, overall the situation of Kenyan Indians gradually improved. Concurrent with a rise in numbers and their average wealth, came increased opportunities for Indian doctors and by 1960 more than 200 Indian private doctors worked in Kenya.[22] Their profile had changed somewhat too. Increasing numbers of these doctors had qualified in the UK and a larger proportion of them were in the possession of specialist medical qualifications. By independence a second, sometimes even third, generation of Indians were being born in Kenya. Indians could not claim to be members of the elite ruling class, but their status and opportunities had grown, and many were energetically prosperous.

***

A history of Indian doctors is important for many reasons. First, it gives a slightly different account of Indian immigration to Africa, above and beyond descriptions of the importation of 'coolie' labour. It also, *inter alia*, shows the tacit reliance the British government put on India, not only in terms of providing models of governance for its other territories, but also in terms of staffing its new African empire. While the importance of Indian personnel has been acknowledged in terms of lower status jobs that required little or usually no qualifications, the role of qualified doctors has been unacknowledged.

As such this story adds essential elements to the colonial history of Kenya. Giving greater richness to studies of colonial society and showing the deep racial rifts that existed within it. The Indians in Kenya lived in their own large communities, and though they were segregated from both the Africans and the Europeans they actually had greater daily contact with Africans than the British did. Their position as the 'middle men' put them in the ambiguous position of being both valuable intermediaries, but also potentially rather suspicious collaborators. Indians were readily lumped together with Africans in as the sites of squalor, disease and corruption, but were simultaneously seen as relatively useful members of the community and of the colonial civil service in particular. They received more privileges than those that were typically granted to Africans, but their social and economic position also meant that they were the targets of more immediate African resentment.

The forgotten history of Indian doctors in Kenya reflects many multifaceted ambiguities. It shows the contradictions of a group's social standing within their own communities, it shows how Indians could be both politically active and politically impotent, it shows their heterogeneous and complex political, national and ethnic allegiances. On one hand, this was a cohort trained in western medicine, relied upon by the British government and providing essential preventative and curative services, yet it could never be fully trusted by a British elite that saw themselves as inherently superior. As such, to forget their story is to lose some of the buried richness within the medical history of colonial Kenya. It is to forget that, far from being an empire of 'blacks ruled by [Oxbridge] blues'[23]; empire in Kenya relied absolutely pivotally upon Indian immigrants. The Indians of this story, at least before 1923, formed the most important backbone of colonial medical governance. Particularly after 1923, but also before, they also constituted a substantial, and not socially and politically un-influential, professional group of private practitioners within colonial society. As such, the story of Indian doctors provides a story of the success of an immigrant community, not only to pragmatically find a niche for itself in not always hospitable circumstances, but also to accommodate, in flexible ways, living in a racially segregated society. Overtime these doctors evidently mustered enough determination not only to contribute, but also to survive and prosper.

Appendix 1: Indians in the Railway Medical Department[1]

	Name	Grade	Year of appointment
1	C.H. Orman1	Assistant Surgeon	1896
2	W. Desmond	Assistant Surgeon	1896
3	Rahmat Ali	Hospital Assistant	1896
4	Ram Saran	Hospital Assistant	1896
5	Ahmad Husseni	Hospital Assistant	1896
6	Balmukund	Hospital Assistant	1896
7	Khanda Baksh	Hospital Assistant	1896
8	Salimuddin	Hospital Assistant	1896
9	Sheikh Nabi Baksh	Hospital Assistant	1897
10	Kanhai Lai	Hospital Assistant	1897
11	Chandrika Prasad	Hospital Assistant	1897
12	J.H. Whittenbury*	Assistant Surgeon	1899
13	A.H. Culpepper*	Assistant Surgeon	1899
14	H.N. Stewart*	Assistant Surgeon	1899
15	W.G. Masterton*	Assistant Surgeon	1899
16	Sheikh Ahmad	Hospital Assistant	1899
17	Buta Mal	Hospital Assistant	1899
18	Dunga Dass	Hospital Assistant	1899
19	Allah Ditta	Assistant Surgeon	1901
20	W. St.J. Hussey*	Assistant Surgeon	1901
21	Muhammad Ali Khan	Hospital Assistant	1900
22	Munshi Basharet Ahmad	Assistant Surgeon	1900
23	Lala Moti Ram	Assistant Surgeon	1900
24	Lala Mathra Das	Assistant Surgeon	1900
25	Muhammad Ibrahim	Hospital Assistant	1901
26	J Doyle*	Assistant Surgeon	1901
27	Lala Ram Dhan Kapur	Assistant Surgeon	1901

*Most probably Anglo-Indians described as Military Assistant Surgeons.'[2]

[1] Information compiled from BL/IOR/L/MIL/7/2177 Collection 48/25 Uganda Railway: Medical Staff for Road to Coast, 1895–1896; BL/IOR/L/MIL/7/2188 Hospital Assistants for the Uganda Railway, 1896; BL/IOR/L/E/7/444 File 626 Revenue and Statistics Department Papers, Notes 12 June 1899, 12 July 1899, 26 January 1900, 29 January 1900, 12 February 1901; BL/IOR/L/MIL/7/14454 Indian Soldiers Invalided to India from British Africa and Rates of Pay, 1899–1900.
[2] BL/IOR/L/E/7/444 File 626 Revenue and Statistics Department Papers, Letter from IMS to Secretary of State India, 12 July 1899.

Appendix 2: Indians in the Colonial Medical Service, 1895–1940

[Broadly chronologically arranged]

1	E.J.H. Oorloff	Assistant Surgeon
2	A.N. Nyss	Assistant Surgeon
3	H.L. Sargent	Assistant Surgeon
4	W.N. Sargent	Assistant Surgeon
5	P.H. Nand	Assistant Surgeon
6	T.D. Nair	Assistant Surgeon
7	G.V. Vinod	Assistant Surgeon
8	Murari Lal	Assistant Surgeon
9	A. Whittle	Assistant Surgeon
10	Imam Buksh	Sub-Assistant Surgeon
11	Dare Khan	Sub-Assistant Surgeon
12	Kushall Chand	Sub-Assistant Surgeon
13	Mahomed Ali Khan	Sub-Assistant Surgeon
14	B.N. Bandoppandhyay	Sub-Assistant Surgeon
15	D.S. Tipnis	Sub-Assistant Surgeon
16	G.V. Patel	Sub-Assistant Surgeon
17	Lal Singh Sandhu	Sub-Assistant Surgeon
18	R. Pillay	Sub-Assistant Surgeon
19	J.A. Karvekar	Sub-Assistant Surgeon 1921; Assistant Surgeon 1939
20	J.P. Andrews	Sub-Assistant Surgeon 1921; Assistant Surgeon 1939
21	Kesar Chand	Sub-Assistant Surgeon
22	Sukram das	Sub-Assistant Surgeon
23	Jaswant Singh	Sub-Assistant Surgeon
24	Milki Ram	Sub-Assistant Surgeon
25	S.F. D'Costa	Sub-Assistant Surgeon
26	Karta Ram	Sub-Assistant Surgeon
27	Wilayat Shah	Sub-Assistant Surgeon
28	C.J. Patel	Sub-Assistant Surgeon
29	Anant Ram Sharma	Sub-Assistant Surgeon
30	R.S. Kibe	Sub-Assistant Surgeon
31	Ali Buksh	Sub-Assistant Surgeon
32	Mahomad Bashir	Sub-Assistant Surgeon
33	Alif Khan	Sub-Assistant Surgeon
34	Chuhar Khan	Sub-Assistant Surgeon
35	H.T. Moorjani	Sub-Assistant Surgeon
36	Wazir Chand	Job title unknown
37	Seyed Asghar Ali	Sub-Assistant Surgeon
38	J.W. Comfort	Job title unknown
39	Bhag Singh	Job title unknown

Continued

Continued

40	Bhagwan Singh	Job title unknown
41	Kartar Singh	Sub-Assistant Surgeon (Hospital Assistant)
42	Munshi Ram Gupta	Sub-Assistant Surgeon
43	Karim Buksh	Job title unknown
44	Gokul Chand	Sub-Assistant Surgeon
45	A.K. Pathrekar	Job title unknown
46	C.D. Rana	Job title unknown
47	Jiwa Ram	Sub-Assistant Surgeon
48	B.J. Parvati	Sub-Assistant Surgeon
49	Shaik Abdul Kadir	Sub-Assistant Surgeon
50	Abdullah Khan	Sub-Assistant Surgeon
51	S.M. Sharfuddin	Sub-Assistant Surgeon
52	H.A. Lewis	Sub-Assistant Surgeon
53	Ferosend Din	Sub-Assistant Surgeon
54	Hira Nand	Sub-Assistant Surgeon
55	S.V. Pantwaidya	Sub-Assistant Surgeon
56	Kisar Singh	Sub-Assistant Surgeon
57	S. Varma	Sub-Assistant Surgeon
58	Thatte	Sub-Assistant Surgeon
59	Das	Sub-Assistant Surgeon
60	Chand	Sub-Assistant Surgeon
61	W.E Cody	Assistant Surgeon
62	D.E. Barrett	Assistant Surgeon
63	H.L. Braganza	Assistant Surgeon
64	Brinda ban	Sub-Assistant Surgeon
65	Shah	Sub-Assistant Surgeon
66	Bhawani Shankar	Sub-Assistant Surgeon
67	J. Doyle	Assistant Surgeon
68	D.D. Stuart	Assistant Surgeon
69	E.W. Rodrigo	Assistant Surgeon
70	Maula Buksh	Sub-Assistant Surgeon (Hospital Assistant)
71	N.Z. Pireleybawa	Sub-Assistant Surgeon (Hospital Assistant)
72	M.D. Dessay	Sub-Assistant Surgeon (Hospital Assistant)
73	Gokalchand	Sub-Assistant Surgeon (Hospital Assistant)
74	Niamat Ulla	Sub-Assistant Surgeon (Hospital Assistant)
75	Bapu Gore	Sub-Assistant Surgeon (Hospital Assistant)
76	Ram Saran	Sub-Assistant Surgeon (Hospital Assistant)
77	Bhushall Chand	Sub-Assistant Surgeon (Hospital Assistant)
78	B.S. Maduram	Sub-Assistant Surgeon (Hospital Assistant)
79	Milton Balm	Sub-Assistant Surgeon (Hospital Assistant)
80	P.N. Patel	Sub-Assistant Surgeon (Hospital Assistant)
81	G. Nahapiet	Assistant Surgeon
82	T.D. Nair	Assistant Surgeon
83	Kishan Singh	Sub-Assistant Surgeon
84	Hasara Singh	Sub-Assistant Surgeon
85	Mahomed Ismail Chaudri	Sub-Assistant Surgeon
86	Sarmukh Singh	Sub-Assistant Surgeon
87	G.P. Joshi	Sub-Assistant Surgeon

Continued

Continued

88	Diwan Singh	Sub-Assistant Surgeon
89	K.B. Saksena	Sub-Assistant Surgeon
90	G.S. Gokhale	Sub-Assistant Surgeon
91	B.D. Neurgaonkar	Sub-Assistant Surgeon
92	Autur Singh	Sub-Assistant Surgeon
93	G.B.S. Thakore	Sub-Assistant Surgeon
94	S.N. Sharma	Sub-Assistant Surgeon
95	M.L. Sarup	Sub-Assistant Surgeon
96	M.A. Dastur	Sub-Assistant Surgeon
97	S.D. Bhardwaj	Sub-Assistant Surgeon
98	A.N. Bowary	Sub-Assistant Surgeon
99	R.L. Seth	Sub-Assistant Surgeon
100	Balwant Rai Nanda	Sub-Assistant Surgeon
101	Jhimal Singh	Sub-Assistant Surgeon
102	Anant Ram	Sub-Assistant Surgeon
103	R.S. Verma	Sub-Assistant Surgeon
104	M.M. Bali	Sub-Assistant Surgeon
106	Mohamaed Anwar Shah	Sub-Assistant Surgeon
107	Mohamed Ashraf	Sub-Assistant Surgeon
108	Jodh Singh	Sub-Assistant Surgeon
109	D.R. Chhabra	Sub-Assistant Surgeon
110	G.B. Singh	Sub-Assistant Surgeon
111	Ali Mohammed	Sub-Assistant Surgeon
112	Bakhtawar Singh	Sub-Assistant Surgeon

Appendix 3: Indian Private Practitioners, 1895–1940

	Name	Qualification	Place of practice
1	Adalja, K.V.*	M.B.B.S. (Bombay)	Nairobi
2	Ahluwalia, L.S.	L.R.C.S, L.R.C.P (Edin.); L.R.F.P.S. (Glas.); D.P.H. (Cantab.)	Unknown
3	Amin, G.C.	M.B.B.S. (Bombay)	Nairobi
4	Amin, J.B.	M.B.B.S. (Bombay)	Nairobi
5	Amin, P.M.*	M.B.B.S. (Bombay)	Nairobi
6	Badruddin, A.	M.B.B.S. (Punjab)	Unknown
7	Campos, J.J.	M.B. (Calcutta)	Unknown
8	Cardozo, L.J.*	M.B.B.S. (Bombay)	Unknown
9	Chiplonkar, T.L.	M.R.C.S., L.R.C.P. (Lon.); L.M.S. (Bombay)	Unknown
10	De Mello, J.P.	M.B.B.S. (Bombay)	Nairobi
11	De Mello, R.N.	M.B.B.S. (Bombay)	Nairobi
12	De Menzes, A.N.J.H.	M.B.B.S. (Bombay)	Unknown
13	De Olivera C.A.*	M.B.B.S. (Bombay)	Unknown
14	De Sousa, A.C.L.*	Lic.Med.Surg (Bombay)	Nairobi
15	De Sousa, M.M.*	Lic.Med.Surg (Bombay)	Nairobi
16	Dev, T.B.	L.R.F.P.S. (Glas.); L.R.C.S., L.R.C.P. (Edin.); L.M. (Dub.)	Unknown
17	Dias, E	L.R.C.P., M.R.C.S. (Edin.)	Nairobi
18	Dotiwala, N.A.*	M.B.B.S. (Bombay)	Mombasa
19	Figueiredo, E.P.*	M.B.B.S. (Bombay)	Mombasa
20	Figueiredo, H.*	M.B.B.S. (Bombay)	Mombasa
21	Gautama, M.D.*	M.B.B.S. (Lahore)	Nairobi
22	Ghandhi, D.T.	M.B.B.S. (Bombay)	Unknown
23	Hamin, M.N.*	M.B.B.S. (Bombay)	Mombasa
24	Ismail, A.H.	M.R.C.S., L.R.C.P (Eng.); D.T.M. (Liverpool)	Nairobi
25	Juvaker, G.V.*	M.B.B.S. (Bombay)	Mombasa
26	Karve, S.V.*	M.B.B.S. (Bombay)	Nairobi/ Mombasa
27	Mandalia, N.C.	M.B.B.S. (Bombay); D.O.M.S.	Mombasa
28	Mascarenhas, I.D.*	M.B.B.S. D.U. (Bombay)	Unknown
29	Mehta, R.N.*	M.B.B.S. (Bombay)	Unknown
30	Menderith, R.M.*	M.B.B.S. (Bombay)	Unknown
31	Merchant, A.M.J.	M.B.B.S. (Bombay)	Unknown
32	Mohamed Shah, M.K.	M.B.B.S. (Punjab)	Unknown
33	Najmudean, G.*	M.B.B.S. (Bombay)	Mombasa
34	Patel, C.S.*	M.B.B.S. (Bombay)	Kisumu
35	Patel, M.S.	M.D., C.M. (Dalhousie, Canada)	Unknown

Continued

Continued

	Name	Qualification	Place of practice
36	Patel, R.P.*	M.B.B.S. (Bombay)	Nairobi
37	Patwardhan, V.V.*	M.B.B.S. (Bombay)	Nairobi
38	Qadros, C.A.	M.B.B.S. (Bombay)	Unknown
39	Rana, M.A.*	M.R.C.S. L.R.C.P. (Lon.)	Nairobi/ Mombasa
40	Ribeiro, M.*	M.B.B.S. (Bombay)	Nairobi
41	Sandhu, B.S.*	M.B. Ch.B. (Edin.)	Nairobi
42	Sheth, A.M.*	M.B.B.S. (Bombay)	Mombasa
43	Sorabjee, E.*	M.R.C.S., L.R.C.P. (Lon.)	Nairobi
44	Sood, F.C.*	M.B.B.S. (Lahore)	Kisumu
45	Shah, M.T.*	M.B.B.S. (Bombay)	Nairobi
46	Shah, N.M.*	M.B.B.S. (Bombay)	Nakuru
47	Singh, B.*	M.B.B.S. (Lahore)	Kisumu
48	Topiwala, H.T.*	M.B.B.S. (Bombay)	Nairobi
49	Vaz, F.C.C.*	M.B.B.S. (Bombay)	Unknown
50	Verma, R.S.	M.B.B.S. (Punjab)	Unknown

* Individuals with ten years or more service in Kenya.

Appendix 4: Statistics Concerning Indian Workers, Uganda Railways (Cumulative)

Year	Imported	Repatriated	Invalided	Casualties	Deaths
1896–7	4,209	N/A	200	321	121
1897–8	7,131	N/A	705	1045	340
1898–9	15,598	773	1,206	2,590	611
1899–0	23,379	2,791	3,424	7,349	1,164
1900–1	31,646	4,109	5,811	11,904	1,984
1901–2	31,983	9,616	6,354	18,337	2,367
1902–3	31,983	16,312	6,454	25,259	2,493

Compiled from *Annual Medical Report, East Africa Protectorate*, 1904

Notes

1 'The Empire is not white': Indian Doctors in Kenya

1. V.S. Srinivasa Sastri, 'The Kenya Question', *The New Age*, 10 May 1923, p. 19.
2. Anna Crozier, *Practising Colonial Medicine: the Colonial Medical Service in East Africa*, London, I.B. Tauris, 2007; Ann Beck, *A History of the British Medical Administration of East Africa, 1900–1950*, Cambridge, Mass, Harvard University Press, 1970; David Hardiman (ed.), *Healing Bodies, Saving Souls: Medical Missions in Asia and Africa*, London, Rodopi, 2006; Megan Vaughan, *Curing Their Ills: Colonial Power and African Illness*, Cambridge, Polity Press, 1991; John Iliffe, *East African Doctors: A History of the Modern Profession*, Cambridge, Cambridge University Press, 1998.
3. M. Reda Bhacker, *Trade and Empire in Muscat and Zanzibar: The Roots of British Domination*, London, Routledge, 1992, p. 11.
4. Government of Gujarat, 'Philanthropic and Nationalistic Activities of Gujaratis in East Africa', p. 84 in *Beyond Boundaries*: http://www.nri.gujarat.gov.in/beyond-boundaries.htm (accessed 18 July 2012).
5. Anonymous, 'Espionage: Indian Publicly Executed', *The Leader*, Nairobi, 13 November 1915, p. 16; R.L. Tignor, *The Colonial Transformation of Kenya*, Princeton, Princeton University Press, 1976, p. 6.
6. But interestingly all the studies of middle level healthcare workers have been of black Africans. See, for example, Anne Digby, 'The Mid-Level Health Worker in South Africa: The In-Between Condition of the "Middle"', in Ryan Johnson and Amna Khalid (eds), *Public Health in the British Empire: Intermediaries, Subordinates, and the Practice of Public Health*, New York & London, Routledge, 2012, pp. 171–92; See also the discussion in Marku Hokkanen, *Medicine and Scottish Missionaries in the Northern Malawi Region, 1875–1930*, Lampeter, Edwin Mellen Press, 2007, pp. 412–20.
7. Adeloya Adeloye, *African Pioneers of Modern Medicine: Nigerian Doctors of the Nineteenth Century*, Ibadan, University Press Limited, 1985; Adell Patton, *Physicians, Colonial Racism and Diaspora in West Africa*, Gainesville, University of Florida Press, 1996; Iliffe, *East African Doctors*, Cambridge University Press, 1998; Anne Digby, 'Early Black Doctors in South Africa', *Journal of African History*, 46, 2005, pp. 427–54.
8. The position of Indians as a 'middlemen minority' has been written about with regards to Tanzania in Eric Burton, '"...what tribe should we call him?" The Indian Diaspora, the State and the Nation in Tanzania since ca.1850', *Stichproben: Weiner Zeitschrift für Kritische Afrikastudien*, 25, 2013, pp. 1–28, p. 1.
9. The tendency to oversimplify and romanticise the subaltern has been pointed out by: Sumit Sarkar, 'The Decline of the Subaltern in Subaltern Studies' in Sumit Sarkar (ed.), *Writing Social History*, 88, Delhi, Oxford University Press, 1997, pp. 82–108.
10. John Lonsdale, 'Kenya: Home Country and African Frontier' in Robert Bickers (ed.), *Settlers and Expatriates: Britons over the Seas*, Oxford University Press, 2014 [First published, 2010], pp. 74–111, p. 86.
11. Nancy Rose Hunt, *A Colonial Lexicon of Birth, Ritual, Medicalization, and Mobility in the Congo*, Durham. N.C., Duke University Press, 1999, p. 23 quoted in Digby, 'The Mid-Level Health Worker', pp. 171–92.

12. For example the origins of Indirect Rule can be found in British India. See Michael H. Fisher, 'Indirect Rule in the British Empire: The Foundations of the Residency System in India (1764–1858)', *Modern Asian Studies*, 18.3, 1984, pp. 393–428.
13. Daniel R. Headrick, *The Tools of Empire: Technology and European Imperialism in the Nineteenth Century*, Oxford University Press, 1981.
14. Robert J. Blythe, *The Empire of the Raj: India, Eastern Africa and the Middle East, 1858–1947*, Cambridge University Press, 2003.
15. Lonsdale, 'Kenya', p. 75.
16. See for example on Tanzania: James R. Brennan, *TAIFA: Making Nation and Race in Urban Tanzania*, Athens, Ohio University Press, 2012; Ronald R. Aminzade, Jack A. Gladstone, Doug McAdam, Elizabeth J. Perry, William Sewell, Jr., Sidney Tarrow, and Charles Tilly, *Silence and Voice in the Study of Contentious Politics*, Cambridge, Cambridge University Press, 2001; Ned Bertz, 'Educating the Nation: Race and Nationalism in Tanzanian Schools' in Sara Dorman, Daniel Hammett, and Paul Nugent (eds), *Making Nations, Creating Strangers: States and Citizenship in Africa*, Leiden, Brill, 2007, pp. 161–80.
17. The authors wish to thank Professor Frederick Cooper, NYU, for his incisive comments on a related paper delivered by Anna Greenwood at University of Nottingham, Ningbo China on 17 May 2014. These comments, and the discussion after, offered valuable insights into redefining ideas of colonial power and its operation. John Darwin, *The Empire Project: The Rise and Fall of the British World-System, 1830–1970*, Cambridge, Cambridge University Press, 2009; John Darwin, *Unfinished Empire: the Global Expansion of Britain*, London, Penguin, 2013.
18. J.R. Gregory, *Under the Sun (A Memoir of Dr R.W. Burkitt of Kenya)*, Nairobi, English Press, 1951.
19. For Africa see particularly: Michael Gelfand, *Tropical Victory: An Account of the Influence of Medicine on the History of Southern Rhodesia, 1890–1923*, Cape Town, Juta, 1953; Philip Manson-Bahr, *History of the School of Tropical Medicine in London: 1899–1949*, London, HK Lewis & Co. Ltd., 1956; Aldo Castellani, *Microbes, Men and Monarchs: A Doctor's Life in Many Lands*, London, Gollancz, 1960; Ralph Schram, *A History of the Nigerian Health Services*, Ibadan, Ibadan University Press, 1971; J.J. McKelvey Jr., *Man Against Tsetse: Struggle for Africa*, London, Cornell University Press, 1973; Michael Gelfand, *A Service to the Sick: A History of the Health Services for Africans in Southern Rhodesia, 1890–1953*, Gweru, Mambo Press, 1976.
20. For one of the first books to famously argue for the intentional exploitation of Africa by British colonials see: Walter Rodney, *How Europe Underdeveloped Africa*, London, Bogle-L'Ouverture and Tanzanian Publishing House, 1972.
21. Headrick, *Tools of Empire*; Roy Macleod and Milton Lewis (eds), *Disease Medicine and Empire*, London, Routledge, 1988; Vaughan, *Curing Their Ills*.
22. Colonial History became particularly influenced by the ideas of bio power elucidated by Michel Foucault, *The History of Sexuality Volume 1: An Introduction*, trans. Robert Hurley, New York, Pantheon, 1978 [French publication: 1976]. Issues remained debated, however. See: Hussein Abdilahi Bulhan, *Frantz Fanon and the Psychology of Oppression*, New York, London, Plenum, 1985.
23. Randall Packard, *White Plague, Black Labor: Tuberculosis and the Political Economy of Health and Disease in South Africa*, Berkeley, University of California Press, 1989; John Farley, *Bilharzia: A History of Imperial Tropical Medicine*, Cambridge, Cambridge University Press, 2008 [first published 1991]; Maryinez Lyons, *The Colonial Disease: A Social History of Sleeping Sickness in Northern Zaire 1900–1940*, Cambridge,

Cambridge University Press, 2002 [first published, 1992]; Mark Harrison, *Public Health in British India: Anglo Indian Preventative Medicine, 1859–1914*, Cambridge, Cambridge University Press, 1994.

24. Ranajit Guha, 'The Small Voices of History' in Shahid Amin and Dipesh Chakrabarty (eds), *Subaltern Studies: Writings on South Asian History and Society*, Vol. IX, Oxford, Oxford University Press, 1988.

25. The idea of the 'thin white line' was put forward by Kirk-Greene. See Anthony M. Kirk-Greene, 'The Thin White Line: The Size of the British Colonial Service in Africa', *African Affairs*, 79, 1980, pp. 25–44.

26. Alan Lester, *Imperial Networks: Creating Identities in Nineteenth-Century South Africa and Britain*, London, Routledge, 2001; Alan Lester, 'Imperial Circuits and Networks: Geographies of the British Empire', *History Compass*, 4, 2006, pp. 124–41; Zoe Laidlaw, *Colonial Connections 1815–1845; Patronage, the Information Revolution and Colonial Government*, Manchester, Manchester University Press, 2005; T.M. Metcalf, *Imperial Connections: India in the Indian Ocean Arena, 1860–1920*, Berkeley and London, University of California Press, 2007; Simon J. Potter, 'Webs, Networks and Systems: Globalization and the Mass Media in the Nineteenth- and Twentieth-Century British Empire', *Journal of British Studies*, 46, 2007, pp. 621–46; Brett M. Bennett and Joseph M. Hodge (eds), *Science and Empire: Knowledge and Networks of Science Across the British Empire, 1800–1970*, Basingstoke, Palgrave Macmillan, 2011; Deborah J. Neill, *Networks in Tropical Medicine: Internationalism, Colonialism, and the Rise of a Medical Specialty, 1890–1930*, Stanford, Stanford University Press, 2012.

27. David Arnold, *Colonizing the Body: State Medicine and Epidemic Disease in Nineteenth Century India*, Berkeley, University of California Press, 1993.

28. Anne Digby and Helen Sweet, 'The Nurse as Culture Broker in Twentieth Century South Africa' in Waltraud Ernst (ed.), *Plural Medicine, Tradition and Modernity*, London, Routledge, 2002, pp. 113–29; Philip D. Morgan and Sean Hawkins (eds), *Black Experience and the Empire*, Oxford, Oxford University Press, 2004; Benjamin N. Lawrance, Emily Lynn Osborn, and Richard L Roberts (eds), *Intermediaries, Interpreters, and Clerks: African Employees in the Making of Colonial Africa*, Madison, University of Wisconsin Press, 2006; Anne-Marie Rafferty, 'The Rise and Demise of the Colonial Nursing Service: British Nurses in the Colonies, 1896–1966', *Nursing History Review*, 15, 2007, pp. 147–54; Johnson and Khalid, *Public Health in the British Empire*, 2012.

29. Douglas Haynes, *Imperial Medicine: Patrick Manson and the Conquest of Tropical Disease*, Philadelphia, University of Pennsylvania Press, 2001.

30. Dane Kennedy and Durba Ghosh (eds), *Decentring Empire: Britain, India, and the Transcolonial World*, Hyderabad, Longman Orient Press, 2006; Kevin Grant, Philippa Levine, and Frank Trentmann (eds), *Beyond Sovereignty, 1880–1950: Britain, Empire and Transnationalism*, London, Palgrave, 2007; Pratik Chakrabarti, *Medicine and Empire, 1600–1960*, London, Palgrave MacMillan, 2013.

31. Projit Mukharji, *Nationalizing the Body: The Market, Print and Healing in Colonial Bengal, 1860–1930*, London, Anthem Press, 2009.

32. Ernst, *Plural Medicine*; Pratik Chakrabarti, *Western Science in Modern India: Metropolitan Methods, Colonial Practices*, Delhi, Permanent Black, 2004; Sanjoy Bhattacharya, Mark Harrison, and Michael Worboys (eds), *Fractured States: Smallpox, Public Health and Vaccination Policy in British India, 1800–1947*, New Delhi, Orient Longman and Sangam Books, 2005; Anne Digby, *Diversity and Division in Medicine: Healthcare in South Africa from the 1800s*, Oxford, Peter Lang, 2006; Guy Attewell, *Refiguring Unani Tibb: Plural Healing in Late Colonial India*, New Delhi,

Orient Longman, 2007; Biswamoy Pati and Mark Harrison (eds), *The Social History of Health and Medicine in Colonial India*, London and New York, Routledge, 2009; Hormoz Ebrahimnejad (ed.), *The Development of Modern Medicine in Non-Western Countries: Historical Perspectives*, London and New York, Routledge, 2009.

33. Anne McClintock, *Imperial Leather: Race, Gender and Sexuality in the Colonial Context*, New York, Routledge, 1995; Karen Jochelson, *The Colour of Disease: Syphilis and Racism in South Africa, 1880–1950*, Basingstoke, Palgrave, 2001; Philippa Levine, *Prostitution, Race and Politics: Policing Venereal Disease in the British Empire*, London, Routledge, 2003; Mary P. Sutphen and Bridie Andrews (eds), *Medicine and the Colonial Identity*, London, Routledge, 2003; Warwick Anderson, *Colonial Pathologies: American Tropical Medicine, Race and Hygiene in the Philippines*, Durham, Duke University Press, 2006.

34. Roger Jeffery, 'Recognizing India's Doctors: The Institutionalization of Medical Dependency, 1918–1939', *Modern Asian Studies*, 13.2, 1979, pp. 301–26; Roger Jeffery, 'Doctors and Congress: The Role of Medical Men and Medical Politics in Indian Nationalism' in Mike Shepperdson and Colin Simmons (eds), *The Indian National Congress and the Political Economy of India, 1885–1985*, Aldershot, Brookfield, USA, Avebury, 1988, pp. 160–73.

35. Deepak Kumar, 'Racial Discrimination and Science in Nineteenth Century India', *Indian Economic and Social History Review*, 19, 1982, pp. 63–82.

36. Karen Blixen, *Out of Africa*, London, Putnam & Co. Ltd, 1937; Isak Dinensen, *Letters from Africa*, London, Weidenfeld and Nicolson, 1981; Elspeth Huxley, *The Sorcers's Apprentice: A Journey Through East Africa*, London, Chatto and Windus, 1949; Elspeth Huxley, *Out in the Midday Sun: My Kenya*, London, Chatto and Windus, 1958; Elspeth Huxley, *The Flame Trees of Thika: Memoirs of An African Childhood*, London, Chatto and Windus, 1959; Elspeth Huxley, *Pioneers' Scrapbook: Reminiscences of Kenya, 1890 to 1968*, London, Evan Brothers Ltd.,1980.

37. Elspeth Huxley, *White Man's Country: Lord Delamere and The Making of Kenya*, London, Chatto and Windus, 1935. For a more critical perspective, see also Margery Perham, *Race and Politics in Kenya: A Correspondence Between Elspeth Huxley and Margery Perham*, London, Faber and Faber, 1944. For the prevalent view of Huxley's work see C. S. Nicolls, *Elspeth Huxley: A Biography*, London, Harper & Collins, 2007. Critiques of Husley came however. See Peter Worsley, 'The Anatomy of Mau Mau', *The New Reasoner*, 1, 1957, pp. 13–25; Chloe Campbell, *Race and Empire: Eugenics in Colonial Kenya*, Manchester University Press, 2007, p. 1.

38. Bethwell Allan Ogot, *History of the Southern Luo: Volume I, Migration and Settlement, 1500–1900*, Nairobi, East African Publishing House, 1967; Bethwell Allan Ogot (ed.), *Politics and Nationalism in Kenya*, Nairobi, East African Publishing House,1972; Bethwell Allan Ogot and J.A. Kieran (eds), *Zamani: A Survey of East African History*, Nairobi, East African Publishing House, 1968; Bethwell Allan Ogot, *My Footprints on the Sands of Time: An Autobiography*, Victoria BC, Trafford Publishing, 2006. See also Ali Mazuri, 'European Exploration and African Self Discovery', *The Journal of Modern African Studies*, 4, 1969, pp. 661–76.

39. M.N. Pearson, *Port Cities and Intruders: The Swahili Coast India and Portugal in the Early Modern Era*, Baltimore, The John Hopkins University Press, 1998, p. 17; Particularly illustrative is the debate between Ogot and Oliver, see Kelly Boyd (ed.), *Encyclopedia of Historians And Historical Writing*, Vol. 2, London and Chicago, Fitzroy & Dearborn, 1999, p. 881.

40. Bruce Berman and John Lonsdale, *Unhappy Valley: Conflict in Kenya and Africa*, London, Nairobi, Athens, J. Currey, Heinemann Kenya, Ohio University Press, 1992. David Anderson, *Histories of the Hanged: The Dirty War in Kenya and the*

End of Empire, London, Weidenfield & Nicolson, 2005; See also Dane Kennedy, 'Constructing the Colonial Myth of Mau Mau', *International Journal of African Historical Studies*, 25.2, 1992, pp. 241–60.

41. Mamdani discusses the role of Indian troops 'as a means of coercion in a colony.' M. Mamdani, *From Citizen to Refugee: Uganda Asians Come to Britain*, London, Francis Pinter Ltd., 1973, p. 14.

42. Gregory, *Under the Sun*; Margaret Trowell, *African Tapestry*, London, Faber and Faber, 1957; John A. Carman, *A Medial History of the Colony and Protectorate of Kenya: A Personal Memoir*, London, Rex Collings, 1976. Wellcome Library (thereafter WL)/CMAC/PP/HCT/A5 Elizabeth Bray, *Hugh Trowell: Pioneer Nutritionist*, unpublished biography and tape transcripts, London, 1988; Rhodes House Library, Oxford (thereafter RHL)/MSS.Brit.Emp.r.4 Peter Alphonsus Clearkin, *Ramblings and Recollections of a Colonial Doctor 1913–1958*, Book I, Durban, 1967; RHL/MSS.Afr.s.1653 Theodore Farnworth Anderson, *Reminiscences by T. Farnworth Anderson*, Book I, Limuru, Kenya, 1973; RHL/MSS.Afr.s.702 Robert Arthur Welsford Procter, 'Random Reminiscences, Mainly Surgical', [n.d.] medical historical articles: Arthur Dawson Milne, 'The Rise of the Colonial Medical Service', *Kenya and East African Medical Journal*, 5, 1928–1929, pp. 50–8; H.A. Bödeker, 'Some Sidelights on Early Medical History in East Africa', *The East African Medical Journal*, 12, 1935–6, pp. 100–7; John Langton Gilks, 'The Medical Department and the Health Organization in Kenya, 1909–1933', *The East African Medical Journal*, 9, 1932–3, pp. 340–54; Clare Aveling Wiggins, 'Early Days in British East Africa and Uganda', *The East African Medical Journal*, 37, 1960, pp. 699–708; Clare Aveling Wiggins, 'Early Days in British East Africa and Uganda: Second Tour—1904–1907', *The East African Medical Journal*, 37, 1960, pp. 780–93.

43. Shula Marks, 'What is Colonial about Colonial Medicine? And what has Happened to Imperialism and Health?' *Social History of Medicine*, 10, 1997, pp. 205–19, p. 205.

44. Beck, *History of the Medical Administration*, Cambridge, MA, Harvard University Press, 1970; Ann Beck, *Medicine, Tradition and Development in Kenya and Tanzania 1920–1970*, Waltham, Mass: Crossroads Press, 1981.

45. Maureen Malowany, *Medical Pluralism: Disease, Health and Healing in the Coast of Kenya, 1840–1940*, Ph.D Thesis, McGill University, Canada, 1997, p. 133; Kenneth Ingram, 'Medicine in East Africa', *The Journal of African History*, 12, 1971, pp. 162–3; C.C. Wrigley, Book Review, *The Journal of African History*, 26.4, 1985, pp. 44–2.

46. As well as her monograph see also Ann Beck, 'Problems of British Medical Administration in East Africa between 1900–1930', *Bulletin of the History of Medicine*, 36, 1962, pp. 275–83; Ann Beck, 'Medical Administration and Medical Research in Developing Countries: Remarks on Their History in Colonial East Africa', *Bulletin of the History of Medicine*, 46, 1972, pp. 349–58; Ann Beck, 'The State and Medical Research: British Government Policy Toward Tropical Medicine in East Africa', *Proceedings of the XXIII International Congress of the History of Medicine, London, 2–9 September 1972*, London, Wellcome Institute for the History of Medicine, 1974, pp. 488–93.

47. Dane Kennedy, *Islands of White: Settler Society and Culture in Kenya and Southern Rhodesia, 1890–1939*, Durham, Duke University Press, 1990; Vaughan, *Curing Their Ills*, 1991.

48. Jock McCulloch, *Colonial Psychiatry and 'The African Mind'*, Cambridge University Press, 1995; Iliffe, *East African Doctors*; George Odour Ndege, *Health, State and Society in Kenya*, Rochester, NY, University of Rochester Press, 2001; O.A. Olluumwallah, *Disease in the Colonial State: Medicine, Society and Social Change Among the Aba*

Nyole of Western Kenya, Westport Connecticut, Greenwood Publishing Group, 2002; Kirk Arden Hoppe, *Sleeping Sickness Control in British East Africa, 1900–1960*, Westport Connecticut, Praeger, 2003; Sloan Mahone, 'The Psychology of Rebellion: Colonial Medical Responses to Dissent in British East Africa', *Journal of African History*, 47.2, 2006, pp. 241–60.

49. Crozier, *Practising Colonial Medicine*.
50. Campbell, *Race and Empire*, pp. 61–71.
51. Robert G. Gregory, *South Asians in East Africa: An Economic And Social History 1890–1990*, Oxford, Westview Press, 1993, pp. 217–27; Cynthia Salvadori, *We Came in Dhows*, Nairobi, Paperchase Kenya Ltd, 1996, has material relating to Dr Dias (Vol I, p. 126); Dr Ribeiro (Vol. II, p. 22), Dr Bowry (Vol. III, p. 91), Dr de Sousa (Vol. III, p. 160), and Dr Bödeker (Vol. II, pp. 6–7).
52. Ralph Schram, *Heroes of Healthcare in Africa*, unpublished folio, Isle of White, 1997.
53. U.K. Oza, *The Rift in the Empire's Lute: Being a History of the Indian Struggle in Kenya*, Bombay, Advocate of India Press, 1931; W. Hollingsworth, *The Asians of East Africa*, London, Macmillan and Company Limited, 1960; George Delf, *Asians in East Africa*, Oxford University Press, 1963; Robert G. Gregory, *India and East Africa: A History of Race Relations within the British Empire, 1890–1939*, Oxford, Clarendon Press, 1971; Christopher P. Youé, 'The Threat of Settler Rebellion and the Imperial Predicament: The Denial of Indian Rights in Kenya, 1923', *Canadian Journal of History*, 12, 1978, pp. 347–60; David Himbara, 'The "Asian" Question in East Africa: The Continuing Controversy on the Role of Indian Capitalists in Accumulation and Development in Kenya, Uganda and Tanzania', *African Studies*, 56.1, 1997, pp. 1–18; Deborah L. Hughes, 'Kenya, India and the British Empire Exhibition of 1924', *Race & Class*, 47, 2006, pp. 66–85.
54. Dharam P. Ghai and Yash P. Ghai (eds), *Portrait of a Minority: Asians In East Africa*, Nairobi, Oxford, Oxford University Press, 1970; D.A. Seidenberg, *Uhuru and the Kenya Indians: the Role of a Minority Community in Kenya Politics, 1939–1963*, New Delhi, Vikas Publishing House, 1983; Christopher A. Bayly, *Indian Society and the Making of the British Empire*, Cambridge, Cambridge University Press, 1988; Metcalf, *Imperial Connections*; Sana Aiyar, 'Empire, Race and the Indians in Colonial Kenya's Contested Public Political Sphere, 1919–1923', *Africa: The Journal of the International African Institute*, 81.1, 2011, pp. 132–54; Neera Kapila, *Race, Rail and Society: Roots of Modern Kenya*, Nairobi, East African Educational Publishers, 2011; Brennan, *TAIFA*; Sana Aiyar, *Indians in Kenya: The Politics of Diaspora*, Cambridge, MA, Harvard University Press, *forthcoming*, 2015.
55. M.G. Visram, *Alidina Visram: The Trail Blazer*, Mombasa, M.G. Visram Publisher, 1990; Makrand Mehta, 'Gujarati Business Communities in the East African Diaspora: Major Historical Trends', *Economic and Political Weekly*, 36.20, 2001, pp. 1738–47; G. Oonk, *Settled Strangers: Asian Business Elites in East Africa: 1800–2000*, London, Sage Publications, 2013.
56. N.K. Mehta, *Dream Half-Expressed: An Autobiography*, Bombay, Vakils, Feffer, and Simons, [c.1966]; Paul Marett, *M.P. Shah: His Life and Achievements*, London, Bharatiya Vidya Bhavan, 1988.
57. Zarina Patel, *Challenge to Colonialism: The Struggle of Alibhai Mulla Jeevanjee for Equal Rights in Kenya*, Nairobi, Publisher Distribution Services, 1997; Zarina Patel, *Unquiet: The Life and Times of Makhan Singh*, Nairobi, Zand Graphics, 2006; Zarina Patel, *Manilal Ambalal Desai: The Stormy Petrel*, Nairobi, Zand Graphics, 2010; J.M. Nazareth, *Brown Man Black Country: A Peep into Kenya's Freedom Struggle*, New Delhi, Tidings Publications, 1981.

58. Agehananda Bharati, *The Asians of East Africa: Jayhind And Uhuru*, Chicago, Nelson Hall, 1972; Vincent Cable, 'The Asians of Kenya', *African Affairs*, 68, 1969, pp. 218–31; Stephen Morris, 'Indians in East Africa: A Study in a Plural Society', *The British Journal of Sociology*, 7, 1956, pp. 194–211.
59. See particularly: M.G. Vassanji, *The Book of Secrets*, London, Picador, 1996; Also: M.G. Vassanji, *The Gunny Sack*, Oxford, Heinemann International, 1989; M.G. Vassanji, *No New Land*, New Delhi, Penguin, 1992.
60. Randolph M.K. Joalahliae, *The Indian as an Enemy: An Analysis of the Indian Question in East Africa*, Bloomington, Authorhouse, 2010.
61. Andrew Mickelburgh, 'Bibliography of The South Asian Diaspora and East Africa: An Annotated Bibliography', http://coombs.anu.edu.au/Biblio/biblio_sasiadiaspora.html (accessed 18 July 2012).
62. More recently, doctors and other professional Indian migrants are also over looked in John C. Hawley (ed.), *India in Africa; Africa in India: Indian Ocean Cosmopolitanisms*, Bloomington, IN, Indiana University Press, 2008.
63. Sometimes 'Kenya' rather than the more cumbersome title is loosely used in the text when referring to events in the early period. The boundary between EAP and Uganda was changed in 1902.
64. Campbell, *Race and Empire*; Lonsdale, 'Kenya', pp. 74–111, pp. 80–1.
65. Antibacterial agents like arsphenamine (Salvarsan; introduced in 1910 for the treatment of syphilis) and sulphonamides (first described in 1935) preceded the introduction of naturally occurring antibiotics like penicillin (not widely available until after World War Two) which were the real game changers in the treatment of bacterial infections. A few effective drugs for the treatment of protozoal infections such as malaria and sleeping sickness were available during the period under study.
66. Chanan Singh, 'Later Asian Protest Movements' in B.A. Ogot (ed.), *Hadith 4: Politics and Nationalism in Colonial Kenya*, Nairobi, East Africa Publishing House, 1972, p. 164.
67. Subaltern Studies adopted this phrase as a new means of reading between the lines of traditional elite source material. The phrase was originally Walter Benjamin's description of a new historical lens.
68. E.g. Errol Trzebinski, *The Kenya Pioneers*, London, Mandarin Paperbacks, 1991, p. 44 BL/IOR/L/E/7/1330:1923–1929, File 529: Indians in Kenya: representation in Municipal Councils for Ribeiro's appointment by the Governor.
69. K.V. Adalja, 'Thirty Two Years in General Practice in Nairobi', *East African Medical Journal*, 36, 1959, pp. 442–8; See also K.V. Adalja, 'The Development of Medical Service in Kenya', *East African Medical Journal*, 39, 1962, pp. 105–14. S.D. Karve, 'An Experiment in Midwifery', *East African Medical Journal*, 10, 1933–4, pp. 358–63; S.D. Karve, 'Some Indian Methods of Midwifery', *East African Medical Journal*, 11, 1934–5, pp. 286–7.
70. Although only a fragment of its full run has been preserved in the British Library at Colindale.
71. British Library (thereafter BL)/IOR/L/E/7/1295, File 191, Indians in Kenya: Representations from Associations and Individual Opinions, EAINAC letter to Viceroy of India, 27 January 1923; see also in same file C.F. Andrews, 'Indian Question in East Africa', October1921, p. 13.
72. The authors are currently in the possession of the private papers of Drs A.C.L. de Souza and H.T. Topiwala. These will be deposited at Rhodes House Library, Oxford (as 'Papers Collected by H. Topiwala Related to Indian Doctors in Kenya', expected deposit date, 2015).

73. For example: BL/IOR/L/MIL/7/2188, 1896, Collection 48/35, Hospital Assistants for Uganda Railway and Uganda. BL/IOR /MIL/7/2189:1897–99, Employment of Indian Officers and Native solders in Africa including military awards to six named individuals including the Indian MO in charge Surgeon Lt. H.M. Masani and Hospital Assistants B. Kasinath, Maula Baksh, Rahim Baksh, Sheikh Ahmed, and Niyamtullah. A little more is available on Indian doctors working for the Uganda Railway. See BL/IOR/MIL/7/2153-2193:1888–1901, Collection 48 Enlistment of natives of India for service in Africa: loan of officers, which mentions Indian medical staff accompanying the Railway Survey in 1891; BL/IOR/MIL/7/2177:1895–1896, *Collection 48/25 Uganda Railway: Medical Staff for Road to Coast*, letter A.D. Mackinnon to F. Jackson, 13 April 1895; BL/IOR/MIL/7/2188: 1896, Collection 48/35, Hospital Assistants for Uganda Railway and Uganda; BL/IOR/MIL/7/14462: 1899–1901, Collection 323/40 Promotion of Uganda Railway Hospital Assistant Rahmat Ali, which records a rare personal account and the testimonials of A.E. Sieveking and E. Whitehouse.
74. For instance, no records of the Indian Medical and Dental Association (IMDA) of Kenya (founded in 1935 and merged with BMA in 1962) can be traced. An exchange of letters between Dr H.T. Topiwala and the Medical Union of Nairobi confirms the Medical Union was also formed in 1935. Its relationship with IMDA is not clear, it might even be possible that they were the same organisation. See RHL/Papers Collected by H. Topiwala Related to Indian Doctors in Kenya, expected deposit date 2015.
75. This mirrors the point made by Metcalf: 'In effect, the late Victorian and Edwardian empire in Africa and Southeast Asia was run by Indians (and by Britons trained in India)', Metcalf, *Imperial Connections*, p. 2 [authors' parantheses].
76. The two tier system within medical provision with European doctors in superior position to Indian doctors has been described in: Jeffery, 'Recognizing India's Doctors'; an additional tier of ethic complexity (that espoused by Goan graduates) is described in: Cristiana Bastos, 'The Inverted Mirror: Dreams of Imperial Glory and Tales of Subalternity from the Medical School of Goa', *Etnográfica*, 6.1, 2002, pp. 59–76.
77. Crozier, *Practising Colonial Medicine*; Iliffe, *East African Doctors*; Kennedy, *Islands of White*; Vaughan, *Curing Their Ills*, 1991.

2   Indians, Migration, and Medicine

1. Sir Harry Johnston, Cd.671, 'Report of His Majesty's Special Commission on the Protectorate of Uganda', 1901, p. 7.
2. C.J. Martin, 'A Demographic Study of an Immigrant Community: The Indian Population of British East Africa', *Population Studies*, 6.3, 1953, pp. 233–47; R.G. Gregory, *South Asians in East Africa: An Economic and Social History, 1890–1990*, Boulder, Westview Press Inc., 1993, p. 13.
3. R. Coupland, *East Africa and its Invaders from the Earliest Times to the Death of Seyyid Said 1856*, London, Clarendon Press, 1938, p. 16; M.N. Pearson, *Port Cities and Intruders: The Swahili Coast, India and Portugal in the Early Modern Area*, Baltimore, John Hopkins University Press, 1998, p. 11.
4. M. Reda Bhacker, *Trade and Empire in Muscat and Zanzibar: The Roots of British Domination*, London, Routledge, 1992, p. 11.
5. Michael N. Pearson (ed.), *The World of Indian Ocean: 1500–1800*, Aldershot, Ashgate, 2005, p. 243.

6. Abdul Sheriff, *Slaves, Spices & Ivory in Zanzibar: Integration of an East African Commercial Empire into the World Economy, 1770–1873*, London, James Curry, 1987; Abdul Sheriff, *Dhow Cultures and the Indian Ocean: Cosmopolitanism, Commerce, and Islam*, Columbia University Press, 2010.

7. E.A. Alpers, 'Gujarat and the Trade of East Africa, c.1500–1800', *The International Journal of African Studies*, 9.1,1976, pp. 22–44; E.A. Alpers, *East Africa and the Indian Ocean*, Princeton, Markus Wiener Publishers, 2009, p. 4.

8. Bhacker, *Trade and Empire in Muscat and Zanzibar*, p. 70; The Hindu concept also used by Khojas after converting to Islam.

9. J.S. Mangat, *A History of the Asians in East Africa, c.1886–1945*, Oxford University Press, 1969, pp. 18–50; W. Hollingsworth, *The Asians of East Africa*, London, Macmillan and Company Limited, 1960, pp. 19–35; Robert G. Gregory, *India and East Africa: A History of Race Relations within the British Empire, 1890–1939*, Oxford, Clarendon Press, 1971, p. 39.

10. Henry Morton Stanley, *Through the Dark Continent*, London, Sampson Low, Marston, Searle & Rivington, 1878, p. 63.

11. Imam Sultan Muhammad Shah (Aga Kahn III), *India in Transition: A Study in Political Evolution*, Bombay, Bennett, Coleman and Co. Ltd., 1918, p. 117.

12. Mangat, *History of the Asians in East Africa*, p. 28.

13. RHL/MSS.Brit.Emp.s.22G5 IBEAC, Report of the Court of Directors to the Annual Shareholders Meeting 27 July 1891, p. 41; Gregory, *India and East Africa*, p. 49.

14. Cited in Mangat, *History of the Asians in East Africa*, p. 12.

15. John Ainsworth address to Indians, *East African Chronicle*, 14 August 1920, p. 14.

16. Harry Johnston, Letter to the Editor, *The Times*, 22 August 1922, p. 4.

17. Frederick D. Lugard, *Rise of Our East African Empire*, London 1893, Volume 1, pp. 488–90 cited in Gregory, *India and East Africa*, pp. 49–50; See also Gregory, *India and East Africa*, p. 40.

18. NA/ FO/107/28, letter from Lloyd William Mathews, 19 January 1894.

19. C.9125 Report by Sir A. Hardinge on the British East Africa Protectorate for the Year 1897–98, London, HMSO, 1899, p. 17.

20. Sir Harry Johnston [n.d. c.1894] cited in R. Oliver, *Sir Harry Johnston and the Scramble for Africa*, London, 1957, p. 254.

21. Lugard, *Rise of Our East African Empire*, pp. 488–90 cited in Gregory, *India and East Africa*, p. 50.

22. RHL/MSS.Brit.Emp.s.22G5 IBEAC, Report of the Court of Directors to the Annual Shareholders Meeting 27 July 1891, p. 4.

23. Robert J. Blyth, *The Empire of the Raj: India, Eastern Africa and the Middle East, 1858–1947*, Cambridge University Press, 2003, p. 10.

24. BL/IO/OIR026.954, Penelope Tuson, *Zanzibar: Sources in the India Office Records*, [n.d. c.1985]; Legal codes were also influenced by those used in Nigeria and South Africa. H.F. Morris, *Government Publications Relating to Kenya*, Microform Academic Publishers, Wakefield, UK, 1976, p. 9; On monetary as well as legal similarities see T.M. Metcalf, *Imperial Connections: India in the Indian Ocean Arena, 1860–1920*, Berkeley and London, University of California Press, 2007, p. 171.

25. Frere is cited in Mangat, *History of the Asians in East Africa*, p. 12; Sir Bartle Frere is cited in Metcalf, *Imperial Connections*, p. 166.

26. R. Coupland, 'Zanzibar: An Asiatic Spice Island, Kirk and Slavery', *The Times*, 5 October 1928, p. 15.

27. Mangat, *History of the Asians in East Africa*, p. 3.

28. See for example: National Archive (thereafter NA)/FO/107/28, letter from H.G. Colvile to R. Owen, 17 November 1893 and letter from R. Rodd to Lord Rosebury, 14 February 1894.
29. Michael H. Fisher, 'Indirect Rule in the British Empire: The Foundations of the Residency System in India (1764–1858)', *Modern Asian Studies*, 18.3, 1984, pp. 393–428.
30. BL/IOR/l/PO/1/1A(iv) letter from Mr Sandback Baker to Mr Jeevanjee, attached to EAINAC submission on 'Kenya Indians' to Viscount Milner Secretary of State, 1920.
31. Gregory, *India and East Africa*, pp. 52, 55; H. Gunston, 'The Planning and Construction of the Uganda Railway', *Transactions of the Newcomen Society*, 74, 2004, pp. 45–71.
32. Johnston, Letter to the Editor, p. 4.
33. John Lonsdale, 'Kenya: Home Country and African Frontier' in Robert Bickers (ed.), *Settlers and Expatriates: Britons over the Seas*, Oxford University Press, 2014 [first published 2010], pp. 74–111, 95.
34. For more on Indian struggles for land and representation in East Africa see: Robert J. Blythe, *The Empire of the Raj: India, Eastern Africa and the Middle East, 1858–1947*, Chippenham and Eastbourne, Palgrave Macmillan, 2003, pp. 93–131.
35. Mangat, *History of the Asians in East Africa*, p. 9; Gregory, *India and East Africa*, p. 36; Gregory, *South Asians in East Africa*, pp. 12, 35.
36. R. Coupland, *Exploitation of East Africa 1856–90: Slave Trade and the Scramble*, London, Faber, 1968, p. 44 refers to 66 Europeans.
37. BL/IOR/L/E/7/1267 'The Position of Indians in Other Parts of Empire', Cabinet Paper Presented by Government of India, 3 May 1921, p. 1.
38. Gijsbert Oonk, 'The Indian Diasporas. A Disputed Empirical and Historical Framework', www.asiansinafrica.com/conference/positionpaperconference.pdf (accessed 9 August 2012).
39. Martin, 'A Demographic Study of an Immigrant Community', p. 234.
40. Makrand Mehta, 'Gujarati Business Communities in East African Diaspora: Major Historical Trends', *Economic and Political Weekly*, 36.20, 2001, pp. 1738–47.
41. Frustration by historians over the lack of sources can be found for example Cynthia Salvadori, *We Came in Dhows*, Nairobi, Paperchase Kenya Ltd, 1996, Vol. I, pp. x–xiv; Zarina Patel, *Challenge to Colonialism: The Struggle of Alibhai Mulla Jeevanjee for Equal Rights in Kenya*, Nairobi, Publisher Distribution Services, 1997, p. 193.This lack of data is particularly frustrating as leading Indian traders kept professional account books, corresponded with distant markets and were known to be proficient in foreign languages. See Bhacker, *Trade and Empire in Muscat and Zanzibar*, p. 11.
42. The word 'banyan' is derived from the word 'vanya' or 'vaishyas' that describes a Hindu caste of merchants.
43. Cynthia Salvadori [Andrew Fedders (ed.)], *Through Open Doors: A View of Asian Cultures in Kenya*, Nairobi, Kenway Publications, 1989, p. 311.
44. Cristiana Bastos, 'The Inverted Mirror: Dreams of Imperial Glory and Tales of Subalternity from the Medical School of Goa', *Etnográfica*, 6.1, 2002, pp. 59–76.
45. Notably Burton did not venture inland. Richard F. Burton, *Zanzibar, City, Island and Coast*, London, Tinsley Brothers, 1872, pp. 5–75, 256–57, 326, 342; James Christie, *Cholera Epidemics in East Africa: An Account of the Several Diffusions of the Disease in that Country from 1821 till 1872*, London, Macmillan, 1876, pp. 299–349.
46. See for example: J.R. Hinnells, *The Zoroastrian Diaspora: Religion and Migration*, London, Oxford University Press, 2005, pp. 245–313; H.M. Amiji, 'The Bhoras of East Africa', *The Journal of Religion in Africa*, 5, 1975, pp. 27–61.

47. G. Oonk, 'The Changing Culture of the Hindu Lohana Community in East Africa', *Contemporary Asian Studies*, 13.1, 2004, pp. 7–23; Salvadori, *Through Open Doors*; Gregory, *South Asians in East Africa*, pp. 9–36.
48. Agehananda Swami Bharati, *The Asians in East Africa: Jayhind and Uhuru*, Chicago, Nelson Hall Company, 1972.
49. Gregory, *India and East Africa*, pp. 204–06; Keith Kyle, 'Gandhi, Harry Thuku and Early Kenya Nationalism', *Transition*, 27, 1966, pp. 16–22; Michael Twaddle, 'Z.K. Sentongo and the Indian Question in East Africa', *History in Africa*, 24, pp. 309–36.
50. Sana Aiyar, 'Empire, Race and the Indians in Colonial Kenya's Contested Public Political Sphere, 1919–1923', *Africa: The Journal of the International African Institute*, 81, 1, 2011, pp. 132–54.
51. Vaughan Megan, *Curing Their Ills: Colonial Power and African Illness*, Cambridge, Polity Press, 1991.
52. Donna Nelson, 'Problems of Power in a Plural Society: Asians in Kenya', *Southwestern Journal of Anthropology*, 23, 1972, pp. 255–64.
53. Mary Parker, 'Race Relations and Political Development in Kenya', *African Affairs*, 50.198, 1951, pp. 41–42; John M. Nazareth, *Brown Man Black Country: A Peep into Kenya's Freedom Struggle*, New Delhi, Tidings Publications, 1981.
54. David Anderson, *Histories of the Hanged: Britain's Dirty War in Kenya and the End of Empire*, London, Weidenfeld and Nicolson, 2005. See also John Lonsdale, 'The Conquest State, 1895–1904' in William R. Ochieng (ed.), *A Modern History of Kenya 1895–1980 in Honour of B.A. Ogot*, Nairobi, London, Evans Brothers (Kenya), 1989, pp. 6–33, p. 6; Christopher P. Youé, 'The Threat of Settler Rebellion and the Imperial Predicament: The Denial of Indian Rights in Kenya, 1923', *Canadian Journal of History*, 12, 1978, pp. 347–60; C.J. Duder, 'The Settler Response to the Indian Crisis of 1923 in Kenya: Brigadier General Philip Wheatly and "Direct Action" ', *Journal of Imperial and Commonwealth History*, 17.3, 1989, pp. 349–73.
55. Dharam P. Ghai and Yash P. *Ghai*, 'Asians in East Africa: Problems and Prospects', *Journal of Modern African Studies*, 3, 1965, pp. 35–51. For specific examples of punishments meted out against Indians, see 'The Public Hanging of Indians', *The Leader*, Nairobi, 13 November 1915, p. 16; As John Lonsdale observed: 'Asians believed that economic survival required political silence' quoted in Anderson, *Histories of the Hanged*, p. 9.
56. Duder, 'The Settler Response to the Indian Crisis of 1923 in Kenya'.
57. Nicola Swainson, *The Development of Corporate Capitalism in Kenya, 1918–77*, London, Heinemann Educational, 1980, pp. 50–54; M. Chandaria personal interview by Harshad Topiwala, Nairobi, 7 May 2006.
58. BL/IOR/L/PJ/8/254 A.W. Pim, Report of the Commission Appointed to Enquire into and Report into the Financial Position and System of Taxation in Kenya, London, HMSO, 1936, p. 6.
59. Parker, 'Race Relations and Political Development in Kenya', pp. 41–52.
60. BL/IOR/L/PJ/8/254 Pim Report, HMSO, 1936, p. 235.
61. Swainson, *The Development of Corporate Capitalism in Kenya*, p. 53; Patel, *Challenge to Colonialism*, p. 76. Crown Ordinance is discussed by M.P.K. Sorrenson, *Origins of European Settlement in Kenya*, Nairobi, Oxford University Press, 1968, p. 174.
62. Swainson, *The Development of Corporate Capitalism in Kenya*, p. 126.
63. For example the Kenya *Official Gazette* in 1911 records all the bankruptcies of which more than 80% were Asian; See also Swainson, *The Development of Corporate Capitalism in Kenya*, p. 51.

64. M. Chandaria personal interview by Harshad Topiwala, Nairobi, 7 May 2006; For example, the Premchand Brothers who established a wattle extraction factory and cotton ginnery in (usually white dominated) Thika. Swainson, *The Development of Corporate Capitalism in Kenya*, p. 125.
65. BL/IOR/L/E/ 7/1330 Nairobi Municipality Report of the Commission, Denham to Governor (Appendix viii), April 1924.
66. BL/IOR/L/E/ 7/1330 Nairobi Municipality Report of the Commission, Denham to Governor (Appendix viii), April 1924.
67. Mangat, *History of the Asians in East Africa*, pp. 172–75; Gregory, *South Asians in East Africa*, p. 192.
68. BL/IOR/L/PJ/8/254 Pim Report, HMSO, 1936, pp. 52–54. NB the cost of salaries included emoluments.
69. D.G. Crawford, *A History of the Indian Medical Service, 1600–1913*, London Thacker and Co., 1914; Robert Heussler, *Yesterday's Rulers: The Making of the British Colonial Service*, New York, Syracuse University Press, 1963; Mark Harrison, *Public Health in British India: Anglo-Indian Preventative Medicine, 1859–1850*, New Delhi, Oxford University Press, 1999; Anthony Kirk-Greene, *On Crown Service: A History of HM Colonial and Overseas Civil Services, 1837–1997*, London, I.B. Tauris, 1999; Anthony Kirk-Green, *Symbol of Authority*, London, I.B. Tauris, 2005; Anna Crozier, *Practising Colonial Medicine: The Colonial Medical Service in East Africa*, London, I.B. Tauris, 2007.
70. BL/IOR/L/PJ/8/254 Pim Report, HMSO, 1936, Appendix XI.
71. BL/IOR/L/PJ/8/254 Pim Report, HMSO, 1936, p. 24.
72. The Pim Report itself provides ample evidence of this. BL/IOR/L/PJ/8/254 Pim Report, HMSO, 1936.
73. Mahmood Mamdani, *From Citizen to Refugee: Uganda Asians Come to Britain*, Oxford, Pambazuka Press, 2011 [first pub.1973], p. 14; R.G. Zwanenburg, 'Robertson and the Kenya Critic' in K. King and A. Salim (eds.), *Kenya Historical Biographies*, Nairobi, East African Publishing House, 1971, p. 152.
74. Gregory, *South Asians in East Africa*, p. 359; Vincent Cable, 'The Asians of Kenya', *African Affairs*, 272, 1969, pp. 218–31, 222.
75. Paul Theroux, 'Hating the Asians', *Transition*, 33, 1967, pp. 46–51.
76. The idea of the Indian as 'The Jew of Africa' is mentioned in D.A. Seidenberg, *Uhuru and the Kenya Indians: The Role of a Minority Community in Kenya Politics, 1939–1963*, New Delhi, Vikas Publishing House, 1983, p. 14.
77. Mangat, *History of the Asians in East Africa*, pp. 104–112; D.P. Ghai and Y.P. Ghai (eds.), *Portrait of a Minority: Asians In East Africa*, Oxford University Press, 1970, p. 5; Gregory, *India and East Africa*, pp. 58, 90, 206; Seidenberg, *Uhuru and the Kenya Indians*, p. 14.
78. Pratik Chakrabarti, *Medicine and Empire: 1600–1960*, Palgrave Macmillan, 2014, pp. 86–7.
79. Hinnells, *The Zoroastrian Diaspora*, p. 268.
80. BL/IOR/L/PJ/6/430 Letter from Hardinge to the Secretary of the Government of India, 6 September 1895; Salvadori, *We Came in Dhows*, Vol. 3, p. 36 mentions two Goans also in Zanzibar.
81. BL/IOR/MIL/7/2177 Collection 48/36 Medical Staff for Road to Coast, 1895–1896, Letter A.D. Mackinnon to Mr Jackson 13 April 1895; BL/IOR/MIL/7/2153 Collection 48 Enlistment of Native of India for Service in Africa 1888–1901; BL/IOR/L/MIL/7/2188 Collection 48/35 Hospital Assistants for Uganda Railway and Uganda, 1896 lists the military awards to six named individuals including

the Indian MO in charge Surgeon Lt. H.M. Masani and Hospital Assistants B. Kasinath, Maula Baksh, Rahim Baksh, Sheikh Ahmed and Niyamtullah.

82. The essential role of Goan clerks in the administration of the Medical Department was also pointed out by Arthur Milne. Arthur Dawson Milne, 'The Rise of the Colonial Medical Service', *Kenya and East African Medical Journal*, 5, 1928–9, pp. 50–8.

83. Mridula Ramanna, 'Indian Doctors, Western Medicine and Social Change, 1845–1885', in Mariam Dossal and Ruby Maloni (eds.), *State Intervention and Popular Response: Western Indian in the Nineteenth Century*, Mumbai, Popular Prakashan Press, 1999, pp. 40–62.

84. Dissection was not a problem after 1880 however. Ramanna, 'Indian Doctors', p. 43

85. BL/IOR/MIL/7/2159, Letter from Military Department India to Foreign Office, 21 December 1891; BL/IOR/MIL/7/2175 Foreign Office Letter 21 August 1895; NA/CO/544/1 Annual Medical Report, 1908, p. 3.

86. Metcalf, *Imperial Connections*, p. 166.

87. Maureen Malowany, *Medical Pluralism: Disease, Health and Healing in the Coast of Kenya, 1840–1940*, PhD Thesis, McGill University, Canada, 1997, p. 12.

88. R. Watermann, 'Medicine and Hospitals along East – African Coasts in 16th Century', in *Actes du XXXIIe Congrès International d'Histoire de la Médecine, Anvers, 3–7 Septembre 1990/Proceedings of the XXXIInd International Congress on the History of Medicine, Antwerp, 3–7 September 1990*, Bruxelles [Belgium], Societas Belgica Historiae Medicinae, pp. 1029–38.

89. Interestingly there are references to the presence of full-time traditional practitioners in Portuguese Mozambique but without a specific names being mentioned. See R. Watermann, 'Medicine and Hospitals along East – African Coasts in 16th Century', pp. 1029–38, 1034. For the influence of Unani and Ayurvedic medicine along the Kenya coast see: Malowany, *Medical Pluralism*, pp. 21–7.

90. Malowany, *Medical Pluralism*, p. 235.

91. Which is far from saying it was not contested or passively received without adaptation. See Andrew Cunningham and Bridie Andrews (eds.), *Western Medicine as Contested Knowledge*, Manchester University Press, 1997.

92. R. Kochhar, 'European Medical Men in India' *Journal of Bioscience*, 24, 1999, pp. 259–68.

93. There are many examples of histories of medical resistance, but for example, David Arnold, *Colonizing the Body: State Medicine and Epidemic Disease in Nineteenth Century India*, University of California Press, 1993; Neshat Quaiser, 'Colonial Politics of Medicine and Popular Unani Resistance', *Indian Horizons*, 47.2, 2000, pp. 29–42.

94. John Iliffe, *East African Doctors: A History of the Modern Profession*, Cambridge University Press, 1998, pp. 7–33.

95. Robert W. Strayer, *The Making of Mission Communities in East Africa: Anglicans and Africans in Colonial Kenya, 1875–1935*, London, Heinemann Educational, 1978, pp. 14–20.

96. Poonam Bala, *Imperialism and Medicine in Bengal: A Socio Historical Perspective*, New Delhi, Sage Publications, 1991; V.R. Muraleedharan, 'Professionalising Medical Practice in Colonial South India', *Economic and Political Weekly*, 27.4, 1992; pp. PE27–30, PE35–37; Mridula Ramanna, *Western Medicine and Public Health in Colonial Bombay, 1845–1895*, Delhi, Orient Longman, 2002; M. Gopal, D. Balasubramanian, P. Kanagarajah, A. Anirudhan, P. Murugan, 'Madras Medical College, 175 Years of Medical Heritage', *The National Medical Journal of India*, 23.2, 2010, pp. 117–20.

97. Gopal et al., 'Madras Medical College', p. 117.

98. Although the Portuguese had run an elementary course in medicine at the Royal Hospital in Goa since 1703 and opened a medical school at Mandavi in Goa in 1842; Bastos, 'The Inverted Mirror'; 1822 is cited as the beginnings of formal western medical education in Calcutta in Roger Jeffery, 'Recognizing India's Doctors: The Institutionalization of Medical dependency, 1918–1939', *Modern Asian Studies*, 13.2, 1979, pp. 301–26, 302.
99. First regular medical establishment was the Bengal Medical Service. Harrison, *Public Health in British India*, p. 7.
100. William Dalrymple, *White Mughals: Love and Betrayal in Eighteenth Century India*, Harper Collins, London, 2002.
101. Roger Jeffery, 'Doctors and Congress: The Role of Medical Men and Medical Politics in Indian Nationalism', in Mike Shepperdson and Colin Simmons (eds.), *The Indian National Congress and the Political Economy of India, 1885–1985*, Aldershot, Brookfield, USA, Avebury, 1988, pp. 160–73, 162.
102. Calcutta Medical College, The Centenary of the Medical College, Bengal, 1835–1934, Calcutta, 1935; Bala, *Imperialism and Medicine in Bengal*; S.N. Sen, *Scientific and Technical Education in India 1781–1900*, New Delhi, Indian National Science Academy, 1991.
103. Gopal et al., 'Madras Medical College'; Abdur Pashid, *History of the King Edward Medical College Lahore, 1860–1960*, King Edward Medical School, Lahore, 1960.
104. Estimate obtained from analyzing the lists of registered and licensed practitioners with Indian names published in the *Official Gazettes* of 1911, 1921, 1930 and 1940.
105. Ramanna, 'Indian Doctors', p. 44.
106. Arnold, *Colonizing the Body*, p. 64.
107. Ramanna, 'Indian Doctors', p. 46.
108. Ramanna, 'Indian Doctors', p. 45.
109. 'Medicine in India', *British Medical Journal* [supplement], 28 January 1939, p. 43
110. Jeffery, 'Recognizing India's Doctors', pp. 317–18.
111. At least never in Kenya, one example of an Indian Medical Officer has been found to be serving in Uganda. See Dr. Karan Singh, Medical Officer, Namale, Uganda, LRCP, LRCS (Ed) and LFPS (Glasgow), *Medical Directory*, 1913, p. 1725.
112. Ramanna, 'Indian Doctors', p. 46.
113. Milne, 'The Rise of the Colonial Medical Service,' p. 58.
114. Requests for more Hospital Assistants presumed that these were qualified and experienced medical personnel. Requests can be found in BL/IOR/MIL/7/2175, Letter Dr A.D. Mackinnon to Mr Jackson, 13 April 1895; BL/IOR/L/MIL/7/14471 Letter, Medical Subordinates for Service in East Africa, 28 January 1907.
115. H.F. Morris, Government Publications Relating to Kenya Including Those Relating to the East Africa High Commission and the East African Common Services Organisation, 1897–1963, Wakefield, EP Microform, 1976, p. 6.
116. NA/CO 822/63/1 1934–1935, East Africa Original Correspondence, Medical Registration, Note 10 May 1935.
117. Deepak Kumar, 'Probing History of Medicine and Public Health in India', *Indian Historical Review*, 37.2, 2010, pp. 259–73, 259.
118. Mark Harrison, 'Medical Experimentation in British India: The Case of Dr Helenus Scott' in Hormoz Ebrahimnejad (ed.), *The Development of Modern Medicine in Non-Western Countries: Historical Perspectives*, Royal Asiatic Society Books, London and New York, Routledge, 2009, pp. 24–40.

119. Pratik Chakrabarti, *Western Science in Modern India: Metropolitan Methods, Colonial Practices*, Delhi, Permanent Black, 2004; Kavita Sivaramakrishnan, *Old Potions, New Bottles: Recasting Indigenous Medicine in Colonial Punjab 1850–1940*, New Delhi, Orient Longman, 2006; Guy N.A. Attewell, *Refiguring Unani Tibb: Plural Healing in Late Colonial India, New Perspectives in South Asian History*, New Delhi, Orient Longman, 2007.
120. Kumar, 'Probing History of Medicine', p. 265.
121. Poonam Bala, ' "Defying" Medical Autonomy: Indigenous Elites and medicine in Colonial India', in Poonam Bala (ed.), *Biomedicine as a Contested Site: Some Revelations in Imperial Contexts*, Lanham, Boulder, New York, Toronto, Plymouth, UK, Lexington Books, 2009, pp. 29–44.
122. Harrison, *Public Health in British India*, p. 32.
123. Although Gandhi was perhaps less anti western medicine than he is sometimes caricatured as being. See Jeffery, 'Doctors and Congress', p. 166.
124. For Gandhi's attitudes to western science and medicine see: Mahatma Gandhi, *India of My Dreams*, Delhi, Rajpal and Sons, 2009, pp. 149–60.
125. For this trend explored in many colonial context see: Ebrahimnejad (ed.), *The Modern Medicine in Non-Western Countries*, London, New York, Routledge, 2009.
126. The complex negotiation of immigrant identity, simultaneously accepting ideologies and cultures from both homeland and hostland has been examined by Sana Aiyar, 'Anticolonial Homelands Across the Indian Ocean: The Politics of the Indian Diaspora in Kenya, ca. 1930–1950', *American Historical Review*, 116.4, 2011, pp. 987–1013.
127. For a description of the role of Indian doctors in the INC in India see: Jeffery, 'Doctors and Congress'.
128. Seidenberg, *Uhuru and the Kenya Indians*, p. 14.
129. Motives identified by Crozier, *Practising Colonial Medicine*, I.B. Tauris, 2007, pp. 46–71.

3   Indians, Western Medicine, and the Establishment of the Protectorate

1. BL/IOR/L/PJ/254 Cmd. 3234 *Hilton Young Commission Report*, 1928–9, p. 27.
2. Winston S. Churchill, *My African Journey*, London, Hodder and Stoughton, 1908, p. 49.
3. Daniel Headrick, 'Technology, Imperialism and History', in Daniel R Headrick, *The Tools of Empire: Technology and European Imperialism in the Nineteenth Century*, Oxford University Press, 1981, pp. 3–14; Michael Worboys, 'Colonial and Imperial Medicine', in Debra Brunton (ed.), *Medicine Transformed: Health, Disease and Society in Europe, 1800–1930*, The Open University and Manchester University Press, 2004, pp. 211–38.
4. David Spurr, *The Rhetoric of Empire: Colonial Discourse in Journalism, Travel Writing and Imperial Administration*, Durham and London, Duke University Press, 1993; Nancy Leys Stepan, *Picturing Tropical Nature*, London, Reaktion Books Ltd., 2001
5. Philip D. Curtin, '"The White Man's Grave": Image and Reality, 1780–1850',*The Journal of British Studies*, 1, 1961, pp. 94–110; Philip Curtin, *Death by Migration: Europe's Encounter with the Tropical World in the Nineteenth Century*, Cambridge, Cambridge University Press, 1989.
6. Patrick Brantlinger, 'Victorians and Africans: The Genealogy of the Myth of the Dark Continent', *Critical Inquiry*, 12, 1985, pp. 166–203.

7. John Langton Gilks, Speech to East Africa Branch of BMA, 'The Annual Dinner of the Kenya Branch of the British Medical Association', *Kenya Medical Journal*, 2.11, 1925–6, pp. 321–5.

8. For the new-found optimism in the benefits of medical science to aid colonization see: L. Westenra Sambon, 'Acclimatization of Europeans in Tropical Lands', *The Geographical Journal* 12, 1898, pp. 589–98; Historians have written widely on the cautious pre-scientific ideas of the disease environment, but see particularly: David Arnold, 'The Place of "The Tropics" in Western Medical Ideas Since 1750', *Tropical Medicine and International Health*, 24, 1997, pp. 303–13; Mark Harrison, *Climates and Constitutions: Health, Race and British Imperialism in India, 1600–1850*, Oxford, Oxford University Press, 2002 [first published 1999].

9. John Farley, *Bilharzia: A History of Imperial Tropical Medicine*, Cambridge, Cambridge University Press, 2003 [first published, 1991]; Douglas M. Haynes, *Imperial Medicine: Patrick Manson and the Conquest of Tropical Disease*, Philadelphia, University of Pennsylvania Press, 2001.

10. For example, Manson's work on filariasis in Amoy, China, Ross's work on malaria in Calcutta, India and the work of Koch on rinderpest in South Africa.

11. Official provision was eventually made legalising mass Indian emigration to East Africa in 1896 with the Indian Emigration Act. See: J.S. Mangat, *A History of the Asians in East Africa, c.1886–1945*, Oxford University Press, 1969, pp. 32–33; Robert G. Gregory, *India and East Africa: A History of Race Relations within the British Empire, 1890–1939*, Oxford, Clarendon Press, 1971, p. 51.

12. NA/FO/107/9 Berkley Memo, 13 September 1893.

13. John Lonsdale, 'The Conquest State 1895–1904', in W.R. Ochieng (ed.), *A Modern History of Kenya*, London, Evans Brothers, 1989, pp. 6–34, p. 10; Robert L. Tignor, *The Colonial Transformation of Kenya: The Kamba, Kikuyu and Masai from 1900–1939*, Princeton, Guildford, Princeton University Press, 1976, p. 6.

14. BL/IOR/L/MIL/7/2153 1876–1888 Collection 48/1 British East Africa Company: requests for 100 Sikhs and 100 Punjabis declined by Secretary of State, letter from Cross to Governor General of India, 27 December 1888; BL/IOR/L/MIL/7/2175 Collection 48/23 1895 Viceroy to Foreign Office, 23 August 1895.

15. BL/IOR/L/MIL/7/2175 Collection 48/23 1895 Foreign Office letter to India Office 21 August 1895.

16. BL/IOR/L/MIL/7/12668 Foreign Office letter (as directed by Lord Salisbury) 20 February 1896.

17. BL/IOR/L/MIL/7/2175 and BL/IOR/L/MIL/7/2153 Collection 48/1 1876–1888 British East Africa Company; NB this file also records that another Parsi, Sorabji Bryce was serving in Harry Johnston's earlier campaign in central Africa when Johnston was exceedingly impressed with Punjabi soldiers, see his letter to Foreign Office, 18 December 1891.

18. C.9125 *Report by Sir A. Hardinge on the British East Africa Protectorate* 1897–98, London, HMSO, p. 17.

19. BL/IOR/L/MIL/7/5356 for awards to 9 solders of the Sikh regiment with descriptions of individual acts of great valour; BL/IOR/L/MIL/7/2188 Dispatch: Mazuri expedition awards, 2 March 1899.

20. Malcolm Page, *A History of the King's African Rifles and East African Forces*, Pen and Sword, Barnsley, 2011, p. 5.

21. Mangat, *History of the Asians*, p. 43.

22. BL/IOR/L/MIL/7/12668, Note of A.R. Badcock, 'A Scheme for the Dispatch of Native Infantry Regiment'. 26 February 1896.

23. W. Lloyd—Jones, *Havash! Frontier Adventures in Kenya*, London, Arrowsmith, 1925, p. 290.
24. W. Lloyd—Jones, *K.A.R.: Being an Unofficial Account of the Origin and Activities of the King's African Rifles*, London, Arrowsmith, 1926, p. 64.
25. Lloyd—Jones, *Havash!*, p. 289; Lloyd–Jones, *K.A.R.*, p. 64; Cynthia Salvadori, *We Came in Dhows*, Nairobi, Paperchase Kenya Ltd., 1996, Vol. II, p. 191.
26. Norman Parsons Jewell, [ed. Tony Jewell], *On Call*, Privately Published, *forthcoming*, 2015 [Chapter 11, not yet paginated].
27. BL/IOR/L/MIL/7/12673, H.D. Masani, Report on the Health of the Mombasa Force, including 24th Bombay Infantry, 3 June 1896.
28. D.G. Crawford, *Roll of the Indian Medical Service: 1615–1930*, London, W.Thacker & Co., 1930, p. 479.
29. BL/IOR/L/MIL/7/12673, Masani Report, 1896.
30. BL/IOR/L/MIL/7/2188 Dispatch: Mazuri Expedition Awards, 2 March 1899 (details the award of Ashanti medals to five Indian hospital assistants).
31. BL/IOR/L/MIL/7/12673 H Masani Report, 1896, p. 3.
32. BL/IOR/L/MIL/7/12673 Masani Report, 1896, p. 2.
33. BL/IOR/L/MIL/7/12673 Masani Report, 1896, p. 3.
34. BL/IOR/L/MIL/7/12673 Masani Report, 1896, p. 3.
35. Lloyd—Jones, *Havash!*, p. 148.
36. Lloyd—Jones, *Havash!*, p. 148.
37. Herbert Henry Austin, *With Macdonald in Uganda: A Narrative Account of the Uganda Mutiny and Macdonald Expedition in the Uganda*, London, Dawsons of Pall Mall, 1973, Appendix A, p. 288.
38. Edward Paice, *Tip and Run: The Untold Tragedy of the Great War in Africa*, London, Weidenfeld and Nicolson, 2007, p. 3; Additionally Paice argues that that Colonial Office suppressed facts surrounding the bloodshed and mortality (and subsequent famines) resultant of these campaigns, p. 393.
39. Paice, *Tip and Run*, p. 3; for special legislation to compel Africans to enlist see 'Our War Legislation: The Native Carriers', *The Leader*, 21 August 1915, p. 14.
40. S.D. Pradhan, *The Indian Army in East Africa, 1914–1918*, New Delhi, National Book Organisation, 1991; BL/L/MIL/7/18448 Major General T.E. Scott, Report on Abnormal Wastage of Indian Troops in East Africa, 1 October 1917.
41. BL/L/MIL/7/18448 Major General T.E. Scott Report, 1917.
42. BL/L/MIL/7/18448 Major General T.E. Scott Report, 1917, pp. 3–5.
43. BL/L/MIL/7/18448 Major General T.E. Scott Report, 1917, p. 5.
44. BL/L/MIL/7/18448 Major General T.E. Scott Report, 1917, p. 5.
45. Aiyar has argued: making common cause with Indian nationalism that was gaining momentum in the Sub-Continent. Sana Aiyar, 'Empire, Race and the Indians in Colonial Kenya's Contested Public Political Sphere, 1919–1923', *Africa: The Journal of the International African Institute*, 81. 1, 2011, pp. 132–54.
46. M.F. Hill, *The Permanent Way: The Story of Kenya and Uganda Railways*, Nairobi, East African Literature Bureau, 1961; Ronald Hardy, *The Iron Snake*, London, Collins, 1965; C. Miller, *The Lunatic Express: An Entertainment in Imperialism*, London, Macdonald, 1971; A. Clayton and D.C. Savage, *Government and Labour in Kenya: 1895–1963*, London, Frank Cass, 1974, pp. 10–14; H. Gunston, 'The Planning and Construction of Uganda Railway', *Transactions of the Newcomen Society*, 74, 2004, pp. 45–71; Satya V. Sood, *Victoria's Tin Dragon: A Railway that Built a Nation*, Cambridge, Vanguard Press, 2007; Neera Kapila, *Race, Rail and Society: Roots of Modern Kenya*, Nairobi, East African Educational Publishers, 2011; Stephen Mills and Brian Yonge, *A Railway to Nowhere: The Building of the Lunatic Line, 1896–1901*,

Nairobi, Mills Publishing, 2012; N. Green, 'African in Indian Ink: Urdu Articulations of Indian Settlement in East Africa', *Journal of African History*, 53, 2012, pp. 131–50. A fascinating article that is based upon on a rare Urdu travelogue written by a railway worker: although no mention is made of health conditions, this is one of the earliest direct evidence of Indian attitudes towards the new environment of East Africa.

47. Quoted in Hill, *The Permanent Way*, p. 235.

48. Cd.769 *Report By HM Commissioner on the East Africa Protectorate*, London, HMSO, August 1901, p. 19.

49. Sir Edward Grigg, quoted by Hardy, *The Iron Snake*, p. 14.

50. Lord Salisbury quoted in E. Bentley, *Handbook of the Uganda Question and Proposed East Africa Railway*, London, Chapman & Hall, 1892, pp. 39, 40. See also Cd. 2164, *Final Report of the Committee of Uganda Railways*, London, HMSO, 1904, p. 4 stated 'HMG became convinced in 1891 that the construction of Railway was the cheapest and most efficient way of stopping Slave Trade'.

51. Miller, *The Lunatic Express*, p. 7.

52. Bentley, *Handbook of the Uganda Question and Proposed East Africa Railway*.

53. BL/IOR/L/PJ/8/254 A.W. Pim, *Report of the Commission Appointed to Enquire into and Report into the Financial Position and System of Taxation in Kenya*, London, HMSO, 1936, pp. 239–47.

54. For example, the meter gauge system, the first rolling stock and wagons selected for the railway were based on Indian railways; See Cd 8435, *Report on the Progress of the Uganda Railway, 1896–97*, London, HMSO, pp. 2,5; and Gunston, 'The Planning and Construction of Uganda Railway', pp. 45–71, 54.

55. Miller, *The Lunatic Express* p. 277 for quote see also Cd. 8435, *Report on the Progress of the Uganda Railway, 1896–97*, London, HMSO, p. 4.

56. C. 9331, G. Molesworth, *Report on the Uganda Railway*, London, HMSO, 1899, p. 20.

57. Cd. 97, *Memorandum on the Uganda Railway*, London, HMSO, 1900, p. 4.

58. BL/IOR/L/E/7/444, File 626, Letter Uganda Railway to Sir C. Bernard, 19 October 1899; the language point was also made by Gracey. See Cd. 670, Correspondence respecting the Uganda Railway, 1901, Report by Colonel Gracey on the Uganda Railway: Subtitle Medical, p. 24.

59. C. 9333 *Report by the Uganda Railway Committee on the Progress of Works, 1898–99*, London, HMSO, p. 3.

60. Cd. 1770, *Report on Progress of the Works, 1902–1903*, London, HMSO, p. 4; Cd. 2164, *Final Report of the Uganda Railway Committee*, London, HMSO, 1904, p. 13; Cd. 674, *Report on the Progress of Works, 1901*, London, HMSO, p. 4 cites over 22,000 workers in total, a figure which includes Africans however. See also John Darwin's excellent study, *Unfinished Empire: The Global Expansion of Britain*, London, Allen Lane, 2012, pp. 218–20.

61. H. Johnson cited in R.A. Oliver, *Sir Harry Johnson and the Scramble for Africa*, London, Chatto and Windus, 1957, p. 293.

62. Mangat, *History of the Asians*, pp. 27–40; Gregory, *India and East Africa*, pp. 50–61.

63. BL/ IOR /L/MIL/7/2153, Foreign Office Letter, 21 December 1891.

64. Hill, *The Permanent Way*, p. 70.

65. J.R.L MacDonald with a new introduction by A.T. Matson, *Soldiering and Surveying in British East Africa, 1891–1894*, London, Dawsons, 1973; Austin, *With Macdonald in Uganda*, pp. 287–89 named: McPherson, McLoughlin, Ferguson, Turner, Standage and Cook in Appendix A.

66. The spelling of Molah Bux varied in different sources, as do many Indian names used in this study. Austin, *With Macdonald in Uganda*, p. 108.

67. H. Tinker, *A New System of Slavery: The Export of Indian Labour Overseas, 1830–1920*, Oxford University Press, 1974 discusses the debate in the Indian legislature regarding an immigration bill in 1883 when concern was expressed about indentured labour in general; See BL/IOR/L/E/7/444 File 626 for more specific correspondence between Uganda Railway and India, 8 May, 9 August and 2 October 1899 about the plight of returning invalids.

68. C.8435, *Report on the Progress of the Mombasa-Victoria Railway, 1896–97*, London, HMSO, 1897, p. 3.

69. BL/IOR/L/MIL/7/2177 Letter to Lord Hamilton, 11 March 1896.

70. The only reference to health conditions along the railway line found is in: Kapila, *Race, Rail and Society*, pp. 39–43.

71. Anna Crozier, *Practising Colonial Medicine: The Colonial Medical Service in East Africa*, London, I.B. Tauris, 2007, p. 6.

72. Cd. 8942, *Report of the Progress of Uganda Railway, 1897–98*, London, HMSO, p. 4; Arthur Dawson Milne, 'The Rise of the Colonial Medical Service', *Kenya and East African Medical Journal*, 5, 1928–9, pp. 50–58, 53 for European MOs enlisting in the campaign to put down the Sudanese Mutiny; Also BL/IOR/L/MIL/7/14462 Promotion of Uganda Railways Hospital Assistant Rahmat Ali, 24 May 1899, p. 1 Ali mentioned Dr Turner's absence for 'weeks and weeks' even when he was with the Railways; NA/FO/403/277 1898, Construction of Uganda Railway for mention of Turner/Brock/Sieveking appointments. Also see O'Callaghan, Memorandum of the Managing Member of the Railway Committee (about the difficulty of replacing the PMO) in Cd. 670 Correspondence Respecting the Uganda Railway, Colonel Gracey's Report, London, HMSO, 1901, p. 57.

73. BL/IOR/L/MIL/7/2177 Letter to Lord Hamilton, 11 March 1896.

74. BL/L/MIL/7/2188 Hospital Assistants for Uganda Railway and Uganda, 23 January 1897. Six hospital assistants and six compounders were requested, letter 28 April1897; BL/IOR/L/MIL/7/14425 Telegram to Viceroy, 12 December 1900, (stating services of medical subordinates 'were urgently needed').

75. BL/IOR/L/MIL/7/2177, Lord Elgin letter to George Francis Hamilton, Secretary of State for India, 11 March 1896.

76. Milne, 'The Rise of the Colonial Medical Service,' p. 53.

77. H.A. Bödeker, 'Some Sidelights on Early Medical History in East Africa', *The East African Medical Journal*, 12, 1935–36, pp. 100–07, p. 105.

78. Bödeker, 'Some Sidelights', p. 105.

79. Cd. 9331, Molesworth, *Report on the Uganda Railway*, p. 21.

80. BL/IOR/L/MIL/7/14462 Rahmat Ali Promotion, 24 May 1899.

81. BL/L/MIL/7/2188 Foreign Office letter, 23 January 1897 [requesting urgent help as SMO attacked with fever]. In 1897 six hospital assistants and six compounders were requested, letter 28 April1897; BL/IOR/L/MIL/7/14454 Telegram to Viceroy, 5 December 1900, [stating services of medical subordinates 'were urgently needed'].

82. BL/IOR/L/MIL/7/2177 Lord Elgin letter to George Francis Hamilton, Secretary of State for India, 11 March 1896; BL/IOR/L/MIL/7/2177; BL/IOR/L//MIL/7/14425, letter from Francis Bertie (under direction of Marquis of Salisbury), 31 August 1897.

83. BL/IOR/L/MIL/7/14454 Letter from Francis Bertie (under direction of Marquis of Salisbury), 19 November 1898.

84. BL/IOR/L/MIL/7/14454 Despatch, 13 December 1900.

85. BL/IOR/L/MIL/7/2177, Lord Elgin letter to George Francis Hamilton, Secretary of State for India, 11 March 1896, p. 2.

86. BL/IOR/L/MIL/7/2177, Lord Elgin letter to George Francis Hamilton, Secretary of State for India, 11 March 1896, pp. 2–3.
87. BL/IOR/L/MIL/7/14462 Rahmat Ali Promotion, 24 May 1899.
88. Hardy, *The Iron Snake*, p. 86.
89. BL/IOR/L/MIL/7/14462 Rahmat Ali Promotion, 24 May 1899, pp. 1–5. In same file see letter by A.R. Sieveking 28 May 1899.
90. BL/IOR/L/MIL/7/14462 Rahmat Ali Promotion, 24 May 1899, p. 1.
91. BL/IOR/L/MIL/7/14462 Rahmat Ali Promotion, 24 May 1899, p. 5.
92. BL/IOR/L/MIL/7/14462 Rahmat Ali Promotion, 24 May 1899, p. 3. E. Whitehouse letter dated 7 June 1899.
93. BL/IOR/L/MIL/7/14462 Rahmat Ali Promotion, 24 May 1899, Despatch from Government of India, 7 December 1899, concerning recommendation by W.S.A. Lockhart, C.E. Dawkins, T. Raleigh and D.C.J. Ibbetson.
94. Cd. 8435, *Report on the Progress of Uganda Railway, 1896–1897*, London, HMSO, p. 5; Cd.2164, *Final Report of the Uganda Railways Committee*, London HMSO, 1904, p. 13.
95. Miller, *Lunatic Express*, p. 294.
96. Cd. 8435 *Report on the Progress of Uganda Railway, 1896–1897*, London, HMSO, p. 5.
97. Cd. 8942 *Report on the Progress of the Uganda Railway, 1897–1898*, London, HMSO, p. 4; See also Kapila, *Race, Rail and Society*, pp. 39–43.
98. Cd. 9331, Molesworth, *Report on the Uganda Railway*, p. 13.
99. Cd 9333 *Report by the Uganda Railway Committee, 1898–1899*, London, HMSO, p. 3; C.9333 *Report on the Progress of Uganda Railway, 1898–1899*, London, HMSO, p. 3, acknowledges Indians worked in bare feet. Gunston, 'Uganda Railway', *Transactions of the Newcomen Society*, 2004, p. 62 for illustrative photograph.
100. Cd. 9331, Molesworth, *Report on the Uganda Railway* p. 21.
101. Miller, *Lunatic Express*, p. 294.
102. Cd.2164 *Final Report on the Progress of the Uganda Railways*, London, HMSO, 1904, p. 13; Also Cd. 9331, Molesworth, *Report on the Uganda Railway*, p. 21; Cd. 674, *Report on the Progress of Works on the Uganda Railway*, London, HMSO, 1900–01, p. 4 (Hospital Returns). Note 8% was an underestimate as deaths of workers repatriated to India were not included time.
103. BL/IOR/L/E/7/444 File 626 Letter from Chief Engineer Uganda Railway to Secretary of the Government of India, 2 October 1899.
104. Tinker, *A New System of Slavery*, p. 175.
105. Tinker, *A New System of Slavery*, p. 175; For other comments regarding returning ships with coolie labour see BL/IOR/L/E/7/444 File 626, Minute 'Alleged Mortality among Uganda Railway Coolies', 8 May 1899. Also BL/IOR/L/E/7/425 has correspondence from India in October 1899 regarding concerning coolies infesting India with jiggers.
106. Cd. 8435 *Report on the Progress of the Uganda Railway*, London, HMSO, 1897, p. 4 mentions plans to erect hospitals. Also Cd. 670, Correspondence respecting the Uganda Railway, London, HMSO, 1901, Report by Colonel Gracey on the Uganda Railway criticises facilities and care in 3 locations, p. 24.
107. Hardy, *The Iron Snake*, p. 86.
108. Cd. 670 Correspondence respecting the Uganda Railway, 1901, Report by Colonel Gracey on the Uganda Railway, p. 24.
109. Cd. 670 Correspondence respecting the Uganda Railway, 1901, Report by Colonel Gracey on the Uganda Railway, pp. 25, 57; For a mention of hospital provision see: C. 9333 *Report on the Progress of the Works, 1898–1899*, London, HMSO, 1899, p. 3.

110. Cd.9331 Molesworth, *Report on the Uganda Railway*, p. 21. A year later the rate increased to 50 per 1000.
111. Bödeker, 'Some Sidelights,' p. 105 (for the 'efficiently carried out' praise); Cd. 670, Correspondence respecting the Uganda Railway, 1901, Report by Colonel Gracey on the Uganda Railway: Subtitle Medical, p. 24; Cd. 9331 Molesworth, *Report on the Uganda Railway*, p. 21; Cd.2164 *Final Report on the Progress of the Uganda Railways*, London, HMSO, 1904, pp. 13, 27.
112. Cd.2164 *Final Report on the Progress of the Uganda Railways*, London, HMSO, 1904, p. 13.
113. Clayton and Savage, *Government and Labour in Kenya*, p. 11.
114. 'Editorial', *The Kenya Medical Journal*, 1.10, 1925, p. 291; Darwin, *Unfinished Empire*, p. 187 refers to a death rate of 76 per 1000 in mining in Southern Rhodesia in 1906 and 16.5 in UK adult males.
115. Milne, 'The Rise of the Colonial Medical Service', p. 53; Bödeker, 'Some Sidelights,' p. 105 also refers to Turner resigning from Railways and volunteering to serve in the military operations and then returning to India and not to Uganda Railways.
116. Cd. 674 *Report on the Progress of the Works, 1900–1901*, London, HMSO, 1901, pp. 2–3.
117. BL/IOR/L/MIL/7/14425 Hospital Assistants for the Uganda Railway, 1897–1898.
118. BL/IOR/ L/MIL/7/1441 File 19, F. Rawson to Foreign Office, 21 June 1900.
119. BL/IOR/MIL/7/2175, Foreign Office to Government of India 1 January 1896 and Government of India to Foreign Office, 11 March 1896 refers to double pay and Lord Salisbury's opposition and instructions to engage on most economic terms. The debate on continued to 1907 when PMO Will complained of pay not being sufficient to attract candidates.
120. Casualties were also very high during the subsequent construction of the extension of the railway to Magadi in 1911. See Clayton and Savage, *Government and Labour in Kenya*, p. 46.
121. 'Editorial', *The Kenya Medical Journal*, 1.10, 1925, p. 291.
122. V.M. Fisher, 'Medical Arrangements for Native Labour on the Thika-Nyeri Railway Construction', *The Kenya Medical Journal*, January 1925, pp. 293–319, 293; R.A.W. Proctor, 'The Health of Labour on the Kisumu-Yala Railway Construction', *Kenya and East African Medical Journal*, 7.10, 1931, pp. 276–81, 276.
123. 'Editorial', *The Kenya Medical Journal*, 1.10, 1925, p. 291.
124. Fisher, 'Medical Arrangements,' p. 293; Proctor, 'The Health of Labour on the Kisumu-Yala Railway Construction', pp. 276–81, 276.
125. BL/IOR/L/PJ/254 Cmd. 3234 *Hilton Young Commission Report*, 1928–29, p. 27.

4   Race and Medicine

1. BL/IOR/L/PJ/6/1718 *Economic Commission Report*, East African Protectorate, Nairobi, Swift Press, 1919, p. 21.
2. E.g. Diane Wylie, 'Confrontation over Kenya: The Colonial Office and Its Critics, 1918–1940', *Journal of African History*, 18.3, 1977, pp. 427–47; Christopher P. Youé, 'The Threat of Settler Rebellion and the Imperial Predicament: The Denial of Indian Rights in Kenya, 1923', *Canadian Journal of History*, 12, 1978; Also see Chapter 8, 'The Indian Question in Kenya' in L.W. Hollingsworth, *The Asians of East Africa*, London, Macmillan, 1960, pp. 76–97.

3. Although no record can be found as to whether he was successful in obtaining these appointments. Hardinge's Administrative Proposals, 6 July 1895 in G.H. Mungeam, *Kenya: Select Historical Documents 1884–1923*, Nairobi, East African Publishing House, 1979, pp. 69–75.
4. M.P.K. Sorrenson, *Origin of European Settlement in Kenya*, London, Oxford University Press, 1968, p. 34.
5. Sorrenson, *Origins of European Settlement*, pp. 34–6. Frere is cited in J.S. Mangat, *A History of The Asians in East Africa, c.1886–1945*, Oxford University Press, 1969, p. 12; Similarly, Ainsworth's comments on the positive role of Indians can be seen in his farewell speech reported in *East African Chronicle*, 14 August 1920, p. 14; See also Harry Johnston, Letter to the Editor, *The Times*, 22 August, 1921, p. 4.
6. Bruce Berman, John Lonsdale, *Unhappy Valley: Conflict in Kenya and Africa [Book 1 State and Class]*, London, James Currey, 1991, p. 34.
7. Cd. 769 *Report by His Majesty's Commissioner on the East African Protectorate*, London, HMSO, 1901, p. 4.
8. Cd. 1626, *Report by His Majesty's Commissioner on the East African Protectorate*, London, HMSO, 1903, p. 13.
9. Cd. 769 *Report by His Majesty's Commissioner on the East African Protectorate*, London, HMSO, 1901, p. 4; Cd. 1626, *Report by His Majesty's Commissioner on the East African Protectorate*, London, HMSO, 1903, p. 7.
10. R.M. Maxon, *John Ainsworth and the Making of Kenya*, Lanham, MD, University Press of America, 1980, p. 408; BL/IOR/L/E/7/1295, File 191, Indians in Kenya: Representations from Associations and Individual Opinions, EAINC, Letter to Viceroy of India, 27 January 1923, p. 1. This letter also discusses the dangers of following the South African model.
11. Roland Anthony Oliver, *Sir Harry Johnston and the Scramble for Africa*, London, Chatto and Windus, 1957, p. 257.
12. Winston Churchill, *My African Journey*, London, Hodder and Stoughton, 1908, p. 45
13. G. H. Mungeam, *British Rule in Kenya, 1895–1912*, Oxford, Clarendon Press, 1966, p. 191.
14. Sorrenson, *Origins of European Settlement*, p. 61, 159, 229. R.M. Maxon, 'The Years of Revolutionary Advance: 1920–1929' in William R. Ochieng, *A Modern History of Kenya 1895–1980*, Nairobi, London, Evans Brothers (Kenya) Ltd., 1989, pp. 74, 86. See also BL/IOR/L/PJ/6/1718 *Economic Commission Report*, Nairobi, Swift Press, 1919, pp. 20–1.
15. Editorial, *The Leader*, 4 December 1914, p. 14.
16. Wylie, 'Confrontation over Kenya', p. 446. Diana Wylie, 'Norman Leys and McGregor: A Case Study in the Conscience of African Empire 1900–39', *The Journal of Imperial and Commonwealth History*, 5.3, 1977, pp. 294–309.
17. Robert M. Maxon, *Struggle for Kenya: The Loss and Reassertion of Imperial Initiative, 1912–1923*, Rutherford, N.J., Farleigh Dickinson University Press, 1993.
18. Maxon, *Struggle for Kenya*, pp. 161, 191. Also BL/IOR/ L/PO/1/1A Milner letter to the Governor, 21 May 1920.
19. The recalls of Governors Girourard (in 1912 over the Masai issue) and Northey (in 1922) were examples of Governors overstepping the boundaries set by London. Maxon, *Struggle for Kenya*, pp. 41, 191.
20. Ronald Hyam, *Britain's Declining Empire: The Road to Decolonisation 1918–1968*, Cambridge, Cambridge University Press, 2006, p. 19. See also Wylie, 'Confrontation over Kenya', p. 445.
21. Mangat, *History of the Asians*, pp. 97–131; Maxon, *Struggle for Kenya*.

22. Keith Kyle, 'Gandhi, Harry Thuku and Early Kenya Nationalism', *Transition*, 27, 1966, pp. 16–22; Harry Thuku, *An Autobiography*, Nairobi, Oxford University Press, 1970. Zarina Patel, *Unquiet: The Life and Times of Makhan Singh*, Nairobi, Zand Graphics, 2006; Channan Singh, 'Manilal Ambalal Desai', pp. 129–41, 136; p. 140; for post 1950 see Margaret Frenz, 'Swaraj for Kenya, 1949–1965. The Ambiguities of Transnational Politics', *Past and Present*, 218, 2013, pp. 151–77.
23. Sorrenson, *Origins of European Settlement*, p. 165.
24. Kyle, 'Gandhi, Harry Thuku', p. 17; Robert J. Blythe, *The Empire of the Raj: India, Eastern Africa and the Middle East, 1858–1947*, Cambridge, Cambridge University Press, 2003, pp. 104–119.
25. C.W. Hobley, enclosure in Mungeam, *Kenya*, p. 464.
26. Churchill, *African Journey*, p. 48.
27. BL/IOR/L/PO/1/A(iv) Official Report of Parliamentary Debate, House of Lords, 14 July 1920, p. 119.
28. Robert L. Tignor, *The Colonial Transformation of Kenya: The Kamba, Kikuyu and Maasai from 1900–1939*, Princeton and Guildford, Princeton University Press, 1976, p. 97; p. 208.
29. Tignor, *Colonial Transformation of Kenya*, p. 97, 208.
30. BL/IOR/L/PJ/6/1718 Appointment of Commission by H.C. Belfield, *Economic Commission Report*, Nairobi, Swift Press, 1919.
31. BL/IOR/L/PJ/6/1718 *Economic Commission Report*, Nairobi, Swift Press, 1919, p. 21; BL/IOR/L/PJ/254 Cmd. 3234 *Hilton Commission Report*, 1928, London, HMSO p. 27
32. BL/IOR/L/PJ/6/1718 *Economic Commission Report*, East African Protectorate, Nairobi, Swift Press, 1919, p. 21.
33. BL/IOR/L/PJ/6/1718 *Economic Commission Report*, Nairobi, Swift Press, 1919, p. 21.
34. BL/IOR/L/PO/1//A(iv) Official Report of Parliamentary Debate, House of Lords, 14 July 1920, p. 161; W.G. Ross, *Kenya from Within: A Short Political History*, London, George Allen, 1927, p. 320; Marjorie Ruth Dilley, *British Policy in Kenya Colony*, London, Frank Cass & Co, 1966, p. 149.
35. Edward Paice, *The Lost Lion of Empire: The Life of 'Cape to Cairo' Grogan*, London, Harper Collins, 2001, pp. 291, 450.
36. 'Asiatic Salaries Cut', *The Leader*, 20 May 1922, p. 8.
37. 'Asiatic Salaries Cut', *The Leader*, 20 May 1922, p. 8; See also 'New Kenya Civil Service Scales', *East African Chronicle*, 4 September 1920, pp. 11–12; NA/CO/544/14, Executive Council Minutes, 10 June 1922.
38. 'A Terrible Accusation', *East African Chronicle*, 19 February 1921, p. 5.
39. 'Mr Desai's Appeal for Funds', *East African Chronicle*, 3 September 1921, p. 10.
40. Robert G. Gregory, *India and East Africa: A History of Race Relations within the British Empire, 1890–1939*, Oxford, Clarendon Press, 1971, p. 443; H. Thuku, M.A. Desai correspondence in *Harry Thuku: An Autobiography*, Nairobi, Oxford University Press, pp. 91–4; U.K. Oza, 'Letter to Editor', *East African Standard*, 1 March 1930, p. 38.
41. Although, because he was denied a translator and also because of conflict of interests Jeevanjee who spoke little English attended very few meetings of the Legislative Council. See Zarina Patel, *Challenge to Colonialism: The Struggle of Alibhai Mulla Jeevanjee for Equal Rights in Kenya*, Nairobi, Publishers Distribution Services 1997, p. 74. Jeevanjee was a very successful self-made business man whose activities within the British Empire were described as 'patriotism with 5 percent' and was said to represent similar characteristics to Cecil Rhodes. See

'Asian Commerce', [Report on the activities of A.M. Jeevanjee], *Leader*, 7 May 1910, cited in Mungeam, *Kenya*, p. 441.
42. Patel, *Challenge to Colonialism*, p. 54.
43. BL/IOR/L/PO/1/1A(iv) 'Kenya Indians', EAINC Submission 1920.
44. Gregory, *India and East Africa*, p. 453.
45. Patel, *Challenge to Colonialism*, p. 54.
46. Patel, *Challenge to Colonialism*, p. 55.
47. Patel, *Challenge to Colonialism*, Publishers Distribution Services 1997, p. 55.
48. For lists of subscribers to the appeal fund see: *East African Chronicle*, 17 September 1921, p.E(23); *East African Chronicle*, 5 November 1921, p. 6.
49. BL/IOR/L/E/7/1339 Education Ordinance, Report of Sub Inspector Bachan Singh enclosed in Dispatch India Office to Governor General of India, 23 February 1928, p. 6.
50. BL/IOR/L/E/7/1330/529(ii) Indians in Kenya: Representation on Municipal Councils, 9 May 1928.
51. BL/IOR /L/E/7/1558/697 Dispatch from the Governor, 14 January 1933.
52. 'Cleavage of Opinion on Indian Affairs', *East African Standard*, 15 November 1930, p. 7; 'Editorial', *East African Standard*, 7 February 1931, p. 43.
53. Channan Singh, 'Manilal Ambalal Desai', in Kenneth King and Ahmed Salim (eds.), *Kenya Historical Biographies*, Nairobi, East African Publishing House, 1971, pp. 129–41, 136, 140.
54. Chloe Campbell, *Race and Empire: Eugenics in Colonial Kenya*, Manchester, Manchester University Press, 2007.
55. James Christie, *Cholera Epidemics in East Africa, from 1821 till 1872*, London, Macmillan & Co., 1876.
56. 'Manners and Customs of the Inhabitants as Affecting the Spread of Epidemic Disease', in Christie, *Cholera Epidemics*, pp. 299, 347–48.
57. Christie, *Cholera Epidemics*, pp. 299, 335, 347–48.
58. Megan Vaughan, *Curing Their Ills: Colonial Power and African Illness*, Cambridge, Polity Press, 1991, p. 39.
59. Warwick Anderson, 'Excremental Colonialism: Public Health and the Poetics of Pollution', *Critical Inquiry*, 1995, 21, pp. 640–69; Anna Crozier, 'Sensationalising Africa: British Medical Impressions of Sub-Saharan Africa 1890–1939,' *Journal of Imperial and Commonwealth History*, 35, 3, 2007, pp. 393–415.
60. K.V. Adalja, 'Thirty Two Years in General Practice in Nairobi', *East African Medical Journal*, 36, 1959, pp. 442–8, 443.
61. Although interestingly, although fully accepted as part of the European establishment, there was some discussion over Bödeker's origins- with some speculation that he was half Burmese. See Cynthia Salvadori, *We Came in Dhows*, Nairobi, Paperchase Kenya Ltd, 1996, Vol. II, pp. 6–7.
62. NA/CO/544/1 'Report by H.A. Bödeker, Medical Officer, Kisumu', East Africa Protectorate Annual Medical Report, 1908, Administration Reports 1903–1909, p. 367; NA/CO 533/1 Kenya Original Correspondence, Dispatches Regarding Kisumu Plague, 1905, 15 February 1905 [where goods from 10 Indian premises burnt].
63. J.R. Gregory, *Under the Sun: A Memoir of Dr. R.W. Burkitt of Kenya*, Nairobi, The English Press, 1952, p. 33; 'European Mass Meeting', *East African Chronicle*, 13 August 1921, p. 8.
64. Adalja, 'Thirty Two Years', p. 447; Gregory, *Under the Sun*, p. 33; 'European Mass Meeting', *East African Chronicle*, 13 August 1921, p. 8.
65. 'European Mass Meeting', *East African Chronicle*, 13 August 1921, p. 9.

66. Gregory, *India and East Africa*, p. 32.
67. 'Dr Burkitt's Divine Doctrine', *East African Chronicle*, 20 August 1921, p. 3.
68. Cited in 'Dr Burkitt's Divine Doctrine', *East African Chronicle*, 20 August 1921, p. 3.
69. Cynthia Salvadori, *Two Indian Travellers: East Africa 1902–1905*, Mombasa, Friends of Fort Jesus, 1997, p. 164.
70. Adalja, 'Thirty Two Years', p. 444.
71. 'A Salutary Example', *East African Chronicle*, 16 October1920, p. 1.
72. Roger Jeffery, 'Recognizing India's Doctors: The Institutionalization of Medical dependency, 1918–1939', *Modern Asian Studies*, 13.2, 1979, pp. 301–26, 313.
73. Mridula Ramanna, 'Indian Doctors, Western Medicine and Social Change, 1845–1885', in Mariam Dossal and Ruby Maloni (eds.), *State Intervention and Popular Response: Western Indian in the Nineteenth Century*, Mumbai, Popular Prakashan Press, 1999, pp. 40–62, 49.
74. BL/IOR/L/PO/1/1A: 1914–1923, Kenya Indians, East African Indian National Congress letter to Viceroy, 22 March 1919.
75. 'An Ordinance to Make Provision for the Registration of Medical Practitioners and Dentists', 27 September 1910, *The Official Gazette*, 1 October 1910, pp. 575–7.
76. A similar point is made in: Pratik Chakrabarti, ' "Sign of the Times": Medicine and Nationhood in British India', *OSIRIS*, 24, 2009, pp. 188–211.
77. 'An Ordinance to Make Provision for the Registration of Medical Practitioners and Dentists', 27 September 1910, *The Official Gazette*, 1 October 1910, pp. 575–7.
78. NA/CO/544/2, Legislative Council Minutes 1907–1909, Minutes of 3 August, 22 November and 4 December 1909; 'Report by the Special Committee', *East African Standard*, 11 December 1909, p. 12.
79. Jeevanjee attended only one session in April 1910 and his term expired in September 1911. 'An Ordinance to Make Provision for the Registration of Medical Practitioners and Dentists, 27 September 1910', *The Official Gazette*, 1 October 1910, p. 575.
80. Patel, *Challenge to Colonialism*, p. 74.
81. BL/IOR/L/E/7/1265 EAINC Resolution No6, 15 November 1919, Enclosures to E.S. Montague from Governor General of India, 3 March 1921. The resolution stated 'that all Indian medical practitioners not below the rank of Sub—Assistant Surgeons be allowed to carry on independent medical practice in British East Africa'.
82. *Hansard*, 66, 17 February 1927, cc.133–56.
83. Deepak Kumar, 'Racial Discrimination and Science in Nineteenth Century India', *Indian Economic and Social History Review*, 19, 1982, pp. 63–82.
84. Michael Worboys, 'Colonial Medicine', in Roger Cooter and John Pickstone (eds.), *Medicine in the Twentieth Century*, Amsterdam, 2000, pp. 67–80, 70.
85. RHL/MSS.Brit.Emp.s.22/G5 W.H. Macdonald, Medical Return 18 August 1890, Imperial British East Africa Company papers.
86. Cd.769, *Report By HM Commissioner, on the East Africa Protectorate Health*, London, HMSO, August 1901, p. 8.
87. 'Appendix C', *Annual Medical Report*, 1928, p. 7.
88. *Annual Medical Report*, 1934, pp. 8–9. The 1932 *Annual Medical Report* made a startling admission that in respect to the bulk of the population (which presumably meant African) 'We have no knowledge whatsoever as to whether the general birth and death rates are increasing or decreasing, and no knowledge as to whether the people as a whole are fitter or less fit than they were last year, or ten, or fifty years ago.' *Annual Medical Report*, 1932, p. 10.
89. *Annual Medical Report*, 1934, pp. 8–9.
90. A. Milne, 'Municipal Committee', *The Leader*, 17 April 1915, p. 1.

91. Thomas R. Metcalf, *Ideologies of the Raj*, Cambridge, Cambridge University Press, 1994, pp. 177–81.
92. This was throughout the British Empire. Chakrabarti has provided illuminating analysis over the discussions to establish a Central Research Institute in India that came to a standstill, as consensus could not be reached as to where to locate it (the British argued for isolation of the proposed facility). Chakrabarti, 'Sign of the Times', *OSIRIS*, 2009.
93. I. Sham-su-deen [letter to the editor] 'Mr Mangaldas Answered', *East African Chronicle*, 13 November 1920, p. 13, where he complained that segregation even applied to Christian worship and graveyards.
94. Cd. 670, Correspondence respecting the Uganda Railway, 1901, Report by Colonel Gracey on the Uganda Railway, 21 April 1901, p. 13.
95. Cd. 670, Correspondence respecting the Uganda Railway, 1901, Report by Colonel Gracey on the Uganda Railway, p. 14.
96. Wellcome Library Archives (thereafter WLA) MS 6807/24 Dr Sieveking, Dr W Radford, Dr D.L. Falkener and Dr A Spurrier, Report on the Sanitary Aspect of the Site of Nairobi Township, 1902.
97. G.B. Williams, *Report on the Sanitation of Nairobi*, London, Waterloo, 1907, pp. 9–15
98. Williams, *Report on the Sanitation of Nairobi*, p. 50.
99. BL/IOR/L/E/7/1264/867 Indians in Kenya, 23 February 1922; A. Jeevanjee Letter and Pamphlet, 22 March 1922.
100. Clare Aveling Wiggins, 'Early Days in British East Africa and Uganda', *The East African Medical Journal*, 37, 1960, pp. 699–708, 701.
101. NA/CO/544/1, *Annual Medical Report*, 1908, p. 28.
102. 'Editorial: Mr. Moynagh's Opportunity', *East African Chronicle*, 18 September 1920, p. 7; W.J. Moynagh, Letter to the Editor 'Housing in Nairobi', *East African Chronicle*, 18 September 1920, p. 14; *Annual Medical Report*, 1913 p. 53; W.J. Simpson, *Report on the Sanitary Matters in the East Africa Protectorate, Uganda and Zanzibar*, London, Colonial Office, African No 1025, February 1915, pp. 4, 23; *Annual Medical Report*, 1922, p. 71.
103. 'Notice to Curtail Movement of Non Europeans under Infectious Disease Ordinance of 1903', *The Official Gazette*, 23 May 1911, p. 1.
104. *Annual Medical Report*, 1911, p. 16.
105. NA/CO/533/171, Kenya Dispatches 1916, Letter from the Indian Association, 24 October 1916.
106. Simpson, *Report*, African No 1025, 1915, p. 79.
107. R.N. Hunter 'Plague in Kenya', *Kenya Medical Journal*, 3.3, 1925, pp. 75–86, p. 75
108. J.H. Cross, 'Sanitary Reforms Urgently Needed', *The Leader*, 26 March 1910, p. 6
109. Annual Medical Report, 1911, p. 12.
110. BL/IOR/L/E/7/1263 A. Jeevanjee, 'Sanitation in Nairobi', May 1915; J.I. Roberts, 'Plague Conditions in an Urban Area of Kenya', *Journal of Hygiene*, 36.4, 1936, pp. 467–84.
111. BL/IOR /L/E/7/1263 Letter from Dr Cherrett to Nairobi Town Clark, 6 October 1913.
112. *Annual Medical Report*, 1911, p. 24. Simpson, *Report*, African No 1025, 1915, p. 78.
113. Report of D.S. Skelton, 'Report on the Epidemic of Plague in Mombasa', *Annual Medical Report*, for year ending 1913, pp. 6–90, 67.
114. *Annual Medical Report*, 1911, p. 12; also see quotation from: C.S. Nicholls, *Red Strangers, The White Tribe of Kenya*', Timewell Press, London, 2005, p. 134, who cites the words of the MO at Eldoret, M. Wetherell, who spoke (c.1920) of 'unrestricted immigration of 50,000 Indians to Kenya would spell the practical

extermination of the native because they would contaminate water with bacillus coli and cause typhoid fever and dysentery'.

115. *Annual Medical Report*, 1913 p. 18.
116. Mark Harrison, *Public Health in British India: Anglo Indian Preventative Medicine, 1859–1914*, Cambridge, Cambridge University Press, 1994, pp. 113–16.
117. Simpson, *Report*, [Appendices] pp. 77–91.
118. Simpson, *Report*, p. 9.
119. 'The Asiatic Question', *The Leader*, 16 January 1915, p. 4; 'Professor Simpson's Report', *The Leader,* 20 February 1915, p. 1; 'Professor Simpson's Scheme', *The Leader*, 17 March 1915, p. 1.
120. NA/CO/533/156 Despatches, September—November 1915, Minutes of the Nairobi Municipal Committee Meeting 14 April 1915.
121. NA/CO/533/156 Despatches, September—November 1915, Minutes of the Nairobi Municipal Committee Meeting 14 April 1915.
122. J.M. Campos, 'Township The Scandal of Nairobi', *East African Chronicle*, 16 October 1920, p. 10.
123. J.M. Campos, 'Township The Scandal of Nairobi', *East African Chronicle*, 16 October 1920, p. 10.
124. 'Showing Their Hand', *East African Chronicle*, 28 August 1920, p. 6; 'Councilor Riddell Supports', *East African Chronicle*, 16 October 1920, p. 11.
125. 'The Malaria Menace in Nairobi', *East African Standard*, 8 May 1935, p. 33.
126. The first Public Health Ordinance in Kenya was in 1921, which was supplemented over time by the Native Authority (Amendment) Ordinance, 1921, The Public Health (Amendment) Ordinance, 1928, Local Government (Municipalities) Ordinance, 1928. For a summary see: Annual Medical Report, 1936.
127. BL/IOR/L/PO/1/1A(iv) Kenya: 1914–1923, Lords debate 14 July 1920.
128. BL/IOR/L/PO/1/1A(iv) Kenya: 1914–1923, Lords debate 14 July 1920.
129. BL/IOR/L/PO/1/1A(iv) Kenya: 1914–1923, Lords debate 14 July 1920.
130. *Annual Medical Report*, 1913, p. 5.
131. 'Notice under Infectious Disease Ordinance of 1903', *The Official Gazette*, 15 August 1909, 354; 15 September 1909, p. 383; 23 May 1911, p. 1 (all of which curtail the movement of non Europeans). See also George Odour Ndege, *Health, State and Society in Kenya*, Rochester, NY, University of Rochester Press, 2001, p. 37.
132. 'Notice under Infectious Disease Ordinance of 1903', *The Official Gazette*, 15 August 1909, 354; 15 September 1909, p. 383; 23 May 1911, p. 1 (all of which curtail the movement of non Europeans).
133. 'Notice under Infectious Disease Ordinance of 1903', *The Official Gazette*, 15 August 1909, 354; 15 September 1909, p. 383; 23 May 1911, p. 1 (all of which curtail the movement of non Europeans).
134. *East African Chronicle*, 29 January 1921, p. 18. (See *Annual Medical Report*, 1911, p. 7 for first mention).
135. W.M. Ross, *Kenya From Within: A Short Political History*, London, George Allen & Unwin, 1927, p. 321.
136. John Langton Gilks, 'The Medical Department and the Health Organization in Kenya, 1909–1933', *The East African Medical Journal*, 9, 1932–3, pp. 340–54, 349; Ross, *Kenya From Within*, pp. 266, 331.
137. Ross, *Kenya From Within*, p. 330.
138. 'Mr Phadke Resigns', *East African Chronicle*, 29 January 1921, p. 5; 'Kenya's Public Health', *East African Chronicle*, 29 January 1921, p.A(17).
139. Ross, *Kenya From Within*, p. 330.

140. 'Mr Phadke Resigns', *East African Chronicle*, 29 January 1921, p. 5; 'Kenya's Public Health', *East African Chronicle*, 29 January 1921, p.A(17).
141. BL/IOR/L/E/7/1265, 'Indians in Kenya', [unidentified newspaper cutting], 29 January 1921; NA/CO 544/1 Minutes of Legco Meeting, 10 February 1921 records the issue and the voting on the decision.
142. Ross, *Kenya From Within*, p. 331.
143. Ross, *Kenya From Within*, p. 331.
144. 'Segregationists Baulked', *East African Chronicle*, 21 May 1921, p. 5 reported that Churchill had refused consent.
145. Ross, *Kenya From Within*, p. 275.
146. NA/CO/533/394/1 Racial Segregation in Towns, February 1930-March 1931.
147. John Maximilian Nazareth, *Brown Man Black Country: A Peep into Kenya's Freedom Struggle*, New Delhi, Tiding Publications, 1981, p. 109.
148. Ethel Younghusband, *Glimpses of East Africa and Zanzibar*, London, John Long, 1910, p. 219.
149. D.D. Puri, Letter to the Editor, 'Legislative Council Education Costs', *East African Standard*, 6 December 1935, p. 35. The only official data confirming the vastly unequal distribution of resources is given in the Appendix of Moyne Report, Cmd. 4093.
150. BL/IOR/L/E/7/1263/867 'Indians in Kenya', A. Jeevanjee Letter and Pamphlet, 22 March 1922.

5   Indians in the Colonial Medical Service

1. British Medical Association Archives (thereafter BMA)/B/1/162/1/10, Dominion Committee Documents, Session 1922–23, Letter from Secretary of State for the Colonies, Winston Churchill response to BMA, 10 August 1922.
2. Colin Baker, 'The Government Medical Service in Malawi: An Administrative History, 1891–1974', *Medical History*, 20,1976, pp. 296–311; A. Bayoumi, *The History of the Sudan Health Services*, Nairobi, Kenya Literature Bureau, 1979; E.B. Van Heyningen, ' "Agents of Empire": The Medical Profession in the Cape Colony, 1880–1910', *Medical History*, 33,1989, pp. 450–71; Heather Bell, *Frontiers of Medicine in the Anglo-Egyptian Sudan, 1899–1940*, Oxford, Clarendon Press, 1999; Crozier, *Practising Colonial Medicine*, I.B. Tauris, 2007.
3. For example, C a, 'The Second Class Doctor and the Medical Assistant in South Africa', *South African Medical Journal*, 31 March 1973, pp. 509–12; Karen Shapiro, 'Doctors or Medical Aids—The Debate over the Training of Black Medical Personnel in the Rural Black Population in South Africa in the 1920s and 1930s', *Journal of Southern African Studies*, 13, 1987, pp. 234–55; Maryinez Lyons, 'The Power to Heal: African Medical Auxiliaries in Colonial Belgian Congo and Uganda', in David Engels and Shula Marks (eds.), *Contesting Colonial Hegemony: State and Society in Africa and India*, London, British Academic Press, 1994, pp. 202–23; Anne Digby, 'The Mid-Level Health Worker in South Africa: The In-Between Condition of the "Middle" ', in Ryan Johnson and Amna Khalid (eds.), *Public Health in the British Empire: Intermediaries, Subordinates, and the Practice of Public Health*, New York & London, Routledge, 2012, pp. 17–92. For East Africa see John Iliffe, *East African Doctors: A History of the Modern Profession*, Cambridge University Press, 1998.
4. Ryan Johnson, 'The West African Medical Staff and the Administration of Imperial Tropical Medicine, 1902–1914', *Journal of Imperial and Commonwealth History*, 38.3, 2010, pp. 419–39; Ryan Johnson, ' "An All White Institution": Defending Private

Practice and the Formation of the West African Medical Staff', *Medical History*, 54.2, 2010, pp. 237–54.

5. Beck makes passing reference to the existence of Indian medical practitioners in her book Ann Beck, *A History of the British Medical Administration of East Africa: 1900–1950*, Cambridge, Massachusetts, Harvard University Press, 1999 (first published, 1970), p. 13. Her other works make no reference at all, see for example Ann Beck, 'Problems of British Medical Administration in East Africa between 1900–1930' *Bulletin of the History of Medicine*, 36,1962, pp. 275–83; Ann Beck, 'Native Medical Services in British East Africa and Native Patterns of Society', in *Verhandlungen des XX Internationalen Kongresses für Geschichte der Medizin, Berlin, 22–27 August 1966*, Hildesheim, Georg Olms Verlagsbuchhandlung, 1968, pp. 870–75.

6. Lord Hailey, *An African Survey: A Study of the Problems Arising in Africa South of the Sahara*, Oxford University Press, 1938, p. 1182.

7. This was, in fact, a description of the Sudan, but the stereotype held for most of British Africa. Richard Hillary, *The Last Enemy*, London, Macmillan and Co. Ltd., 1950, p. 15.

8. BL/IOR L/MIL/7/14626, Collection 324A/122 Proposal for grant of pensionable status to subordinate medical staff of East Africa and Uganda Protectorates, 1915–1925; COD No.855 'Revised Rules for the Employment of Assistant Surgeons, Sub-Assistant Surgeons and Compounders in the British East Africa and Uganda Protectorates Recruited From Sources Outside the Service of the Government of India', 9 November 1917, p. 1. This states that no formal qualifications were required for Compounders unlike Sub-Assistant Surgeons.

9. Crozier, *Practising Colonial Medicine*, p. 52.

10. The three operated throughout the colonial period in India, with a few amendments and additions, such as the Royal Army Medical Corps. See 'The Framework of Medicine in India', *The Lancet*, 16 October, 1937, pp. 933–5.

11. M.S. Rao, 'The History of Medicine in India and Burma', *Medical History*, 12.1, 1968, pp. 52–61, 56; D.G. Crawford, *A History of the Indian Medical Service, 1600–1913* (two vols.), London, W. Thacker, 1914, p. 649; Mark Harrison, *Public Health in British India: Anglo Indian Preventative Medicine, 1859–1914*, Cambridge, Cambridge University Press, 1994, p. 32.

12. 'Exclusion of Natives of India From the Army Medical Service', *The Times*, 25 February 1861, p. 5; Also see Roger Jeffery, 'Recognising India's Doctors: The Institutionalisation of Medical dependency, 1918–1939', *Modern Asian Studies*, 13.2, 1979, pp. 301–26, 304.

13. 'Indian Medical Service', *The Times*, 24 August 1880, p. 3.

14. For the campaigns to improve Indian access to the IMS see: Roger Jeffery, 'Doctors and Congress: The Role of Medical Men and Medical Politics in Indian Nationalism', in Mike Shepperdson and Colin Simmons (eds.), *The Indian National Congress and the Political Economy of India, 1885–1985*, Aldershot, Brookfield, USA, Avebury, 1988, pp. 160–73.

15. 'The Framework of Medicine in India', *The Lancet*, 16 October, 1937, pp. 933–5, 934.

16. The first four European doctors were Dr A.D. Mackinnon, Dr W.H. MacDonald, Dr J.S. MacPherson, Dr R.U. Moffat. See Arthur Dawson Milne, 'The Rise of the Colonial Medical Service', *Kenya and East African Medical Journal*, 5, 1928–9, pp. 50–8, 53.

17. BL/Microfilm/Government Publications Relating to Kenya, 1897–1963, East African Protectorate Blue Book, 1901/1902.

18. H.A. Bödeker, 'Some Sidelights on Early Medical History in East Africa', *The East African Medical Journal*, 12, 1935–36, pp. 100–7, 101, 105.

19. Milne, 'The Rise of the Colonial Medical Service', p. 53.
20. Crozier, *Practising Colonial Medicine*, p. 95.
21. BL/IOR/ MIL/7/14471,1907–1909, Collection 323/49 Medical Subordinates for Service in East Africa and Uganda, letter from Dr. Will, Principal Medical Officer, 28 January 1907.
22. BL/IOR/ MIL/7/14471,1907–1909, Collection 323/49 Medical Subordinates for Service in East Africa and Uganda, letter from Dr Will, Principal Medical Officer, 28 January 1907.
23. BL/IOR/ MIL/7/14471,1907–1909, Collection 323/49 Medical Subordinates for Service in East Africa and Uganda, letter from India Office, 29 March 1907, letter from Colonial Office, 5 July 1907, letter from Colonial Office, 11 September1907, letter from Colonial Office, 25 September, Letter from Military Department, India Office, 26 October 1907.
24. NA/CO/544/1 Administration Reports 1903–1909, *Annual Medical Report*, 1908, p. 3.
25. NA/CO/544/1 Administration Reports 1903–1909, *Annual Medical Report*, 1908, p. 3.
26. BL/IOR/MIL/7/2175 A.D. Mackinnon to Mr. Jackson, 13 April 1895; Milne, 'The Rise of the Colonial Medical Service', p. 58.
27. *Annual Medical Report*, 1921, p. 21.
28. BL/IOR L/MIL/7/2188 Collection 48/35 Hospital Assistants for Uganda Railway and Uganda, Foreign Office letter 23 January 1897.
29. Beck, *History of the Medical Administration*, p. 13.
30. Lord Hailey, *An African Survey: A Study of the Problems Arising in Africa South of the Sahara Revised 1956*, London, New York, Toronto, Oxford University Press, 1957, p. 1081; For more detail on the history of African medical education in East Africa see Iliffe, *East African Doctors*.
31. BL/IOR/L/MIL/7/14626, Collection 324A/122 1915–25, Proposal for Grant of Pensionable Status to Subordinate Medical staff of East Africa and Uganda Protectorates, COD No.855 'Revised Rules for the Employment of Assistant Surgeons, Sub-Assistant Surgeons and Compounders in the British East Africa and Uganda Protectorates Recruited From Sources Outside the Service of the Government of India', 9 November 1917; See also in the same collection Bonar Law letter, 28 October 1915.
32. BL/IOR/L/MIL/7/14626, Collection 324A/122 1915–25, Proposal for Grant of Pensionable Status to Subordinate Medical staff of East Africa and Uganda Protectorates, COD No.855 'Revised Rules for the Employment of Assistant Surgeons, Sub-Assistant Surgeons and Compounders in the British East Africa and Uganda Protectorates Recruited From Sources Outside the Service of the Government of India', 9 November 1917.
33. BL/IOR/L/MIL/7/14626, Collection 324A/122 1915–25, Proposal for Grant of Pensionable Status to Subordinate Medical staff of East Africa and Uganda Protectorates, COD No.855 'Revised Rules for the Employment of Assistant Surgeons, Sub-Assistant Surgeons and Compounders in the British East Africa and Uganda Protectorates Recruited From Sources Outside the Service of the Government of India', 9 November 1917.
34. *Annual Medical Report*, 1914, Promotions, p. 11.
35. BL/IOR/L/MIL/7/14441, 1898–1901 Collection 323/19 Medical Subordinates for Uganda Railway: Extension of Service and Replacement, F. Ranson letter 21 June 1900; BL/IOR/L/MIL/7/14471, 1907–1909, Collection 323/49 Medical Subordinates for Service in East Africa and Uganda, Colonial Office note 4 November 1907.

36. KNA/MOH/1/716, Appointment of Government Asian Doctors, Memo from Director of Medical Services, 11 January 1955; See also NA/CO/822/107/22 1942–43, Conditions of Service: Non-Europeans Appointed to Posts Ordinarily Assigned to Europeans; NA/CO/822/34/8 Governor's note to Colonial Office 23 July 1931.

37. RHL/Papers Collected by H. Topiwala Related to Indian Doctors in Kenya, expected deposit date 2015, *Dr A.C.L de Souza, In Memorium*, Nairobi, Majestic Printing Works, 1958, pp. 10, 28.

38. 'Dr Nair back from England', *East African Chronicle*, 16 October 1920, p. 18; 'Dr T.D. Nair promoted' *East African Chronicle*, 27 November 1920, p. 11.

39. RHL/Papers Collected by H. Topiwala Related to Indian Doctors in Kenya, expected deposit date 2015, Manmeet Singh, *The Life and Times of late Dr Bakhtawar Singh*, 6 May 2008.

40. Indian Sub- Assistant Surgeon, *The Lancet*, 19 October 1867, pp. 503–4; BL/IOR/L/MIL/7/19041, Note by R.H. Charles 2 September 1921; BL/IORL/MIL/7/14626, Coryndon letter dated 1 April 1921 [recommends improving the pay of Sub-Assistant Surgeons to stimulate recruitment].

41. BL/IOR/MIL/7/2175:1895, Collection 48/23 East Africa Protectorate: Raising of Indian Contingent of 300 men for Mombasa, Memo to Lord Hamilton, Secretary of State, 11 March 1896; John Iliffe states that Medical Officers got two or three times higher pay than Assistant surgeons. Iliffe, *East African Doctors*, p. 78. For European Medical Officer salaries see Crozier, *Practising Colonial Medicine*, pp. 27–8.

42. BL/IOR/L/MIL/7/14626 Collection 324A/122 1915–1925 Proposal for Grant of Pensionable Status to Subordinate Medical Staff of East Africa and Uganda Protectorates, COD No 855: Revised Rules for the Employment of Assistant Surgeons and Compounders in the British East Africa and Uganda Protectorates Recruited from Sources Outside of the Service of the Government of India, 9 November 1917.

43. BL/IOR/L/MIL/7/14626 Collection 324A/122 1915–25 Proposal for Grant of Pensionable Status to Subordinate Medical Staff of East Africa and Uganda Protectorates, Bonar Law letter, 28 October 1915; Robert Coryndon, letter, 1 April 1921.

44. 'New Kenya Civil Service Scales: Comparison between Asian and European Emoluments', *East African Chronicle*, 4 September 1920, p. 12.

45. BL/IOR/L/MIL/7/ Collection 324A/122 1915–1925 Proposal for Grant of Pensionable Status to Subordinate Medical Staff of East Africa and Uganda Protectorates, letter from Robert Coryndon, 1 April 1921.

46. Crozier, *Practising Colonial Medicine*, p. 27.

47. NA/CO/822/34/8 Appointment of Indian men and Women to Higher Grade Posts, Governor's note to Colonial Office, 23 July 1931; NA/CO/822/284 Salaries in Relation to Race and Domicile, 1951–1953, Note of E.L. Scott, 27 June 1951.

48. NA/CO/544/14 1916–1922, Executive Council Minutes, 1918, p. 363.

49. NA/CO/544/14 1916–1922, Executive Council Minutes (Confidential), 1916, pp. 42, 109.

50. Crozier, *Practising Colonial Medicine*, p. 30.

51. NA/CO/822/38/4 1930–1938, East African Medical Service Regulations: Private Practice, letter, Governor Gowers to Secretary of State, 10 July 1931.

52. BMA/B/162/1/9, Dominions Committee: 1921–22, Meeting 30 June 1922, p. 3; See also see *Annual Medical Report*, 1919, p. 19 which mentions government paying full time salaries to Medical Officers whose work, to a large part, consists in the treatment of private patients; debate in the press spoke explicitly about the

threat that Indian doctors posed to European private practitioners if they engaged in private practice on the side. See: 'Licensing Indian Doctors', *The Leader* [supplement], 25 March 1922, p. 4; more generally: 'Geddes Considerations', *The Leader*, 22 April 1922, p. 14 stated that government doctors (presumably both European and Indian) had 'practically wiped out' the private practitioner in the provinces and outstations.

53. NA/CO/691/89/9 1926–1927, Dewan Chand, Sub-Assistant Surgeon: Medical Retirement; Gratuity Award, letter from Governor to L. Amery, 31 January 1927, pp. 4, 7.

54. NA/CO/691/123/15 Petition about retrenchment by Sant Ram, former Sub-Assistant Surgeon, Medical Department, 1 January 1932–31 December 1932, letter, Governor to Secretary of State 18 June 1932.

55. NA/CO/536/184/5 1935, Dr A.V.S Rao, Former Sub-Assistant Surgeon, Medical Department: Petition Submitted in Regard to his Dismissal from Post, and Applying for the Grant of a Licence to Practice again in Uganda, 28 August 1935–30 November 1935.

56. BMA/B/162/1/12, Dominions Committee Meeting, Memorandum on the Medical and Sanitary Services of Kenya from BMA Kenya Branch, 6 March 1925, p. 3.

57. NA/CO/822/34/8 Appointment of Indian Men and Women to Higher Grade Posts, Colonial Office note, 16 January 1931 [here the complaints of the Indian Overseas Association are discussed].

58. NA/CO/822/34/8 Appointment of Indian Men and Women to Higher Grade Posts, Colonial Office note, 16 January 1931.

59. NA/CO/877/1/43122 Minute: Furse, Method of Dealing with Applications for Colonial Employment on the Part of Natives of India', 29 August 1921.

60. NA/CO/877/1/43122 Minutes: 'Applications by Natives of India for Appointments in the Colonial Service', 5 October 1921, Discussion of Letter by P.L. Gupta, 17 August 1923.

61. NA/CO 822/34/8 1931, Note to Bottomley, 16 January 1931.

62. NA/CO 822/284, 1951–3 Note, E.L. Scott, Salaries in Relation to Race and Domicile, 27 June 1951.

63. KNA/MOH/1/716, Appointment of Asian Medical Officers, Memo 11 January 1955 (also confirms that the salary of each Asian was based on 3/5ths of that which a European received- colloquially known as 'the 3/5th rule').

64. KNA/MOH/1/716, Appointment of Asian Medical Officers, Memo 11 January 1955.

65. *Annual Medical Report*, 1921, p. 21; John Langton Gilks, 'The Medical Department and the Health Organization in Kenya, 1909–1933', *The East African Medical Journal*, 9, 1932–3, pp. 340–54, 342.

66. BL/IOR/L/E/7/1295 1921–1924, Industries and Overseas Department Papers, Notes of the Meeting between Sec of State for the Colonies and the Indian Delegation from Kenya, 4 May 1923, p. 16.

67. *Annual Medical Report*, 1924 p. 1; Iliffe, *East African Doctors*, p. 24.

68. *Annual Medical Report*, 1936, p. 7.

69. *Annual Medical Report*, 1933, p. 1.

70. *Annual Medical Report*, 1932, p. 2; *Annual Medical Report*, 1936, p. 1 In both cases Paterson excluded Indian subordinate doctors from the count of medically qualified staff and made no references to their work.

71. *Annual Medical Report*, 1937, p. 7.

72. *Annual Medical Report*, 1937, p. 5 lists 24 Asian Sub-Assistant Surgeons and 2 Asian Assistant Surgeons.

73. BL/IOR/L/MIL/7/14471 1907–1909 Collection 323/49 Memo: Medical Subordinates for Service in East Africa and Uganda, Government of India note, 17 June 1909; 10 September 1909. See also BL/IOR/L/MIL/7/2188 Government of India note 24 June 1897 (The case of Hospital Assistant R.F. Holder who refused to board the boat in Bombay when he learnt his family was refused free to passage to Mombasa).

74. Crozier, *Practising Colonial Medicine*, p. 112. (It should be noted that incomplete data does not allow for the computation of an average term of service for Indian subordinate doctors over the whole period. However, a comparison of information from the 1911 Blue Book with the information subsequently printed in the *Official Gazettes* of 25 February 1920 (p. 138), 14 January 1930, 19 January 1932, p. 59; 24 January 1939, p. 114 indicates only 15% of those serving in 1911 had been in the jobs for more than 10yrs. Similarly, about 20% those serving in 1920 had had careers longer than 10 years; and about a third of those serving in 1930 and half of those serving in 1939.

75. *The Official Gazette*, 24 January 1939; RHL/Papers Collected by H. Topiwala Related to Indian Doctors in Kenya, expected deposit date 2015, email communication between Harshad Topiwala and Anwar Sheikh (son of Sayyid Wilayat Shah) 5 November 2007.

76. RHL/MSS.Afr.s.702 Robert Arthur Welsford Procter, 'Random Reminiscences, Mainly Surgical' [no date]; Crozier, *Practising Colonial Medicine*, pp. 87–88.

77. Syracuse University (SU), Kenya National Archive Records (KNA), Mircrofilm Number 2801, Annual and Quarterly Reports (Provincial and District) Reel 15: Provincial Medical Report, E.B. Horne, Meru, 1915 and 1916; Provincial Medical Report, Abdulla Khan, Meru, 1918–20; Provincial Medical Report, Ali Baksh, Meru, 1922; Reel 21: Provincial Medical Report, Gokul Chand, Kabarnet, 1921; Reel 56: Provincial Medical Report Maula Buksh, Kilifi, 1918–19; Reel 59: Provincial Medical Reports, Maula Buksh, Malindi, 1918–19.

78. A.N. Nyss, *Annual Medical Report*, 1919, p. 14; *Annual Medical Report*, 1921, p. 22.

79. SU/KNA/Mircrofilm Number 2801, Annual and Quarterly Reports (Provincial and District), Reel 21: Provincial Medical Report, Gokul Chand, Kabarnet, 1921, Provincial and District Reports, Gokul Chand, Kabarnet Medical Report, p. 30; See also WL/CMAC/PP/HCT/A5 Elizabeth Bray, *Hugh Trowell: Pioneer Nutritionist*, unpublished biography, London 1988.

80. Christine Stephanie Nicholls, *Red Strangers: The White Tribe of Kenya*, London, Timewell Press, 2005, p. 209.

81. To be fair European Medical Officers were also given very little on the job training: typically just a few days orientation in either Nairobi or Mombasa. They usually were required to have undertaken the Diploma of Tropical Medicine and Hygiene however at either the London or Liverpool schools. See Crozier, *Practising Colonial Medicine*, p. 26.

82. *The Official Gazette*, 24 January 1939; RHL/Papers Collected by H. Topiwala Related to Indian Doctors in Kenya, expected deposit date 2015, email communication between Harshad Topiwala and Anwar Sheikh (son of Sayyid Wilayat Shah) 5 November 2007.

83. Cynthia Salvadori, *We Came in Dhows*, Nairobi, Paperchase Kenya Ltd, 1996, Vol. III, p. 90 [Contribution by Usha Dupta].

84. Not least as the editor of the EAMJ between 1922–28 was Christopher Wilson, who held explicitly white supremacist view. See C.J. Wilson, *Before The Dawn in Kenya*, Nairobi, The English Press, 1952; C.J. Wilson, *Kenya Warning: The Challenge to White Supremacy in Our British Colony*, Nairobi, The English Press, 1954.

85. It might be possible to speculate here whether Indian researchers were denied the same journal space as Europeans through editorial decisions however.
86. *Annual Medical Report*, 1923, p. 1; BMA/B/162/1/9 Dominions Committee Documents, Session 1921–1922, Meeting 30 June 1922, p. 1, hh.
87. A.N. Nyss, *Annual Medical Report*, 1913, pp. 77–80.
88. T.D. Nair, 'A Tana River Yaws Campaign', *Kenya and East Africa Medical Journal*, 4.7, 1927–28, pp. 201–7.
89. Salvadori, *We Came in Dhows*, Paperchase Kenya, 1996, Vol. III, p. 140.
90. *Annual Medical Report*, 1913, p. 77.
91. *Annual Medical Report*, 1919, p. 14.
92. Nair, 'Tana River Campaign', p. 201.
93. Nair, 'Tana River Campaign', pp. 205–7.
94. Salvadori, *We Came in Dhows*, pp. 140–1 [A summary of the 29 page medical report from Dastur while he was in Baringo in 1935].
95. Salvadori, *We Came in Dhows*, p. 141.
96. As evidenced by the absence of reports by Maula Baksh, Abdulla Khan, Gopal Chand and Ali Baksh in the *Annual Medical Reports* though they submitted reports regularly at district level.
97. 'The Indian and Cocaine', *The East African Standard*, 18 April 1931, p. 32.
98. 'Echo of the Murder at Meru', *The East African Standard*, 18 July 1931, p. 42.
99. SU/KNA Mircrofilm Number 2801, Annual and Quarterly Reports (Provincial and District) Reel 15: Provincial Medical Report, E.B. Horne, Meru, 1915 and 1916; Provincial Medical Report, Abdulla Khan, Meru, 1918–20; Provincial Medical Report, Ali Baksh, Meru, 1922; Reel 21: Provincial Medical Report, Gokul Chand, Kabarnet, 1921; Reel 56: Provincial Medical Report Maula Buksh, Kilifi, 1918–19; Reel 59: Provincial Medical Reports, Maula Buksh, Malindi, 1918–19.
100. SU/KNA Mircrofilm Number 2801, Annual and Quarterly Reports (Provincial and District) Reel 15: Provincial Medical Report, E.B. Horne, Meru, 1915 and 1916; Provincial Medical Report, Abdulla Khan, Meru, 1918–20; Provincial Medical Report, Ali Baksh, Meru, 1922; Reel 21: Provincial Medical Report, Gokul Chand, Kabarnet, 1921; Reel 56: Provincial Medical Report Maula Buksh, Kilifi, 1918–19; Reel 59: Provincial Medical Reports, Maula Buksh, Malindi, 1918–19.
101. SU/KNA Mircrofilm Number 2801, Annual and Quarterly Reports (Provincial and District) Reel 21: Provincial Medical Report, Gokul Chand, Kabarnet, 1921; See in the same report comments of District Commisioner Bamber on the performance of Gokul Chand, p. 1.
102. Milne, 'The Rise of the Colonial Medical Service', p. 58.
103. Milne, 'The Rise of the Colonial Medical Service', p. 58.
104. RHL/MSS.Afr.s. 1872/75, Robert Samuel Hennessey, Memorandum on Experiences in the Colonial Medical Service in Uganda, 192–955.
105. RHL/MSS.r.4.I, P.A. Clearkin, Rambling and Recollections of a Colonial Doctor, 1913–58, unpublished manuscript, p. 126.
106. Norman Parsons Jewell, [ed. Tony Jewell], *On Call*, Privately Published, *forthcoming*, 2015 [Chapter 4; Chapter 11, not yet paginated].
107. 'Lord Ellenborough and the Indian Medical Service', *The Lancet*, 14 December 1861, p. 581.
108. 'Indian Medical Degrees: The General Medical Council and Indian Qualifications: History of a Ten Year Controversy', *British Medical Journal*, 15 March 1930, pp. 505–11.

109. 'The Framework of Medicine in India', *The Lancet*, 16 October 1937, pp. 933–5, p. 934.
110. BL/IOR L/MIL/7/14157 'Rules Governing Admission –BMA Memorandum on the Present Position and Future Prospects of the Indian Medical Service', 20 February 1914, p. 1; also see Cd.7900, HMSO, 1915 Minutes of the evidence relating to Medical Services, p188, R.W. Lyon, Surgeon General ' I consider the maintenance of a preponderance of European element essential....'
111. 'The Future of Medicine in India: Sir Gerald Gifford's Address' *British Medical Journal*, 22 November 1924, p. 966; Jeffery, 'Recognizing India's Doctors', p. 324.
112. 'The Future of Medicine in India' *British Medical Journal*, 22 November 1924, p. 966.
113. Jeffery, 'Recognizing India's Doctors', p. 307; Jeffery, 'Doctors and Congress', p. 161.
114. BMA/B/162/1/12 Dominions Committee Documents, Session 1924–5, 6 March 1925, p. 3; Memorandum on Medical and Sanitary Services from Kenya BMA Branch, 5 March 1925, p. 3.
115. KNA/MOH/File 1/716, Memo from Director of Medical Services, 'Appointment of Asian Doctors', 11 January 1955.

6 Squeezing Indians Out of Government Medicine

1. BL/IOR/L/PJ/6/1718 *Economic Commission Report*, East Africa Protectorate, 1919 Nairobi, Swift Press, 1919, p. 20 NB: no copy of the Bowring Committee Report has been found, so second hand reporting of its findings in other sources has been used throughout.
2. W.S. Churchill, *My African Journey*, London, Hodder and Stoughton, 1908, p. 49.
3. *Annual Medical Report*, 1937, p. 7. Table III shows a third of the hospitals in the reserves were still run by Indians.
4. 'Colonial Medicos', *The East African Chronicle*, 25 September 1920, p. 4; 'Call the Doctor', *The East African Chronicle*, 2 October 1920, pp. 3–4; BL/IOR/L/PO/1/1A(iii) Letter Indian Association to Viceroy, 22 March 1919, and EAINC submission to Viscount Milner, (not dated), 1920; NA/CO/544/29 Legislative Council Minutes, 3 March 1922; BL/IOR/L/E/7/1265, Enclosures, E.S. Montague to Governor General of India, 3 March 1921, EAINC Resolution No 6., 15 November 1919.
5. BL/IOR/L/PO/1/1A(iii) Letter Indian Association to the Viceroy, 22 March 1919.
6. BL/IOR/L/PO/1/1A(iii) Letter Indian Association to the Viceroy, 22 March 1919.
7. BL/IOR/L/PO/1/1A(iii) Letter Indian Association to the Viceroy, 22 March 1919.
8. BL/IOR/L/PO/1/1A (iii) Milner to Officer administrating the Government of East Africa, 21 May 1920.
9. 'Interview with Mr. M.A. Desai', *The East African Chronicle*, 21 August 1920, p. 9.
10. BMA/B/162/1/9 Dominion Committee Documents, Session 1921–22, Meeting 30 June 1922.
11. BMA/B/162/1/9, Dominion Committee Documents, Session 1921–22, Meeting 30 June 1922, p. 1.
12. BMA/B/162/1/9, Dominion Committee Documents, Session 1921–22, Meeting 30 June 1922, p. 2.
13. BMA/B/162/1/9, Dominion Committee Documents, Session 1921–22, Meeting 30 June 1922, p. 3.
14. BMA/B/162/1/10 Dominion Committee Documents, Session 1921–22, Meeting 6 October 1922, p. 1.

15. NA/CO/533/276 Despatches: March – April 1922; Also see R.M. Maxon, *John Ainsworth and the Making of Kenya*, Lanham, MD, University Press of America, 1980, pp.199,. 231–3.
16. *Hansard*, 66, 1927, pp. 133–56, 141.
17. NA/CO/533/288, Colonial Office note, 'Kenya Expenditure', Bottomley minute, p. 1, 8 February 1922.
18. NA/CO/533/288, Colonial Office note, 'Kenya Expenditure', Bottomley minute, comparative table, 8 February 1922.
19. NA/CO/533/288 Churchill to Governor Northey 14 February 1922.
20. NA/CO/544/29 Kenya Legislative Council Minutes, 1921–1929, 3 January 1922, p. 162.
21. NA/CO/544/29 Kenya Legislative Council Minutes, 1921–1929, 3 January 1922, p. 163.
22. W.M. Ross, *Kenya from Within: A Short Political History*, London, George Unwin, 1927, p. 158.
23. 'Economic and Financial Committee' extract from 'Report of Proceedings of Economic and Financial Committee', in G.H. Mungeam (ed.), *Kenya: Select Historical Documents 1884–1923*, Nairobi, East African Publishing House, p. 449; Legislative Council Proceedings 27 March 1922 as quoted in Ross, *Kenya from Within*, p. 158.
24. 'Economic and Financial Committee'; extract from 'Report of Proceedings of Economic and Financial Committee', p. 449.
25. Ross, *Kenya from Within*, p. 160; (there is no record of Shamsuddun in any reports of the committee after Sept 1922).
26. 'Kenya's Economic Problems: How Expenditure has Grown', *The Leader*, 24 June 1922, pp. 11–12.
27. 'Geddes Task', *The Leader*, 17 June 1922, p. 3.
28. 'Geddes Task', *The Leader*, 17 June 1922, p. 3;'Geddes Considerations', *The Leader*, 22 April, p. 14 argues the case for subsidizing private practitioners stating that the government doctor 'has practically wiped out the private practitioner'.
29. Ross, *Kenya from Within*, p. 266.
30. Ross, *Kenya from Within*, pp. 266–75.
31. Ross was eventually forced into resignation after claims of his mismanagement of his colonial department failed but his salary was reduced on economic grounds. Claims that he always felt had been manufactured in order to get rid of him. Ross, *Kenya from Within*, pp. 266–75.
32. BL/IOR/L/PJ/6/1718 *Economic Commission Report*, Nairobi, Swift Press, 1919, p. 20.
33. BL/IOR/L/PJ/6/1718 *Economic Commission Report*, Nairobi, Swift Press, 1919, p. 21.
34. BL/IOR/L/E/7/1295 File 191, 'Indians in Kenya', Letter B.S. Varma letter to Secretary of State for India, 25 May 1923.
35. Ross, *Kenya from Within* pp. 150, 160.
36. BL/IOR/L/E/7/1264 'Indians in Kenya', The Devonshire Declaration, White Paper, Cmd. 1922, July 1923; See also BL/IOR/L/E/1295 'Indians in Kenya', May-September 1923, for Indian responses, of which the minutes of the London 4 May Meeting between the Indian delegation (led by M.A. Desai) and the Colonial Office, are the most revealing; see also Robert G. Gregory, *India and East Africa: A History of Race Relations within the British Empire, 1890–1939*, Oxford, Clarendon Press, 1971, p. 251.
37. There is an extensive literature on the Devonshire Declaration. For example, Christopher P. Youé, 'The Threat of Settler Rebellion and the Imperial

Predicament: The Denial of Indian Rights in Kenya, 1923', *Canadian Journal of History*, 12, 1978; Robert J. Blythe, *The Empire of the Raj: India, Eastern Africa and the Middle East, 1858–1947*, Cambridge University Press, 2003, pp. 93–131; Randolph M. K. Joalahliae, *The Indian as an Enemy: An Analysis of the Indian Question in East Africa*, Bloomington, Author house, 2010; Sana Aiyar, *Empire, Race and the Indians in Colonial Kenya's Contested Public Political Sphere, 1919–1923*, Africa: *The Journal of the International African Institute*, 81.1, 2011, pp. 132–154.

38. 'The Political Outlook in Kenya', *East African Standard*, 12 April 1935, p. 27.

39. U.K. Oza, 'Correspondence: Indian Claims in Kenya', *East African Standard*, 26 April 1935, p. 30.

40. Shrinivasa Shastri in India, C.F. Andrews and H.S.L. Polak in London, Lord Islington in the House of Lords and Colonel Josiah Wedgewood in the House of Commons.

41. W.C. Bottomley, senior colonial official, pointed out the 'danger of Kenya being South Africanised', cited in Maxon, *John Ainsworth*, p. 407.

42. See for example, C.J. Wilson, *Before the Dawn in Kenya*, Nairobi, The English Press, 1952; C.J. Wilson, *Kenya Warning: The Challenge to White Supremacy in Our British Colony*, Nairobi, The English Press, 1954.

43. Andrew Hicks, 'Forty Years of the British Medical Association (Kenya Branch)', *East African Medical Journal*, 38.1, 1961, pp. 43–53, 51.

44. Hicks, 'Forty Years of the British Medical Association', p. 49. Interestingly, the pro-settler *East African Standard* funded the publication costs of the journal for a few years.

45. RHL/Papers Collected by H. Topiwala Related to Indian Doctors in Kenya, expected deposit date 2015, F.J. Wright, Letter Secretary Kenya BMA to H.T. Topiwala, 12 June 1943.

46. *Annual Medical Report*, 1922, p.11 mentions a reduction of 29 Sub-Assistant Surgeons and 5 Assistant Surgeons. In fact the reduction was more dramatic as from 1920 the replacement of Indian doctors to establishment numbers was already reduced.

47. Mr. Ormsby-Gore in response to question by Mr. Short, Parliamentary Debates, 13 February-23 March 1923, Commons, Volume 160, p 812.

48. *Annual Medical Report*, 1922, p. 11.

49. *Annual Medical Report*, 1923, p. 1. (Also an earlier hint can be found in the *Annual Medical Report* 1921, p. 18 when it is reported that the 'inclinations' of Indian doctors 'were not as a rule directed towards Africans').

50. BL/IOR/L/E/7/1295 Note by Walton quoting EAINC complaints, 12 February 1924.

51. BL/IOR/L/E/7/1295 Note by Walton quoting EAINC complaints, 12 February 1924.

52. BL/IOR/L/E/7/1295 Memo from EAINC, 30 January 1924.

53. John Langton Gilks, 'The Medical Department and the Health Organization of Kenya, 1909–1933', *The East African Medical Journal*, 9, 1932–3, pp. 340–54, 350.

54. *Annual Medical Report* 1923, Returns, p. 96. *Annual Medical Report*, 1932, p. 2; BL/IOR/L/PJ/8/254, A.W. Pim, *Report on the Financial Position and System of Kenya*, HMS Stationery Office, 1936, Appendix XI shows numbers of Medical Officers went up from 52 in 1921 to 71 in 1928.

55. Gilks, 'The Medical Department and the Health Organization of Kenya', p. 350.

56. BL/IOR/L/PJ/8/254 Pim Report, HMSO, 1936, p. 24 and Appendix XI.

57. Gilks, 'The Medical Department and the Health Organization of Kenya', p. 347.

58. BL/IOR/L/PJ/8/254 Pim Report, HMSO, 1936, p. 195; The policy of 'Europeanisation' was also stressed in 1932 *Annual Medical Report*, 1932, p. 6; *Annual Medical Report*, 1933, p. 1.

59. BL/IOR/L/PJ/8/254 Pim Report, HMSO, 1936, p. 24.
60. Roger Jeffery, 'Doctors and Congress: The Role of Medical Men and Medical Politics in Indian Nationalism', in Mike Shepperdson and Colin Simmons (eds.), *The Indian National Congress and the Political Economy of India, 1885–1985*, Aldershot, Brookfield, USA, Avebury, 1988, pp. 160–73, 164.
61. *Annual Medical Report*, 1932, p. 2.
62. *Annual Medical Report*, 1932, p. 2.
63. BMA/B/162/1/12 Dominions Committee, Memo on the Medical and Sanitary Services of Kenya Colony, 6 March 1925, p. 3; BMA/B/162/162/1/9 Meeting 30 June 1922: 'the medical practitioner was registered not only because he was qualified but also because he was honourable and understood the ethics of the profession, whilst the Indian was not honourable but seditious and had no affection for the natives'.
64. BMA/B/162/1/12 Dominions Committee, Memo on the Medical and Sanitary Services of Kenya Colony, 6 March 1925, p. 3.
65. John Iliffe, *East African Doctors: A History of the Modern Profession*, Cambridge University Press, 1998, p. 45.
66. *Annual Medical Report*, 1921, p. 21.
67. Dr Likimani who graduated from Makerere in 1939 was the first African doctor to work in Kenya, 'News: Dr Jason C. Likimani', *East African Medical Journal*, 56, 1979, p. 467.
68. *Annual Medical Report*, 1932, p. 2; The others could have been nurses, but also personnel in other subordinate roles, such as gardeners, sweepers etc.
69. *Annual Medical Report*, 1936, p. 7.
70. A typical positive assessment was that the new staff consisted of 'a fairly highly trained, literate and efficient African staff of about one thousand strong have been brought into being.' *Annual Medical Report*, 1932, p. 2.
71. NA/CO/533/426/8 'Native Medical Service', Memo from Gilks concerning Deputy Governor's letter to Colonial Office, 6 August 1932.
72. WL/CMAC/PP/HCT/A5 Elizabeth Bray, *Hugh Trowell: Pioneer Nutritionist*, unpublished biography, London 1988, Chapter 6, pp. 1–2.
73. 'Hospital and Dispensary Policy', *Annual Medical Report*, 1933, p. 7.
74. *Annual Medical Report*, 1921, p. 21; Gilks speech to BMA Kenya Branch, *Kenya Medical Journal*, February 1926, 2.11, pp. 321–25, 323; F.C. Trowell, 'The Medical Training of Africans', *East African Medical Journal*, 2.2, 1935, pp. 338–53, 346.
75. WL/CMAC/PP/HCT/A5 Elizabeth Bray, *Hugh Trowell: Pioneer Nutritionist*, unpublished biography, London 1988, Chapter 6, p. 1.
76. WL/CMAC/PP/HCT/A5 Bray, *Hugh Trowell*; Iliffe, *East African Doctors*; In Kenya there were no schools for Africans for full secondary education, see NA/CO/822/90/8 1938, 'Report: Higher Education in East Africa', 1932.
77. BL/IOR/L/PJ/8/254, Pim, *Report on the Financial Position*, p. 193; *Annual Medical Report*, 1932, p. 6 mentions the first batch of trained Africans to appear in 1933.
78. NA/CO/533/426/8 Native Medical Service, Memo from Gilks concerning Deputy Governor's letter to Colonial Office, 6 August 1932.
79. *Annual Medical Report*, 1933, p. 5; See WL/CMAC/PP/HCT/A5 Elizabeth Bray, *Hugh Trowell*, p. 13 where he states the doctor refilled the drug bottles in the dispensaries 'every two or three months on his arrival'. The lack of interest in his SMO in seeing patients was also noted in chapter 4, p. 17. Also a similar point about getting to outstations was made by Carman, who mentioned that many roads became inaccessible and were closed off during the rainy seasons. See John A. Carman, *A Medial History of the Colony and Protectorate of Kenya: A Personal Memoir*, London, Rex Collings, 1976, p. 20.

80. *Annual Medical Report*, 1933, p. 5.
81. NA/CO/533/426/8 Native Medical Service, 'Memorandum on the Training of African Staff of the Medical Department' from Gilks contained in Deputy Governor's letter to Colonial Office, 6 August 1932.
82. NA/CO/533/426/8 Native Medical Service, Memo from Gilks contained in Deputy Governor's letter to Colonial Office, 6 August 1932, p. 2.
83. NA/CO/533/426/8 Native Medical Service, Memo from Gilks contained in Deputy Governor's letter to Colonial Office, 6 August 1932, p. 6.
84. *Annual Medical Report*, 1932, p. 3. Compare this with *Annual Medical Report*, 1932, p. 2; p.3 which talked of highly trained, literate, and efficient staff' with comments in the ten page NA/CO/533/426/8 Native Medical Service, Memo from Gilks concerning Deputy Governor's letter to Colonial Office, 6 August 1932.
85. *Annual Medical Report*, 1932, p. 4; Iliffe, *East African Doctors*, p. 54.
86. *Annual Medical Report*, 1919, p. 9; *Annual Medical Report*, 1936, p. 7.
87. *Annual Medical Report*, 1936, p. 9.
88. A.B. Patel, Legislative Council minutes 28 November 1935, p. 744.
89. 'Obituary: J.L. Gilks', *British Medical Journal*, 28 August 1971, p. 538.
90. Suzanne Fisher, *We Lived On the Verandah*, Bognor Regis, New Horizon, 1980, p. 94.
91. *Annual Medical Report*, 1921, pp. 18, 21.
92. *Annual Medical Report*, 1921, p. 18.
93. NA/CO/544/29 1921–1929, Amendment to Medical Ordinance, Legislative Council Minutes, 24 March 1922, p. 3; see also statement made by Lord Arnold, 17 February 1927 vol. 66 cc.133–56, regarding Delamere's attempt to wreck the Bill to license Indian Sub-Assistant Surgeons, who had provided meritorious service, to practice in the Colony after retirement from the Service by introducing the word European instead of person. Earlier in 1920 concerns had been expressed in the House of Commons about the treatment of Indian doctors; HC Deb 24 March 1920 vol. 127 cc450–2W question from Mr Waterson to Under-Secretary of State for the Colonies ('Whether Indian medical practitioners in British East Africa are forbidden to carry on independent medical practice; and, if so, on what grounds?').
94. NA/CO/544/14, Executive Council Minutes, Vote on Common Franchise, 7 May 1921, p.640.
95. NA/CO/544/14 Executive Council Minutes, Meeting 10 June 1922, p. 761.
96. *Annual Medical Report*, 1911, p. 10; also see 'The Report by the Sub Committee', *East African Standard*, 11 December 1909, p. 12. This provides a very flattering portrait of the quality of the Sub-Assistant Surgeon based on Milne's input.
97. He mentions their number, but never explains why Indians suffered worse than Europeans in the cutbacks. Gilks, 'The Medical Department and the Health Organization of Kenya', p. 350.
98. Gilks, 'The Medical Department and the Health Organization of Kenya', p. 341
99. Our parenthesis. NA/CO/544/29 Kenya Legislative Council Minutes, 27 March 1922.
100. Ann Beck, 'Native Medical Services in British East Africa and Native Patterns of Society,' in *Verhandlungen des XX Internationalen Kongresses für Geschichte der Medizin Berlin, 22–27 August 1966*, Hildesheim, Georg Olms Verlags buchhandlung, 1968, pp. 870–5, 871.
101. K.V. Adalja, 'The Development of Medical Service in Kenya', *East African Medical Journal*, 39, 1962, pp. 105–14, 108; Adalja praised both Gilks and Paterson in the Kenya Legislative Council Minutes, 28 October1960, p. 56.

102. Carman, *A Medial History of Kenya*, p. 17.
103. Carman, *A Medial History of Kenya*, p. 17,. 60.
104. RHL/MSS.Brit.Emp, r.4 Peter Alphonsus Clearkin, *Ramblings and Recollections of a Colonial Doctor, 1913–58*, Book 1, Durban, 1967, p. 149; RHL/MSSAf.s.1653 F. Anderson who succeeded Paterson as the PMO also pointed out that Gilks and Paterson had not been to outstations in his unpublished memoir.
105. RHL/MSS.Brit.Emp, r.4 Clearkin, *Ramblings and Recollections*, p. 149.
106. RHL/MSS.Brit.Emp, r.4 Clearkin, *Ramblings and Recollections*, p. 152.
107. RHL/MSS.Brit.Emp, r.4 Clearkin, *Ramblings and Recollections*, p. 152.
108. RHL/MSS.Brit.Emp, r.4 Clearkin, *Ramblings and Recollections*, p. 153.
109. P.A. Clearkin, 'Typhus in the Tropics', *British Medical Journal*, 8 September 1934, p. 487.
110. NA/CO/544/29 Kenya Legislative Council Minutes 1921–29, Debate on Medical Department 29 October 1923.
111. NL/CO/544/29:1922–27, Mr B.S.Varma, Legislative Council Minutes, 29 October 1923.
112. 'Obituary: J.L. Gilks', *British Medical Journal*, 28 August 1971, p. 538.
113. Chloe Campbell, *Race and Empire: Eugenics in Colonial Kenya*, Manchester University Press, 2007, p. 61.
114. Campbell, *Race and Empire*.
115. NA/CO/822/55/1 East African Native: Brains, Structure and Mental Capacity, Fleet's handwritten note, 8 March 1934, p.8.
116. John L. Gilks, Letter to the Editor, 'The Native Brain', *The Times*, 30 December 1933, p. 6.
117. *Annual Medical Report*, 1932, p. 1 gives fulsome credit to Gilks work. NA/CO/822/55/1:1933 Paterson, 'East African Native Brains', 1 November 1933.
118. NA/CO/822/53/6:1933, Flood memo, East African Medical Service Reorganisation, 21 August 1933. This talks of 'interesting but overloaded with unnecessary verbiage'.
119. John Carman said some of Paterson's schemes were grandiose and unrealistic. See Carman, *A Medial History of Kenya*, p. 60.
120. R. Smyth, 'The Development of British Colonial Film Policy, 1927–1939, with Special Reference to East and Central Africa', *Journal of African History*, 20.3, 1979, pp. 437–50.
121. For an example see Arthur Rutherford Paterson, *The Book of Civilization: Part 1, On Cleanliness and Health. The Care of Your Children, Food, and How to Get Rid of Flies*, London, New York, Toronto, Longmans, Green & Co., 1934.
122. NA/CO/822/105/17 Conference of the Directors of Medical Services, 19 July 1939, Item 7.
123. Dr de Sousa, Legislative Council Minutes 4 December 1935 p. 872; N.S. Mangat, Legislative Council Minutes, 6 December 1933, p.831.
124. In *Annual Medical Report*, 1932, pp. 2–4 Paterson surveyed the history of the medical department praising European doctors and nurses together with the African staff but omitted to mention the contribution of Indian doctors; In the Legislative Council debate on government expenditure when Paterson was pressurized by Indian members as to why medical costs were not reduced by greater use of Indian subordinate doctors, he maintained that they 'could not carry out that type of work' which European MOs could. Legislative Council Minutes, 4 December 1935, p. 882.
125. KNA/MOH/1/716 Memo 'Appointment of Asian Medical Officers', 11 January 1955.

126. *Annual Medical Report*, 1932, p.2, p. 3.
127. *Annual Medical Report*, 1932, p. 7.

7 Indian Private Doctors in Kenya

1. RHL/Papers Collected by H. Topiwala Related to Indian Doctors in Kenya, expected deposit date 2015, H.T. Topiwala, Speech Indian Medical and Dental Association, Nairobi, 25 March 1955.
2. The only mentions that have been found in print are in Robert G. Gregory, *South Asians in East Africa: An Economic and Social History 1890–1990*, Oxford, Westview Press, 1993, pp. 217–27; Cynthia Salvadori, *We Came in Dhows*, Nairobi, Paperchase Kenya Ltd, 1996, has material relating to Dr Dias (Vol. I, p. 126); Dr Ribeiro (Vol. II, p. 22), Dr Bowry (Vol. III, p. 91), Dr de Sousa (Vol. III, p. 160). These cannot be counted as substantial analyses.
3. K.V. Adalja, 'Thirty Two Years in General Practice in Nairobi', *East African Medical Journal*, 36, 1959, pp. 442–8, 443.
4. For example see unpublished material has been uncovered for Drs Adalja, Singh, Topiwala, de Sousa, Ribeiro, Sorabjee and Dotiwala.
5. Regular correspondent to the local newspapers included Drs Adalja, Topiwala, de Sousa and Karve.
6. Robert G. Gregory, *India and East Africa: A History of Race Relations within the British Empire, 1890–1939*, Oxford, Clarendon Press, 1971, p. 217.
7. Colonial Office, Blue Book, 1910, Nairobi, East African Protectorate, 1910–11, p. Q1; Margaret Frenz, 'Representing the Portuguese Empire. Goan Consuls in British East Africa, c.1910–1963', in Eric Morier-Genoud and Michel Cahen (eds.), *Imperial Migrations: Colonial Communities and Diaspora in the Portuguese World*, New York, Palgrave Macmillan, 2012, pp. 193–212- for the use of Goan doctors, including Ribeiro, as representatives of the Portuguese government.
8. Errol Trzebinski, *The Kenya Pioneers*, London, Mandarin Paperbacks, 1991, p. 44
9. *The Official Gazette*, 1 December 1911, p. 605. According to Ribeiro's daughter medical qualification from Panjim, Goa. See Salvadori, *We Came in Dhows*, p. 22.
10. A reference has been made to two other Goan doctors who had worked in Zanzibar between 1870 and 1880 but they did not practice on the mainland. Salvadori, *We Came in Dhows*, p. 36.
11. Trzebinski, *The Kenya Pioneers*, p. 44; Photograph of Dr Ribeiro growing grapes in Nairobi, *East African Standard*, 5 April 1935, p. 25.
12. *Official Government Gazette*, 1 December 1911, p. 605; *Official Government Gazette*, 25 February 1920, p. 137.
13. Salvadori, *We Came in Dhows*, p. 127 and Vol. III, p. 104 [for photo].
14. *The Official Gazette*, 1 January 1932, p. 59.
15. *The Official Gazette*, 9 January 1940, p. 22.
16. K.V. Adalja, 'The Development of Medical Service in Kenya', *East African Medical Journal*, 39, 1962, pp. 105–14, 114.
17. Other sources consulted include: Obituary: R. Ribeiro, *The East African Medical Journal*, 28.1, 1951, p.91; No obituaries found for Lobo and Rodrigues. There is a very short reference to Lobo in Gregory, *South Asians*, p. 217 but no other material has been uncovered.
18. Salvadori, *We Came in Dhows*, p. 174.
19. For example: M.D. Gautama.
20. For example: J.J. Campos.
21. For example: M.S. Patel.

22. Anna Crozier, *Practising Colonial Medicine: The Colonial Medical Service in East Africa*, London, I.B. Tauris, 2007, p. 111.
23. Mridula Ramanna, 'Indian Doctors, Western Medicine and Social Change, 1845–1885', in Mariam Dossal and Ruby Maloni (eds.), *State Intervention and Popular Response: Western Indian in the Nineteenth Century*, Mumbai, Popular Prakashan Press, 1999, pp. 40–62.
24. Freni Sorabjee, interview in Salvadori, *We Came in Dhows*, p. 122. The only other possible exception is that of Edward Dias, it is unlikely, but possible that he was born in East Africa, where his father worked as pharmacist on Zanzibar at the end of the nineteenth century.
25. Crozier, *Practising Colonial Medicine*, pp. 97–101.
26. Mary de Sousa, Obituary, *The Goan Voice*, 18 July 1953, (personal clipping, no page number).
27. Mary de Sousa, Obituary, *The Goan Voice*, 18 July 1953, (personal clipping, no page number).
28. Nairobi Municipal Medical Report, 1930, p. 8, reports an Asian population of 16,000 in Nairobi; Cynthia Salvadori [Andrew Fedders (ed.)], *Through Open Doors: A View of Asian Cultures in Kenya*, Nairobi, Kenway Publications, 1989, p. 11 indicates the total population of Indians Kenya, according to the census, was 30,000 in 1926 and 44,000 in 1931; no separate figures for Mombasa are available.
29. Bashir Mauladad interview, Nairobi, 10 May 2006.
30. Teresa Albuquerque, *Goans of Kenya*, Mumbai, Michael Lobo Publishers, 1999, p. 37; email, Kersi Rustomji to Harshad Topiwala 21 January 2007 regarding the River Road Clinic; Email Dr Serosh Sorabjee to Harshad Topiwala, 24 April 2007.
31. Email, Kersi Rustomjee to Harshad Topiwala 21 January 2007.
32. Email, B. Sood to Harshad Topiwala, 18 January 2007.
33. Ramanna, 'Indian Doctors', p. 47.
34. Adalja, 'Thirty Two Years', *East African Medical Journal*, 1959, p. 442.
35. 'Obituary of G.V. Juvekar', *East African Medical Journal*, 41.10,1964, p. 496.
36. They met Adalja- with whom they formed a three-way partnership on arrival in Kenya. When Patel died, Topiwala and Adalja practised together. Hansa Topiwala interview, Seattle, 3 August 2007. The partnership included the ownership of a residential property in Blenheim Road is registered in *The Official Gazette*, 1 February 1938, p.108.
37. Gregory, *South Asians*, p. 18.
38. Freni Sorabjee, interview in Salvadori, *We Came in Dhows*, Paperchase Kenya, 1996, Vol.I, p. 122; Isak Dinesen, *Letters from Africa 1914–1931*, Chicago, University of Chicago Press, 1981, p. 313.
39. Edward Dias was the first to have this opportunity. Salvadori, *We Came in Dhows*, p. 126.
40. Angela Ribeiro, 'Doctor on a Zebra', in Salvadori (Ed.), *We Came in Dhows*, Paperchase Kenya, 1996, Vol.II, p. 22.
41. Albuquerque, *Goans of Kenya*, p. 38.
42. 'Obituary of G.V. Juvekar', *East African Medical Journal*, 41,1964, p. 496.
43. Email, Dr Manmeet Singh to Harshad Topiwala, 6 May 2008 (containing an unpublished tribute 'The Life and Times of the Late Dr Bakhtawar Singh').
44. Hansa Topiwala interview, Seattle, 3 August 2007.
45. RHL/Papers Collected by H. Topiwala Related to Indian Doctors in Kenya, expected deposit date 2015, privately published family tribute, *Dr A.C.L. de Sousa: In Memorium*, Nairobi, Majestic Printing works, 1959, pp. 10, 28; Gregory, *India and East Africa*, p. 218.

46. RHL/Papers Collected by H. Topiwala Related to Indian Doctors in Kenya, expected deposit date 2015, privately published family tribute, *Dr A.C.L. de Sousa: In Memorium*, Nairobi, Majestic Printing works, 1959, p. 28.
47. A.C.L. de Sousa, Legislative Council Minutes, 4 December, 1935, pp. 870–2.
48. Blue Book, 1903 and 1913, Nairobi, Kenya Government; 1903 List of Non European Medical Department staff confirms he joined in 1897, p. N105.
49. *The Official Gazette*, 11 January 1922, p.53, lists him Assistant Surgeon and *The Official Gazette*, 16 January 1923, p. 31 shows him being licensed as a Medical Practitioner.
50. Salvadori, *We Came in Dhows*, p. 174.
51. Email, Manmeet Singh to Harshad Topiwala, 6 May 2008.
52. S.D. Karve, *Who's Who in East Africa*, 1963–64, Nairobi, Marco Surveys, p. 24.
53. Anon, 'M.B. Gautma: Appreciation', *East African Medical Journal*, 36.3, 1959, p. 178; For information on Rana see Gregory, *South Asians*, p. 219.
54. Sana Aiyar, 'Anticolonial Homelands across the Indian Ocean: The Politics of the Indian Diaspora in Kenya, ca. 1930–1950', *American Historical Review*, 116.4, 2011, pp. 987–1013, 989.
55. Crozier, *Practising Colonial Medicine*, pp. 53–55.
56. Charles Jeffries, *Partners for Progress: The Men and Women of the Colonial Service*, London, 1949, p. 101.
57. RHL/Papers Collected by H. Topiwala Related to Indian Doctors in Kenya, expected deposit date 2015, privately published family tribute, *Dr A.C.L. de Sousa: In Memorium*, Nairobi, Majestic Printing works, 1959 includes several photographs and tributes from prominent individuals including a Minister and a former Mayor of Nairobi; see also Freni Sorabjee, interview in Cynthia Salvadori, *We Came in Dhows*, p. 122.
58. A.C.L. de Sousa was said to have loved to share his patient's joys and sorrows and, according to Dr J.R. Gregory, held a detailed knowledge of family history and illnesses in the local community. RHL/Papers Collected by H. Topiwala Related to Indian Doctors in Kenya, expected deposit date 2015, privately published family tribute, *Dr A.C.L. de Sousa: In Memorium*, Nairobi, Majestic Printing works, 1959, p.6, p.10; Adalja, 'Thirty Two Years', p. 445.
59. Hansa Topiwala interview, Seattle, 3 August 2007.
60. 'An Ordinance to Make Provision for the Registration of Medical Practitioners and Dentists', 27 September 1910, *The Official Gazette*, 1 October 1910, pp. 575–77, 577, Clause 18.
61. de Sousa attacked the behaviour of hakims. A.C.L. de Sousa, Legislative Council Minutes, 27 July, 1937, Vol. 1, p. 71; Adalja, 'Thirty Two Years', p. 444; John Iliffe, *East African Doctors: a History of the Modern Profession*, Cambridge University Press, 1998, p. 121.
62. NA/CO/822/38/4 Letter from Governor of Uganda to Secretary of State for the Colonies, 10 July 1931; 'Licensing Asiatic Doctors', *The Leader*, [Supplement], 25 March 1922, p.4 mentioned 'a European Legco member, Mr Conway Harvey, speaking strongly against a motion to allow Indian subordinate doctors licensing rights in the reserves as they kept "European doctors out of the country" '.
63. 'Geddes Considerations', *The Leader*, 22 April 1922, p.14, claimed 'the official doctor had practically wiped out the private practitioner' [in the provinces and outstations].
64. BMA/B/162/3/1/1 Minutes of the Dominions Committee, 1909–10, K.M. Dubas, 'Report of Other Action by Profession in Bombay' 26 January 1910.

65. BMA/B/162/1/12 Minutes Dominions Committee, 5 March 1925; see also: 'Medical Service in East Africa: BMA Stance to Stop the Withdrawal of Private Practice', *The Times*, 14 January 1926, p. 11.

66. NA/CO/822/38/4 Letter from Governor of Uganda to Secretary of State for the Colonies, 10 July 1931; NA/CO/533/681 Note regarding private practice from Governor of Kenya to Secretary of State, 30 October 1926.

67. RHL/Papers Collected by H. Topiwala Related to Indian Doctors in Kenya, expected deposit date 2015, privately published family tribute, *Dr A.C.L. de Sousa: In Memorium*, p. 37; 'Dr A.U. Sheth, Obituary', *East African Medical Journal*, 46.7, 1969, p. 448.

68. RHL/Papers Collected by H. Topiwala Related to Indian Doctors in Kenya, expected deposit date 2015, Privately published family tribute, *Dr A.C.L. de Sousa: In Memorium*, Nairobi, Majestic Printing works, 1959, p. 24.

69. Albuquerque, *Goans of Kenya*, p. 38.

70. 'Dr A.U. Sheth, Obituary', *East African Medical Journal*, 46.6, 1969, p. 448.

71. Gregory, *South Asians*, p. 218; Anon 'Mombasa Indian Girls', *The East African Standard*, 8 April 1933, p.31.

72. Mary de Sousa, Obituary, *The Goan Voice*, 18 July 1953, [Private newspaper cutting [no Page no.].

73. Adalja, 'Obituary', *East African Standard*, 21 July 1964, p. 387.

74. RHL/Papers Collected by H. Topiwala Related to Indian Doctors in Kenya, expected deposit date 2015, photo: 'Adarsha Vidyalaya-Nairobi: Members of the School Committee and Staff', 1956.

75. BL/IOR/L/PJ/6/295 File 206 contains a note of 1891 from Euan–Smith to Salisbury about Indian merchants philanthropic actions regarding construction of school and hospital; Iliffe, *East African Doctors*, p. 30; Salvadori, *We Came in Dhows*, p. 103

76. Iliffe, *East African Doctors*, p. 30; Salvadori, *We Came in Dhows*, p. 103.

77. Iliffe, *East African Doctors*, p. 30; Salvadori, *We Came in Dhows*, p. 103.

78. *Annual Medical Report*, 1941 p. 3.

79. *Annual Medical Report*, 1941, p. 4; post 1945 Karve played a pivotal role in the establishment of Pandya Memorial Hospital in Mombasa.

80. Rattansi Family Interview in Salvadori, *We Came in Dhows*, p. 44.

81. P.D. Patel, Correspondence 'Social Services League', *East African Standard*, 28 December 1935, p. 31; See Salvadori, *Through Open Doors*, p. 339 mentions the founding dates of SSL; Also see SSL dispensary photograph and information in R.O. Preston, *Oriental Nairobi*, Nairobi, Preston Publications, 1938, p. 318.

82. Patel, Correspondence 'Social Services League', p. 31.

83. Patel, Correspondence 'Social Services League', p. 31.

84. 'Obituary: Dr Krishnalal Vithadas on Adalja', *East African Medical Journal*, 41.8, 1964, p. 387; Hansa Topiwala interview, Seattle, 3 August 2007.

85. Patel, Correspondence 'Social Services League', p. 31.

86. (For Dotiwala) Email, K Rustomji to Harshad Topiwala, 21 January 2007; (for M.N.Hamin) see 'Obituary,' *The East African Medical Journal*, 36.1, 1959, p. 56.

87. Gregory, *South Asians*, p. 219.

88. KNA/GH/7/3 'Indian Association and Indian Political Movements' speech by Dr Adalja regarding the history of the Indian Maternity Home, 5 December 1942.

89. RHL/MSS. Brit.Emp.s 400/Box 136/1 East Africa 'Nursing in Kenya', The Lady Grigg Welfare League pamphlet, Nairobi, January 1927.

90. Mary de Sousa, Obituary, *The Goan Voice*, 18 July 1953, [personal newspaper clipping, no page number].

91. KNA/GH/7/3 'Indian Association and Indian Political Movements', Correspondence Related to Lady Grigg Welfare League, December 1942, pp. 160–71.
92. RHL/Papers Collected by H. Topiwala Related to Indian Doctors in Kenya, expected deposit date 2015, privately published family tribute, *Dr A.C.L. de Sousa: In Memorium*, Nairobi, Majestic Printing works, 1959, p. 21.
93. Ribeiro, 'Doctor on a Zebra', p. 22.
94. Crozier, *Practising Colonial Medicine*, p. 88.
95. RHL/Papers Collected by H. Topiwala Related to Indian Doctors in Kenya, expected deposit date 2015, privately published family tribute, *Dr A.C.L. de Sousa: In Memorium*, Nairobi, Majestic Printing works, 1959, pp. 26–7.
96. Adalja, 'Thirty Two Years', p. 444.
97. Letter from Hansa Topiwala to Harshad Topiwala, 26 June 2007.
98. Letter from Hansa Topiwala to Harshad Topiwala, 26 June 2007.
99. Mina Madhani, interview, Seattle, 4 August 2007.
100. 'Obituary of G.V. Juvekar', *East African Medical Journal*, 41.10, 1964, p. 496.
101. Email, Dr Manmeet Singh to Harshad Topiwala, 6 May 2008 (containing an unpublished tribute 'The Life and Times of the Late Dr Bakhtawar Singh').
102. Letter from Hansa Topiwala to Harshad Topiwala, 26 June 2007.
103. Email, K.Rustomji to Harshad Topiwala, 21 January 2007.
104. Ribeiro, 'Doctor on a Zebra', p. 22.
105. Ribeiro, 'Doctor on a Zebra', p. 22; Trzebinski, *The Kenya Pioneers*, p. 44.
106. 'BMA Branch Meetings', *East Africa Medical Journal*, 11.3, 1934, p. 103.
107. RHL/Papers Collected by H. Topiwala Related to Indian Doctors in Kenya, expected deposit date 2015, K.V. Adalja, 'Jain Juvkona' (in Gujarati), *Jyotna Magazine*, Nairobi, Regal Press, c.1940 p. 29; RHL/Papers Collected by H. Topiwala Related to Indian Doctors in Kenya, expected deposit date 2015, H.T. Topiwala, 'Mulykno' (in Gujarati), *Jyotna Magazine*, c.1940 p. 33. Dr Adalja also referred to contributing articles to the magazine *Orient*: Adalja, 'Thirty Two Years', p. 444.
108. Letter Sudha Young to Harshad Topiwala, 15 July 2007.
109. D. Seidenberg, *Mercantile Adventurers*, Delhi, New Age International Ltd., 1996, p.54.
110. Mary de Sousa, 'The Treatment of Erosion of Cervix with Picric Acid', *Kenya Medical Journal*, Vol.1.9, 1924, p. 281.
111. S.D. Karve, 'An Experiment in Midwifery', *East African Medical Journal*, 10, 1933–34, pp. 358–63, 358; S.D. Karve, 'Some Indian Methods of Midwifery', *East African Medical Journal*, 11, 1934–35, pp. 286–7, 286.
112. K.V. Adalja, 'Intramuscular versus Intravenous Quinine', *The Indian Medical Gazette*, September 1935, p. 538; K.V. Adalja, 'Ayurvedic Midwifery', *East African Medical Journal*, 17, 1940, p. 142; [for reference to Adalja's arguments about the benefits of Ayurvedic medicine] Andrew Hicks, 'Forty Years of the British Medical Association (Kenya Branch)', *East African Medical Journal*, 38.1, 1961, pp. 43–53, 46]; K.V. Adalja, 'Acquired Haemophilia', *British Medical Journal*, 10 April 1937, V1, p. 785.
113. G. Najmudean, 'Ocular Manifestations in Syphilis', *East African Medical Journal*, 8, 1931–2, pp. 316–22, 316.
114. K.V. Adalja, 'Haematuria following M and B 693', *British Medical Journal*, 10 April 1937, p. 643.
115. Email, K. Rustomji to Harshad Topiwala, 21 January 2007.
116. Letter, Sudha Young to Harshad Topiwala, 15 July 2007.
117. Adalja, 'Thirty Two Years', p. 445.
118. Email, B. Sood to Harshad Topiwala, 18 January 2007.

119. Ayres Ribeiro, 'Multicultural General Practice in Nairobi', pp. 979–81, 979.
120. Ribeiro, 'Multicultural General Practice', p. 979.
121. Karve, 'An Experiment in Midwifery', p. 358.
122. Karve, 'An Experiment in Midwifery', p. 358.
123. Adalja, 'Thirty Two Years', p. 444.
124. Karve, 'Some Indian Methods of Midwifery', p. 286.
125. Karve, 'Some Indian Methods of Midwifery', p. 286.
126. Hicks, 'Forty Years', p. 46.
127. Ribeiro, 'Multicultural General Practice', p. 979.
128. Karve, 'An Experiment in Midwifery', p. 358 explains his midwifery experiment was given a trial by the 'progressive and go-ahead' community of Khojas; Dr Ribeiro talks of the 'westernised' Goans and distinguishes between the 'medically well off' Khojas and the not-so-advanced 'orthodox Muslims' whose women folk were mostly confined indoors, See Ribeiro, 'Multicultural General Practice', p. 971.
129. Ribeiro, 'Multicultural General Practice', p. 979.
130. Hansa Topiwala interview, Seattle, 3 August 2007.
131. Mary de Sousa, Obituary, *The Goan Voice*, 18 July 1953, [personal newspaper clipping, no page number].
132. Kalavati Topiwala, interview Seattle, 3 September 1992.
133. Crozier, *Practising Colonial Medicine*, pp. 115–32.
134. www.lodge of Nairobi.org (Accessed 15 June 2013) Website of Nairobi Free Mason Lodge No 7187 (established 1952).
135. RHL/Papers Collected by H. Topiwala Related to Indian Doctors in Kenya, expected deposit date 2015, Neta Peal, 'Mombasa Honours Late Dr S.D. Karve', *Coastweek* [Mombasa newspaper], 24 March 2001 [no page number, newspaper clipping].
136. Gregory, *India and East Africa*, p. 219.
137. Salvadori, *Through Open Doors*, p. 269, 313, 340.
138. Email S. Sorabjee to Harshad Topiwala, 24 January 2007 refers to Topiwala and E. Sorabjee at the S.V Gymkhana Club; RHL/Papers Collected by H. Topiwala Related to Indian Doctors in Kenya, expected deposit date 2015, photograph of tennis players with trophies, untitled shows Adalja, Topiwala and others.
139. Salvadori, *We Came in Dhows*, p. 123 ref to a 1977 cutting from a special anniversary magazine of the Aero Club of East Africa.
140. Email, B. Sood to Harshad Topiwala, 18 January 2007.
141. A.D. Charters, 'Correspondence: Dr K.V. Adalja', *East African Medical Journal*, 41.10, 1964, p. 496; Hicks, 'Forty Years', pp. 43–53, 46; RHL/Papers Collected by H. Topiwala Related to Indian Doctors in Kenya, expected deposit date 2015, privately published family tribute, *Dr A.C.L. de Sousa: In Memorium*, Nairobi, Majestic Printing works, 1959, pp. 10, 26 (for Dr Gregory and Anderson comments on de Sousa).
142. 'R.A. Ribeiro, Obituary', *East African Medical Journal*, 28.1, 1951, p. 91.
143. Da Gama in Salvadori, *We Came in Dhows*, p. 127 refers to 3 October 1903 newspaper article 'International Dispensary mentions energy and affability'; 'Obituary of G.V. Juvekar', *East African Medical Journal*, 41.10, 1964, p. 496; [Reference to Dotiwala], Email, K. Rustomjee to Harshad Topiwala 21 January 2007; RHL/Papers Collected by H. Topiwala Related to Indian Doctors in Kenya, expected deposit date 2015, privately published family tribute, *Dr A.C.L. de Sousa: In Memorium*, Nairobi, Majestic Printing works, 1959, pp. 19, 26; 'Dr A.U. Sheth, Obituary', *East African Medical Journal*, 46.7, 1969, p. 448; Adalja, 'Obituary', *East African Standard*, 21 July 1964, p. 387; RHL/Papers Collected by

H. Topiwala Related to Indian Doctors in Kenya, expected deposit date 2015, Neta Peal, 'Mombasa Honours Late Dr S.D. Karve', *Coastweek* [Mombasa newspaper], 24 March 2001 [no page number, newspaper clipping]; Anon, 'M.B. Gautama: Appreciation', *East African Medical Journal*, 36.3, 1959, p. 178.

144. 'Dr Rozendo Ayres Ribeiro, Obituary', *The East African Medical Journal*, 28.1, 1951, p. 91; Sood tribute in Gregory, *South Asians*, Westview Press Inc., 1993, p. 219.
145. 'Dr M.N. Hamin, Obituary', *The East African Medical Journal*, 36.1, 1959, p. 56.
146. *Who's Who in East Africa 1963–64*, Nairobi, Marco Surveys, 1964, p. 52.
147. Email, K. Rustomji to Harshad Topiwala, 21 January 2007.
148. RHL/Papers Collected by H. Topiwala Related to Indian Doctors in Kenya, expected deposit date 2015, privately published family tribute, *Dr A.C.L. de Sousa: In Memorium*, Nairobi, Majestic Printing works, 1959.
149. Email, B. Sood to Harshad Topiwala, 18 January 2007.
150. 'M.B. Gautama: Appreciation', *The East African Medical Journal*, 36.3, 1959, p. 178.
151. Salvadori, *We Came in Dhows*, Paperchase Kenya Ltd, 1996, Vol.III, p.175 contains the excerpt from the book Vivian Prince, *Kenya, the Years of Change*.
152. Hansa Topiwala interview, Seattle, 3 August 2007.
153. Email, B. Sood to Harshad Topiwala, 18 January 2007; 28 January 2007.
154. RHL/Papers Collected by H. Topiwala Related to Indian Doctors in Kenya, expected deposit date 2015, privately published family tribute, *Dr A.C.L. de Sousa: In Memorium*, Nairobi, Majestic Printing works, 1959, p.29; RHL/Papers Collected by H. Topiwala Related to Indian Doctors in Kenya, expected deposit date 2015, Neta Peal, 'Mombasa Honours Late Dr S.D. Karve', *Coastweek* [Mombasa newspaper], 24 March 2001 [no page number, newspaper clipping]; Email, B. Sood to Harshad Topiwala, 18 January 2007; 28 January 2007; Email, K. Rustomji to Harshad Topiwala, 28 January 2007.

8 Private Doctors: Practising Medicine in a Segregated World

1. 'With Regard to Europeans', *Annual Medical Report*, Kenya Colony, 1934, p. 9.
2. 'Racial Discrimination and Science in Nineteenth Century India', *Indian Economic Social History Review*, 19, 1982, pp. 63–82.
3. R.N. Shah, 'Nairobi Hospitals', *East African Standard*, 2 August 1930, p. 34; also see B.S. Varma in Legislative Council Minutes, 23 October 1922, p. 11, for complaint of inadequate medical relief despite the high tax burden for Indians.
4. *Annual Medical Report*, 1908, p. 39.
5. M.A. Desai, Minutes of the Kenya Legislative Council, 16 December 1925, p.1106. I. Sham-su-deen, Legislative Council Minutes, 2 July 1935, p.151.
6. Quoted in John Iliffe, *East African Doctors: A History of the Modern Profession*, Cambridge University Press, 1998, p. 30.
7. WL/CMAC/PP/HCT/A5 Elizabeth Bray, *Hugh Trowell: Pioneer Nutritionist*, unpublished biography, London 1988, p. 4.
8. J.M. Campos, 'The Scandal of Nairobi Township', *East African Chronicle*, 16 October 1920, p. 10.
9. 'Nairobi, European Hospital', *The East African Standard* 18 April 1931, p. 49.
10. Shah, letter 'Nairobi Hospitals', p. 34.
11. J.H. Sequeira, 'Grouped Hospitals, Attitude of the Medical Profession' rebuts the charge of risk of infection due to contamination', [letter to Editor], *The East African Standard*, 3 June 1930, p.42.
12. *Annual Medical Report*, 1937, p. 8.
13. *Annual Medical Report*, 1911, p. 13; *Annual Medical Report*, 1913, p. 62.

14. Shah, 'Nairobi Hospitals', p. 34.
15. Angela Ribeiro, 'Doctor on a Zebra' in Cynthia Salvadori, *We Came in Dhows*, Nairobi, Paperchase Kenya Ltd, 1996, Vol. II, p. 23.
16. Salvadori, *We Came in Dhows*, Paperchase Kenya Ltd, 1996, Vol. III, p.175.
17. A. Hicks, 'Historical Note', *Nairobi Hospital Proceedings*, Vol. 1, 1997, p.69.
18. Hicks, 'Historical Note', p.71.
19. 'Hospitals and Dispensaries', *Annual Medical Report*, 1911, p. 13; *Annual Medical Report*, 1937, p. 3.
20. Shah, 'Nairobi Hospitals', p. 34.
21. Shah, 'Nairobi Hospitals'0, p. 34.
22. Robert G. Gregory, *South Asians in In East Africa: An Economic And Social History 1890–1990*, Oxford, Westview Press, 1993, p. 35.
23. A.R. Paterson, PMO Response to Group Hospital Questions, Legislative Council Minutes 4 March 1937, 1, p. 42.
24. J.B. Pandya and M.A. Desai, Kenya Legislative Council Minutes, 15 December 1925, pp. 1083–85.
25. K.V. Adalja, 'Thirty Two Years in General Practice in Nairobi', *East African Medical Journal*, 36, 1959, pp. 442–8, 443.
26. K Adalja, 'Thirty Two Years', p. 443; Also RHL/Papers Collected by H. Topiwala Related to Indian Doctors in Kenya, expected deposit date 2015, Topiwala unpublished speech to Indian Medical Association, 1943.
27. Adalja, 'Thirty Two Years', p.444.
28. RHL/Papers Collected by H. Topiwala Related to Indian Doctors in Kenya, expected deposit date 2015, privately published family tribute, *Dr A.C.L. de Sousa: In Memorium*, Nairobi, Majestic Printing works, 1959, p. 21.
29. RHL/Papers Collected by H. Topiwala Related to Indian Doctors in Kenya, expected deposit date 2015, privately published family tribute, *Dr A.C.L. de Sousa: In Memorium*, Nairobi, Majestic Printing works, 1959, p. 21.
30. J.R. Gregory, *Under the Sun (A Memoir of Dr R.W. Burkitt of Kenya)*, Nairobi, English Press, 1951, p. 84; Adalja, 'Thirty Two Years', p. 447.
31. Isak Dinesen, *Letters from Africa 1914–1931*, Chicago, University of Chicago Press, 1981, p. 314. It should be noted that Dr Sorabjee was an UK educated Parsee.
32. Hansa Topiwala interview, Seattle, 3 August 2007.
33. Email, K Rustomji to Harshad Topiwala, 21 January 2007.
34. Adalja, 'Thirty Two Years', p. 444.
35. Hansa Topiwala interview, Seattle, 3 August 2007.
36. RHL/Papers Collected by H. Topiwala Related to Indian Doctors in Kenya, expected deposit date 2015, privately published family tribute, *Dr A.C.L. de Sousa: In Memorium*, Nairobi, Majestic Printing works, 1959, p. 27; Isak Dinesen, *Letters from Africa 1914–1931*, Chicago, University of Chicago Press, 1981, p. 314; Email, Manmeet Singh to Harshad Topiwala, 6 May 2008; Angela Ribeiro, 'Doctor on a Zebra' in Salvadori, *We Came in Dhows*, p. 23; Ayres Ribeiro, 'Multicultural General Practice in Nairobi', *British Medical Journal*, 23 October 1954, pp. 979–81, 979.
37. Hansa Topiwala interview, Seattle, 3 August 2007. In this interview she recalled how a couple of European patients during the thirties depression period had paid their medical fees in paintings and ivory carvings.
38. Ribeiro, 'Multicultural General Practice', p. 981.
39. Hansa Topiwala interview, Seattle, 3 August 2007; There is evidence that some Indian doctors indulged in unfair/unprofessional ways to make money-Sood, for example, mentions the use of touts to meet prospective patients at Kisumu bus stops (Emails, B. Sood to Harshad Topiwala, 18 January 2007; 28 January 2007)

and Trowell talks of promoting 'single injection cures' and how doctors (both Indian and European) competed for patients. See H.C. Trowell, 'Presidential Address of the Conference of Physicians', *East African Medical Journal*, 1955, p. 433

40. Hicks, 'Historical Note', p. 69.

41. Ribeiro, 'Multicultural General Practice', p. 980; RHL/Papers Collected by H. Topiwala Related to Indian Doctors in Kenya, expected deposit date 2015, privately published family tribute, *Dr A.C.L. de Sousa: In Memorium*, Nairobi, Majestic Printing works, 1959, p. 27; Hansa Topiwala interview, Seattle, 3 August 2007.

42. Email, B. Peirara to Harshad Topiwala, 6 January 2007 with extracts from unpublished manuscript E.H. Pereira, *Trail Blazer's Century*.

43. Ribeiro, 'Multicultural General Practice', p. 980; RHL/Papers Collected by H. Topiwala Related to Indian Doctors in Kenya, expected deposit date 2015, privately published family tribute, *Dr A.C.L. de Sousa: In Memorium*, Nairobi, Majestic Printing works, 1959, p. 27; Hansa Topiwala interview, Seattle, 3 August 2007.

44. Ribeiro, 'Multicultural General Practice', p. 980.

45. Examples include: John A. Carman, *A Medial History of the Colony and Protectorate of Kenya: A Personal Memoir*, London, Rex Collings, 1976; Gregory, *Under the Sun*

46. RHL/Papers Collected by H. Topiwala Related to Indian Doctors in Kenya, expected deposit date 2015, letters, H. Trowell to H.T. Topiwala, 13 April and 26 April 1953 [concerning treatment of Mr Vaz who suffered from Rheumatoid Arthritis].

47. Letter, Sudha Young to Harshad Topiwala, 15 July 2007.

48. Roger Jeffery, 'Doctors and Congress: The Role of Medical Men and Medical Politics in Indian Nationalism', in Mike Shepperdson and Colin Simmons (eds.), *The Indian National Congress and the Political Economy of India, 1885–1985*, Aldershot, Brookfield, USA, Avebury, 1988, pp. 160–73, 160.

49. Aiyar would argue that Indian nationalism also called for common cause with African nationalist movements. Sana Aiyar, 'Empire, Race and the Indians in Colonial Kenya's Contested Public Political Sphere, 1919–1923', *Africa: The Journal of the International African Institute*, 81, 1, 2011, pp. 132–54; Cross-racial anti-colonial nationalism became even more evident between 1930–1950. See Sana Aiyar, 'Anticolonial Homelands Across the Indian Ocean: The Politics of the Indian Diaspora in Kenya, ca. 1930–1950', *American Historical Review*, 116.4, 2011, pp. 987–1013.

50. Zarina Patel, *Challenge to Colonialism: The Struggle of Alibhai Mulla Jeevanjee for Equal Rights in Kenya*, Nairobi, Publisher Distributon Services, 1997, p. 152, 181; D.A. Seidenberg, *Uhuru and the Kenya Indians: The Role of a Minority Community in Kenya Politics, 1939–1963*, New Delhi, Vikas Publishing House, 1983, p. 19; also see the EAINAC and the CMA Muslim Central Association split with Dr Rana: S. Aiyar, 'Anticolonial Homelands across the Indian Ocean: The Politics of the Indian Diaspora in Kenya, ca. 1930–1950', *American Historical Review*, 116.4, pp. 987–1013.

51. Salvadori, *We Came in Dhows*, p. 104.

52. Gregory, *South Asians*, p. 222; RHL/Papers Collected by H. Topiwala Related to Indian Doctors in Kenya, expected deposit date 2015, privately published family tribute, *Dr A.C.L. de Sousa: In Memorium*, Nairobi, Majestic Printing works, 1959, p. 29. *Who's Who in East Africa, 1963–64*, Nairobi, Marco Surveys, pp. 24, 52 has entries for Karve and Patwardhan.

53. Salvadori, *We Came in Dhows*, p. 105.

54. Anna Crozier, *Practising Colonial Medicine: The Colonial Medical Service in East Africa*, London, I.B. Tauris, 2007, pp. 89–90.

55. A.R. Paterson, PMO response to Group Hospital questions, Legislative Council minutes, 4 March 1937, Vol. 1, p. 42; see also Legislative Council Minutes 30

October 1922 for the treasurers evasive response to B.S. Varma's question on relative medical expenditure on different races. Nairobi Council debate on Simpson Report recommendation on segregation, 'The Asiatic Question', *The Leader*, 16 January 1915 p. 4.

56. 'Nairobi Municipal Committee', *The Official Gazette*, 17 January 1917, p. 62; BL/IOR/L/E/7/1330 Nairobi Municipality Committee, Letter 15 April 1924; Notably, Phadke, was evidently not quite the governments man that the governor had hoped for. He resigned in 1921, a couple of months after Visram, from the Legislative council over the policy of township segregation, *The East African Chronicle*, 29 January 1929, p. 5.

57. Isher Dass and Pandya asked for an inquiry into why Dr Sheth had been charged. See Legislative Council Minutes, 27 June 1935, p. 299.

58. *Official Gazette*, 9 December 1925, p. 1228.

59. 'Karve -Unofficial Member of the Executive Council', *Official Gazette*, 9 February 1937, p. 131.

60. Gregory, *South Asians*, p.219; Karve nomination to the Legislative Council reported in: *The London Gazette*, 13 June 1952; 'Karve, S.D', *Who's Who in East Africa*, 1963–64, Nairobi, Marco Surveys Ltd., p. 24.

61. 'List of Members', Minutes of Legislative Council debates, 2 June 1931.

62. *Official Gazette*, 26 June 1934, p. 846; *Official Gazette*, 8 June 1948, p. 377; Dr A.U. Sheth, 'Appreciation', *East African Medical Journal*, 46.7, 1969, p. 448.

63. M.A. Rana, *Who's Who in East Africa*, 1967–68, Nairobi, Marco Surveys Ltd, p. 148

64. K.V. Adalja, Obituary, *East African Medical Journal*, 41.8, 1964, p. 387; J. Stuart, *A Jubilee History of Nairobi*, Nairobi, East African Standard, 1950, p.24.

65. Gregory, *South Asians* p. 219; Stuart, *A Jubilee History of Nairobi*, p. 24.

66. *The Official Gazette*, 19 July 1939 records the Municipality Council appointment; 'BMA Meeting' [recorded congratulations to Ismail and Adalja on being nominated to serve on Kenya Legislative Council] *The East African Medical Journal*, 36.6, 1959, p. 338; *Who's Who in East Africa: 1963–64*, Nairobi, Marco Surveys Ltd., p. 21.

67. M.D. Gautama, 'Obituary', *The East African Medical Journal*, 36.3, 1959, p. 178.

68. 'Patwardhan, V.V.', *Who's Who in East Africa, 1963–64*, Nairobi, Marco Surveys Ltd, p. 52; Gregory, *South Asians*, p. 222.

69. 'R.A. Ribeiro, 'Obituary', *East African Medical Journal*, 28.1, 1951, p. 91.

70. 'The Asiatic Question', *The Leader*, 16 January 1915, p. 4.

71. NA/CO/533/156 Letter from the Governor to Secretary of State, 22 September 1915

72. BL/IOR/L/E/7/1263 A. Jeevanjee, 'Sanitation in Nairobi', May 1915.

73. RHL/Papers Collected by H. Topiwala Related to Indian Doctors in Kenya, expected deposit date 2015, privately published family tribute, *Dr A.C.L. de Sousa: In Memorium*, Nairobi, Majestic Printing works, 1959, p. 29.

74. RHL/Papers Collected by H. Topiwala Related to Indian Doctors in Kenya, expected deposit date 2015, privately published family tribute, *Dr A.C.L. de Sousa: In Memorium*, Nairobi, Majestic Printing works, 1959, p. 29.

75. J.S. Mangat, *A History of the Asians in East Africa, c.1886–1945*, Oxford University Press, 1969, p. 20.

76. RHL/Papers Collected by H. Topiwala Related to Indian Doctors in Kenya, expected deposit date 2015, privately published family tribute, *Dr A.C.L. de Sousa: In Memorium*, Nairobi, Majestic Printing works, 1959, p. 34.

77. A.C.L. de Sousa, 'Questions and Comments', Minutes of the Legislative Council, 2, 1937, p.71, p.252, p.597.

78. RHL/Papers Collected by H. Topiwala Related to Indian Doctors in Kenya, expected deposit date 2015, privately published family tribute, *Dr A.C.L. de Sousa: In Memorium*, Nairobi, Majestic Printing works, 1959, p.20.
79. Legislative Council Minutes, 4 December 1935, p. 871; 8 January 1936, pp. 1128–30.
80. RHL/Papers Collected by H. Topiwala Related to Indian Doctors in Kenya, expected deposit date 2015, privately published family tribute, *Dr A.C.L. de Sousa: In Memorium*, Nairobi, Majestic Printing works, 1959, p. 22.
81. RHL/Papers Collected by H. Topiwala Related to Indian Doctors in Kenya, expected deposit date 2015, privately published family tribute, *Dr A.C.L. de Sousa: In Memorium*, Nairobi, Majestic Printing works, 1959, p. 11.
82. M. Nazareth, *Brown Man, Black Country*, Delhi, Tidings Publication, 1981, p. 45.
83. BL/IOR/L/E/7/1330 'Indians in Kenya', Karve, 'Report of the Kenya Local Government Commission', 12 August 1927.
84. BL/IOR/L/E/7/1330 'Indians in Kenya', Karve, 'Report of the Kenya Local Government Commission', 12 August 1927.
85. BL/IOR L/E/7/1339 Enclosures of the letter from Birkenhead to Governor General of India, Police inspector report, Coryndon to Secretary of State, 22 January 1925.
86. Patel, *Challenge to Colonialism*, Publishers Distribution Services 1997, p. 148; Robert G. Gregory, *India and East Africa: A History of Race Relations within the British Empire, 1890–1939*, Oxford, Clarendon Press, 1971, p. 453 [also describes Karve as a moderate]; See also D.A. Seidenberg, *Uhuru and the Kenya Indians: The Role of a Minority Community in Kenya Politics, 1939–1963*, New Delhi, Vikas Publishing House, 1983, p. 19.
87. BL/IOR/L/E/7/1330 cutting from *The East African Standard*, 14 January1929 [on the moderates position as per J.B. Pandya, on participation in Municipalities opposed by B.S. Varma].
88. BL/IOR/L/E/7/1330 Letter from A.B. Patel to EAINC, 8 February 1931.
89. 'The Commissions of Inquiry Ordinance', *The Kenya Gazette*, 22 August 1939, p. 1148.
90. BL/IOR/L/E/7/1330 'Indians in Kenya', Karve, 'Report of the Kenya local Government Commission', 12 August 1927 as Secretary of EAINC to the Secretary of State. See also his contribution in a 1939 Legislative Council session on the white highlands, Legislative Council Minutes, 21 April 1939, p. 302.
91. BL/IOR/L/E/7/1330 'Indians in Kenya', Karve, 'Report of the Kenya Local Government Commission', 12 August 1927.
92. BL/IOR/L/E/7/1330 'Indians in Kenya', Karve, 'Report of the Kenya Local Government Commission', 12 August 1927.
93. Maureen Malowany, *Medical Pluralism: Disease, Health and Healing in the Coast of Kenya, 1840–1940*, Ph.D Thesis, McGill University, Canada, 1997, p. 195.
94. D.G. Tendulkar, *Mahtma: Life of Mohandas Karamchand Gandhi*, Bombay, V.K. Jhaveri and D.G Tendulkar, 1951, Vol. 2, p. 13.
95. RHL/Papers Collected by H. Topiwala Related to Indian Doctors in Kenya, expected deposit date 2015, H.T. Topiwala, 'Indians in Kenya: The Highlands', Letter to the Editor, *The East African Standard*, 12 May 1939, p. 9; RHL/Papers Collected by H. Topiwala Related to Indian Doctors in Kenya, expected deposit date 2015, H.T. Topiwala, draft letter to editor *East African Standard*, 16 March 1937 [cutting, no page no.].
96. For example see collection at RHL/Papers Collected by H. Topiwala Related to Indian Doctors in Kenya, expected deposit date 2015, including: H.T. Topiwala, 'State Medical Service', *The East African Standard*, 21 June 1943; H.T. Topiwala, 'State Medical Service', 6 July 1943; H.T. Topiwala, 'State Medical, 26 July 1943;

H.T. Topiwala, 'Gymkhana Fete and Aftermath', *Colonial Times*, 30 September 1943; H.T. Topiwala, 'Indian Politics', *East African Standard*, 15 August 1943 [cuttings with no page no.].

97. RHL/Papers Collected by H. Topiwala Related to Indian Doctors in Kenya, expected deposit date 2015, H.T. Topiwala, letter to editor, *The East African Standard*, 9 July 1943 [cutting no page number]. This letter refers to an earlier letter by another correspondent that made uncomplimentary personal remarks.

98. K.V. Adalja, 'The Nairobi Swamp', *East African Standard*, 29 September 1934, p. 34.

99. Adalja, 'The Nairobi Swamp', p. 34.

100. A.C.L. de Sousa, 'Malaria in Nairobi', *The East African Standard*, 1 September 1934, p. 36; See also K.V. Adalja comments within 'Report of the Emergency Meeting of the Town Council: The Malaria Menace to Nairobi', *The East African Standard*, 3 May 1935, p.33 which reports on the emergency meeting of the council where Adalja blames the government's prior agreement with the owners of the land for the inaction.

101. Adalja, 'The Nairobi Swamp', p. 34.

102. R. Ward, 'The Nairobi Swamp', *The East African Standard*, 10 May 1935, p. 9; see also Ward, 'The Nairobi Swamp', p. 34.

103. 'Report of the Emergency Meeting of the Town Council: The Malaria Menace to Nairobi', *The East African Standard*, 3 May 1935, p.33.

104. 'Report of the Emergency Meeting of the Town Council: The Malaria Menace to Nairobi', *The East African Standard*, 3 May 1935, p.33.

105. E. Grogan, 'Malaria in Nairobi', 3 May 1935, p.38.

106. RHL/Papers Collected by H. Topiwala Related to Indian Doctors in Kenya, expected deposit date 2015, privately published family tribute, *Dr A.C.L. de Sousa: In Memorium*, Nairobi, Majestic Printing works, 1959, p.15; p.27.

107. RHL/Papers Collected by H. Topiwala Related to Indian Doctors in Kenya, expected deposit date 2015, privately published family tribute, *Dr A.C.L. de Sousa: In Memorium*, Nairobi, Majestic Printing works, 1959, p.24; Notably not all Goans were united behind the Goan institute. The Goan community itself was split between those who identified their origins with the Brahamin or the Chaddo villages of Goa. Members of Goan Institute tended to come from Chaddo, whereas Brahamins joined the rival Goan Gymkhana instead. See Donna Nelson, 'Problems of Power in a Plural Society: Asians in Kenya', *Southwestern Journal of Anthropology*, 28.3, 1972, pp. 255–62, 258.

108. RHL/Papers Collected by H. Topiwala Related to Indian Doctors in Kenya, expected deposit date 2015, privately published family tribute, *Dr A.C.L. de Sousa: In Memorium*, Nairobi, Majestic Printing works, 1959, pp. 9, 25.

109. Email, K Rustomji to Harshad Topiwala, 21 January 2007.

110. Gregory, *South Asians*, p. 219; KNA/GH/7/3 'Indian Association and Indian Political Movements', letter from Dr Rana to Chairman of the Indian Man-Power Committee, Mombasa, 26 May 1942.

111. Gregory, *South Asians*, p. 222.

112. RHL/Papers Collected by H. Topiwala Related to Indian Doctors in Kenya, expected deposit date 2015, 'New Building of the Surat District Association', *The East African Standard*, 13 July 1943 [cutting no page number].

113. http://unitedkenyaclub.com/ukc/history.htm (accessed on 8 August 2013).

114. For the congress resolution of 1920 that recommended returning honours and titles see: Tendulkar, *Mahtma*, Jhaveri and Tendulkar, 1951, Vol. 2, p. 13.

115. Salvadori, *We Came in Dhows*, p. 23.

116. Dr A.U. Sheth, 'Appreciation', *East African Medical Journal*, 46.7, 1969, p. 448.

117. Gregory, *South Asians*, p. 219; M.A. Rana, *Who's Who in East Africa*, 1967–68, Nairobi, Marco Surveys Ltd, p. 148.
118. 'Nairobi Doctor Dies at 63', *The East African Standard*, 21 July, 1964, p. 5.
119. *The Kenya Gazette*, 15 June 1948, p. 384.
120. 'The Association and the Colonies', *British Medical Journal*, i, 1905, p. 1394.
121. Andrew Hicks, 'Forty Years of the British Medical Association (Kenya Branch)', *The East African Medical Journal*, 38.1,1961, pp. 43–53.
122. Hicks, 'Forty Years', p. 44.
123. Crozier, *Practising Colonial Medicine*, I.B. Tauris, 2007, pp. 96–7.
124. 'BMA Branch Meetings Report', *East African Medical Journal*, 11.3, 1934, p. 103; 'BMA, Kenya Branch Annual General Meeting', *East African Medical Journal*, 11.10, 1935, pp.365–367, 365.
125. BMA/B/162/1/4 Dominions Committee Minutes, Alteration of Article IV, discussion related to letter [24 November 1913] from Dr Watt 5 January 1913, p. 2.
126. BMA/B/162/1/4 Dominions Committee Minutes, Alteration of Article IV, discussion related to letter [24 November 1913] from Dr Watt 5 January 1913, p. 2.
127. BMA/B/162/1/4 Dominions Committee Minutes, Alteration of Article IV, discussion related to letter [24 November 1913] from Dr Watt 5 January 1913, p. 2.
128. BMA/B/162/1/4 Dominions Committee Minutes, Alteration of Article IV, discussion related to letter [24 November 1913] from Dr Watt 5 January 1913, p. 2.
129. BMA/B/162/1/9 Minutes of Dominions Committee, 30 June 1922, p. 2; Iliffe, *East African Doctors*, p. 124.
130. BMA/B/162/1/4, Dominions Committee Minutes, Alteration of Article IV, discussion related to letter [24 November 1913] from Dr Watt, 5 January 1913, p. 2.
131. BMA/B/162/1/22/ 16 May 1935, pp. 23–5; For background also see BMA/B/162/1/4 Dominions Committee Minutes, Alteration of Article IV, discussion related to letter [24 November 1913] from Dr Watt, 5 January 1913, p. 2.
132. 'British Medical Association, Branch Meetings', *East African Medical Journal*, 11.3, 1934, p. 103.
133. 'British Medical Association, Kenya Branch Annual General Meeting', *East African Medical Journal*, 11.10, 1935, p. 365. Hicks states that Indians were admitted in 1928 and John Iliffe states 1934 neither statement is consistent with the records. See Hicks, 'Forty Years'; Iliffe, *East African Doctors*, p. 124.
134. 'British Medical Association, Kenya Branch, Annual General Meeting', *East African Medical Journal*, 11.10, 1935, pp. 365–367, 365.
135. Iliffe, *East African Doctors*, Cambridge University Press, 1998, p.124.
136. 'British Medical Association, Kenya Branch, Annual General Meeting', *East African Medical Journal*, 11.10, 1935, pp. 365–367, 365.
137. [For Adalja] 'British Medical Association Meeting', *East African Medical Journal*, 16.10, 1940, p. 441.
138. [For Gautama and Adalja] Hicks, 'Forty Years', p. 46.
139. Hicks, 'Forty Years', p. 46.
140. RHL/Papers Collected by H. Topiwala Related to Indian Doctors in Kenya, expected deposit date 2015, H.T Topiwala, Note to Indian Medical Union [n.d. c.1943?].
141. Hicks, 'Forty Years', p. 47.
142. Hicks, 'Forty Years', p. 47.
143. 'BMA Meeting: Kenya Branch', *East African Medical Journal*, 3.11, 1962, pp. 665–6, 666.
144. For the establishment of IMDA see R.D. Shah, 'The Kenya Medical Association', *East African Medical Journal*, 53.12, 1976, pp. 724–9, 725. However the source

of the information is not referenced and some of the claims appear to be speculative.
145. RHL/Papers Collected by H. Topiwala Related to Indian Doctors in Kenya, expected deposit date 2015, H.T. Topiwala personal note [n.d.c.1943?].
146. Iliffe, *East African Doctors*, p.124.

9 Conclusion

1. Crozier has estimated, in the same period, that 194 European medical officers came to Kenya, to which should to be added the privately practising doctors that also arrived from Europe, which have never been formally enumerated. Anna Crozier, *Practising Colonial Medicine: The Colonial Medical Service in East Africa*, London, I.B. Tauris, 2007, pp. 142–51.
2. Crozier, *Practising Colonial Medicine*, Appendix 2.
3. Although this was said in the context of India, it holds true for Kenya too. 'British Doctors Leaving India', *The Times*, 25 January 1921, p. 7.
4. Syracuse University (SU), Kenya National Archive Records (KNA), Norman Leys, Letter to the Provincial Commissioner, Quarterly Provincial Report, 9 April 1912 details how Leys took the Railway Administration to court for ignoring 'elementary sanitary principles' in its treatment of native staff.
5. RHL/Papers Collected by H. Topiwala Related to Indian Doctors in Kenya, expected deposit date 2015, F.J. Wright, Letter Secretary Kenya BMA to H.T. Topiwala, 12 June 1943.
6. *Annual Medical Report*, 1913, p. 52; For diverting attention n away from other responsibilities see Lesley Doyle, *The Political Economy of Health*, London, Pluto Press, 1979, p. 243.
7. W.J. Moynagh, *East African Chronicle*, 18 September 1922, p. 14; also see *East African Chronicle*, 2 October 1920, p. 5.
8. *Annual Medical Report*, 1928, p.42.
9. Medical Report of the Nairobi Municipality Council, Municipality of Nairobi, 1930, p. 10; *Annual Medical Report* 1934, p. 8. N.B. We by no mean wish to imply that African health conditions were any better. Indeed, we fully acknowledge that Africans remained firmly at the bottom of the colonial hierarchy and were (despite calls for Africanisation) the least pressing colonial priority in terms of the agenda to improve health conditions.
10. H.C. Trowell, 'The Medical Training of Africans', *East African Medical Journal*, 11.10, 1935, pp. 338–53, 346.
11. NA/CO/533/426/8 Native Medical Service, Memo from Gilks concerning Deputy Governor's letter to Colonial Office, 6 August 1932.
12. BL/IOR/L/PJ/8/254 A.W. Pim, *Report on the Financial Position and System of Kenya*, HMS Stationery Office, London, 1936, p. 196.
13. *Annual Medical Report*, 1932, p. 2; notably numbers then dipped between 1930–1933 as part of the worldwide economic depression; *Annual Medical Report*, 1937, pp. 4–5.
14. BL/IOR/L/PJ/8/254 Pim Report, HMSO, 1936, p.195.
15. Dr K.V. Adalja, 'The Development of Medical Services in Kenya', *East African Medical Journal*, 39.3, 1962, pp. 105–14, 111–112.
16. Adalja, 'Medical Services in Kenya', p. 111–112.
17. H.W. Tilling, 'Medical Report of the Municipal Council of Nairobi', Nairobi, Nairobi Municipality, 1939, p. 1.

18. Tilling, 'Medical Report', p. 6.
19. *Annual Medical Report*, 1930 p. 28.
20. Sanjiva Rai, 'Memorandum to Nairobi City Council Commission', *Daily Chronicle*, 10 February 1956, p. 28.
21. Rai, 'Memorandum', *Daily Chronicle*, 1956, p. 28.
22. *The Kenya Gazette*, 19 February 1960, pp. 200–7.
23. Quotation was famously originally used in the context of the Sudan. 'The Men who Ran Empire with a Stony British Stare', *The Economist*, 3 September 2011 at http://www.economist.com/node/21528220 (accessed 8 August 2014).

Bibliography

Abdilahi Bulhan, Hussein, *Frantz Fanon and the Psychology of Oppression*, New York, London, Plenum, 1985

Adalja, K.V., 'Intramuscular Versus Intravenous Quinine', *The Indian Medical Gazette*, September 1935, p. 538

Adalja, K.V., 'Haematuria Following M and B 693', *British Medical Journal*, 10 April 1937, p. 643

Adalja, K.V., 'Acquired Haemophilia', *British Medical Journal*, 10 April 1937,V1, p. 785

Adalja, K.V., 'Ayurvedic Midwifery', *East African Medical Journal*, 17, 1940, p. 142

Adalja, K.V., 'Jain Juvkona' (in Gujarati), *Jyotna Magazine*, Nairobi, Regal Press, c.1940 p. 29

Adalja, K.V., 'Thirty Two Years in General Practice in Nairobi', *East African Medical Journal*, 36, 1959, pp. 442–48

Adalja, K.V., 'The Development of Medical Service in Kenya', *East African Medical Journal*, 39, 1962, pp. 105–14

Adeloye, Adeloya, *African Pioneers of Modern Medicine: Nigerian Doctors of the Nineteenth Century*, Ibadan: University Press Limited, 1985

Aiyar, Sana, *Indians in Kenya: The Politics of Diaspora*, Cambridge, MA, Harvard University Press, 2015

Aiyar, Sana, 'Anticolonial Homelands Across the Indian Ocean: The Politics of the Indian Diaspora in Kenya, ca. 1930–1950', *American Historical Review*, 116.4, 2011, pp. 987–1013

Aiyar, Sana, 'Empire, Race and the Indians in Colonial Kenya's Contested Public Political Sphere, 1919–1923', *Africa: The Journal of the International African Institute*, 81.1, 2011, pp. 132–54

Albuquerque, Teresa, *Goans of Kenya*, Mumbai, Michael Lobo Publishers, 1999

Alpers, E.A., *East Africa and the Indian Ocean*, Princeton, Markus Wiener Publishers, 2009

Alpers, E.A., 'Gujarat and the Trade of East Africa, c.1500–1800', *The International Journal of African Studies*, 9.1, 1976, pp. 22–44

Amiji, H.M., 'The Bhoras of East Africa', *The Journal of Religion in Africa*, 5, 1975, pp. 27–61

Aminzade, Ronald R., Jack A. Gladstone, Doug McAdam, Elizabeth J. Perry, William Sewell, Jr, Sidney Tarrow, Charles Tilly, *Silence and Voice in the Study of Contentious Politics*, Cambridge, Cambridge University Press, 2001

Anderson, David, *Histories of the Hanged: The Dirty War in Kenya and the End of Empire*, London, Weidenfield & Nicolson, 2005

Anderson, Warwick, 'Excremental Colonialism: Public Health and the Poetics of Pollution', *Critical Inquiry*. 1995, 21, pp. 640–69

Anderson, Warwick, *Colonial Pathologies: American Tropical Medicine, Race and Hygiene in the Philippines*, Durham, Duke University Press, 2006

Anon, 'Lord Ellenborough and the Indian Medical Service', *The Lancet*, 14 December 1861, p. 581

Anon, 'Indian Sub-Assistant Surgeon', *The Lancet*, 19 October 1867, pp. 503–4

Anon, 'The Association and the Colonies', *British Medical Journal*, i, 1905, p. 1394

Anon, 'The Future of Medicine in India: Sir Gerald Gifford's Address' *British Medical Journal*, 22 November 1924, p. 966

Anon, 'Editorial', *The Kenya Medical Journal*, 1.10, 1925, p. 291

Anon, 'Indian Medical Degrees: The General Medical Council and Indian Qualifications: History of a Ten Year Controversy', *British Medical Journal*, 15 March 1930, pp. 505–11

Anon, 'BMA Branch Meetings', *East Africa Medical Journal*, 11.3, 1934, p. 103

Anon, 'BMA, Kenya Branch Annual General Meeting', *East African Medical Journal*, 11.10, 1935, pp. 365–67

Anon, 'The Framework of Medicine in India', *The Lancet*, 16 October 1937, pp. 933–35

Anon, 'Medicine in India', *British Medical Journal* [supplement], 28 January 1939, p. 43

Anon, 'R.A. Ribeiro, Obituary', *East African Medical Journal*, 28.1, 1951, p. 91

Anon, 'Dr M.N. Hamin, Obituary', *The East African Medical Journal*, 36.1, 1959, p. 56

Anon, 'M.B. Gautama: Appreciation', *East African Medical Journal*, 36.3, 1959, p. 178

Anon, 'Obituary: Dr Krishnalal Vithadas on Adalja', *East African Medical Journal*, 41.8, 1964, p. 387

Anon, 'Obituary of G.V. Juvekar', *East African Medical Journal*, 41.10, 1964, p. 496

Anon, 'Dr A.U. Sheth, Obituary', *East African Medical Journal*, 46.7, 1969, p. 448

Anon, 'Obituary: J.L. Gilks', *British Medical Journal*, 28 August 1971, p. 538

Arden Hoppe, Kirk, *Sleeping Sickness Control in British East Africa, 1900–1960*, Westport Connecticut, Praeger, 2003

Arnold, David, *Colonizing the Body: State Medicine and Epidemic Disease in Nineteenth Century India*, University of California Press, 1993

Arnold, David, 'The Place of "The Tropics" in Western Medical Ideas Since 1750', *Tropical Medicine and International Health*, 24, 1997, pp. 303–13

Attewell, Guy N.A., *Refiguring Unani Tibb: Plural Healing in Late Colonial India, New Perspectives in South Asian History*, New Delhi, Orient Longman, 2007

Austin, Herbert Henry, *With Macdonald in Uganda: A Narrative Account of the Uganda Mutiny and Macdonald Expedition in the Uganda*, London, Dawsons of Pall Mall, 1973

Baker, Colin 'The Government Medical Service in Malawi: An Administrative History, 1891–1974', *Medical History*, 20, 1976, pp. 296–311

Bala, Poonam, *Imperialism and Medicine in Bengal: A Socio Historical Perspective*, New Delhi, Sage Publications, 1991

Bala, Poonam, '"Defying" Medical Autonomy: Indigenous Elites and medicine in Colonial India', in Poonam Bala (ed.), *Biomedicine as a Contested Site: Some Revelations in Imperial Contexts*, Lanham, Boulder, New York, Toronto, Plymouth, UK, Lexington Books, 2009, pp. 29–44

Bastos, Cristiana, 'The Inverted Mirror: Dreams of Imperial Glory and Tales of Subalternity from the Medical School of Goa', *Etnográfica*, 6.1, 2002, pp. 59–76

Bayly, Christopher A., *Indian Society and the Making of the British Empire*, Cambridge University Press, 1988

Bayoumi, A., *The History of the Sudan Health Services*, Nairobi, Kenya Literature Bureau, 1979

Beck, Ann, *A History of the British Medical Administration of East Africa, 1900–1950*, Cambridge, Mass, Harvard University Press, 1970

Beck, Ann, *Medicine, Tradition and Development in Kenya and Tanzania 1920–1970*, Waltham, Mass.: Crossroads Press, 1981

Beck, Ann, 'Problems of British Medical Administration in East Africa between 1900–1930' *Bulletin of the History of Medicine*, 36, 1962, pp. 275–83

Beck, Ann, 'Native Medical Services in British East Africa and Native Patterns of Society' in *Verhandlungen des XX Internationalen Kongresses für Geschichte der Medizin,*

Berlin, 22–27 August 1966, Hildesheim, Georg Olms Verlagsbuchhandlung, 1968, pp. 870–75

Beck, Ann, 'Medical Administration and Medical Research in Developing Countries: Remarks on Their History in Colonial East Africa', *Bulletin of the History of Medicine*, 46, 1972, pp. 349–58

Beck, Ann 'The State and Medical Research: British Government Policy Toward Tropical Medicine in East Africa', *Proceedings of the XXIII International Congress of the History of Medicine, London, 2–9 September 1972*, London, Wellcome Institute for the History of Medicine, 1974, pp. 488–93

Bell, Heather, *Frontiers of Medicine in the Anglo-Egyptian Sudan, 1899–1940*, Oxford, Clarendon Press, 1999

Bennett, Brett M. and Joseph M. Hodge (eds.), *Science and Empire: Knowledge and Networks of Science Across the British Empire, 1800–1970*, Basingstoke, Palgrave Macmillan, 2011

Bentley, E., *Handbook of the Uganda Question and Proposed East Africa Railway*, London, Chapman & Hall, 1892

Berman, Bruce and John Lonsdale, *Unhappy Valley: Conflict in Kenya and Africa*, London, Nairobi, Athens, OH, Ohio University Press, 1992

Bertz, Ned, 'Educating the Nation: Race and Nationalism in Tanzanian Schools', in Sara Dorman, Daniel Hammett, Paul Nugent (eds.), *Making Nations, Creating Strangers: States and Citizenship in Africa*, Leiden, Brill, 2007

Bethwell, Allan Ogot, *History of the Southern Luo: Volume I, Migration and Settlement, 1500–1900*, Nairobi, East African Publishing House, 1967

Bethwell, Allan Ogot, (ed.), *Politics and Nationalism in Kenya*, Nairobi, East African Publishing House, 1972

Bethwell, Allan Ogot, *My Footprints on the Sands of Time: An Autobiography*, Victoria BC, Trafford Publishing, 2006

Bethwell Allan Ogot, and J.A. Kieran (eds.), *Zamani: A Survey of East African History*, Nairobi, East African Publishing House, 1968

Bhacker, M. Reda, *Trade and Empire in Muscat and Zanzibar: The Roots of British Domination*, London, Routledge, 1992

Bharati, Agehananda Swami, *The Asians in East Africa: Jayhind and Uhuru*, Chicago, Nelson Hall Company, 1972

Bhattacharya, Sanjoy, Mark Harrison and Michael Worboys (eds.), *Fractured States: Smallpox, Public Health and Vaccination Policy in British India, 1800–1947*, New Delhi, Orient Longman and Sangam Books, 2005

Blixen, Karen, *Out of Africa*, London, Putnam & Co. Ltd, 1937

Blyth, Robert J., *The Empire of the Raj: India, Eastern Africa and the Middle East, 1858–1947*, Cambridge, Cambridge University Press, 2003

Bödeker, H.A., 'Some Sidelights on Early Medical History in East Africa', *The East African Medical Journal*, 12, 1935–36, pp. 100–7

Boyd, Kelly, (ed.), *Encyclopedia of Historians And Historical Writing*, Vol. 2, London and Chicago, Fitzroy & Dearborn, 1999

Brantlinger, Patrick, 'Victorians and Africans: The Genealogy of the Myth of the Dark Continent', *Critical Inquiry*, 12, 1985, pp. 166–203

Brennan, James R., *TAIFA: Making Nation and Race in Urban Tanzania*, Athens, Ohio University Press, 2012

Burton, Eric "…what Tribe Should We Call Him?" The Indian Diaspora, the State and the Nation in Tanzania since ca.1850', *Stichproben: Weiner Zeitschrift für Kritische Afrikastudien*, 25, 2013, pp. 1–28

Burton, Richard F., *Zanzibar, City, Island and Coast*, London, Tinsley Brothers, 1872

Cable, Vincent, 'The Asians of Kenya', *African Affairs*, 68, 1969, pp. 218–31

Calcutta Medical College, *The Centenary of the Medical College, Bengal, 1835–1934*, Calcutta, Medical College of Bengal, 1935

Campbell, Chloe, *Race and Empire: Eugenics in Colonial Kenya*, Manchester, Manchester University Press, 2007

Carman, John A., *A Medial History of the Colony and Protectorate of Kenya: A Personal Memoir*, London, Rex Collings, 1976

Castellani, Aldo, *Microbes, Men and Monarchs: A Doctor's Life in Many Lands*, London, Gollancz, 1960

Chakrabarti, Pratik, *Western Science in Modern India: Metropolitan Methods, Colonial Practices*, Delhi, Permanent Black, 2004

Chakrabarti, Pratik, *Medicine and Empire, 1600–1960*, London, Palgrave MacMillan, 2013

Chakrabarti, Pratik, '"Sign of the Times": Medicine and Nationhood in British India', *OSIRIS*, 24, 2009, pp. 188–211

Charters, A.D., 'Correspondence: Dr K.V. Adalja', *East African Medical Journal*, 41.10, 1964, p. 496

Christie, James, *Cholera Epidemics in East Africa: An Account of the Several Diffusions of the Disease in that Country from 1821 till 1872*, London, Macmillan, 1876

Churchill, Winston S., *My African Journey*, London, Hodder and Stoughton, 1908

Clayton A. and D.C. Savage, *Government and Labour in Kenya: 1895–1963*, London, Frank Cass, 1974

Clearkin, P.A.,'Typhus in the Tropics', *British Medical Journal*, 8 September 1934, p. 487

Cooper, Frederick, *Colonialism in Question: Theory, Knowledge, History*, Berkely, University of California Press, 2005

Coupland, R., *East Africa and its Invaders from the Earliest Times to the Death of Seyyid Said 1856*, London, Clarendon Press, 1938

Coupland, R., *Exploitation of East Africa 1856–90: Slave Trade and the Scramble*, London, Faber, 1968

Crawford, D.G., *A History of the Indian Medical Service, 1600–1913*, London Thacker and Co., 1914

Crawford, D.G., *Roll of the Indian Medical Service: 1615–1930*, W.Thacker & Co., London, 1930

Crozier, Anna, *Practising Colonial Medicine: The Colonial Medical Service in East Africa*, London, I.B. Tauris, 2007

Crozier, Anna, 'Sensationalising Africa: British Medical Impressions of Sub-Saharan Africa 1890–1939,' *Journal of Imperial and Commonwealth History*, 35, 3, 2007, pp. 393–415

Cunningham, Andrew and Bridie Andrews (eds.), *Western Medicine as Contested Knowledge*, Manchester, Manchester University Press, 1997

Curtin, Philip D., *Death by Migration: Europe's Encounter with the Tropical World in the Nineteenth Century*, Cambridge, Cambridge University Press, 1989

Curtin, Philip D., '"The White Man's Grave:" Image and Reality, 1780–1850', *The Journal of British Studies*, 1, 1961, pp. 94–110

Dalrymple, William, *White Mughals: Love and Betrayal in Eighteenth Century India*, London, Harper Collins, 2002

Darwin, John, *The Empire Project: The Rise and Fall of the British World-System, 1830–1970*, Cambridge, Cambridge University Press, 2009

Darwin, John, *Unfinished Empire: the Global Expansion of Britain*, London, Penguin, 2013

de Sousa, Mary, 'The Treatment of Erosion of Cervix with Picric Acid', *Kenya Medical Journal*, 1.9, 1924, p. 281

Delf, George, *Asians in East Africa*, Oxford, Oxford University Press, 1963

Digby, Anne, *Diversity and Division in Medicine: Healthcare in South Africa from the 1800s*, Oxford, Peter Lang, 2006

Digby, Anne and Helen Sweet, 'The Nurse as Culture Broker in Twentieth Century South Africa' in Waltraud Ernst (ed.), *Plural Medicine, Tradition and Modernity*, London, Routledge, 2002, pp. 113–29

Digby, Anne, 'Early Black Doctors in South Africa', *Journal of African History*, 46, 2005, pp. 427–54

Digby, Anne, 'The Mid-Level Health Worker in South Africa: The in-between Condition of the "Middle"' in Ryan Johnson and Amna Khalid (eds.), *Public Health in the British Empire: Intermediaries, Subordinates, and the Practice of Public Health*, New York & London, Routledge, 2012, pp. 171–92

Dilley, Marjorie Ruth, *British Policy in Kenya Colony*, London, Frank Cass & Co, 1966

Dinesen, Isak, *Letters from Africa 1914–1931*, Chicago, University of Chicago Press, 1981

Doyle, Lesley, *The Political Economy of Health*, London, Pluto Press, 1979

Duder, C.J., 'The Settler Response to the Indian Crisis of 1923 in Kenya: Brigadier General Philip Wheatly and "Direct Action"', *Journal of Imperial and Commonwealth History*, 17.3, 1989, pp. 349–73

Ebrahimnejad, Hormoz, (ed.), *The Development of Modern Medicine in Non-western Countries: Historical Perspectives*, London and New York, Routledge, 2009

Ernst, Waltraud, (ed.), *Plural Medicine, Tradition and Modernity*, London, Routledge, 2002

Farley, John, *Bilharzia: A History of Imperial Tropical Medicine*, Cambridge, Cambridge University Press, 2008 [first published 1991]

Fisher, Michael H., 'Indirect Rule in the British Empire: The Foundations of the Residency System in India (1764–1858), *Modern Asian Studies*, 18.3, 1984, pp. 393–428

Fisher, Suzanne, *We Lived on the Verandah*, Bognor Regis, New Horizon, 1980

Fisher, V.M., 'Medical Arrangements For Native Labour on the Thika-Nyeri Railway Construction', *The Kenya Medical Journal*, January 1925, pp. 293–319

Foucault, Michel, *The History of Sexuality Volume 1: An Introduction* translated by Robert Hurley, New York, Pantheon, 1978 [French publication: 1976]

Frenz, Margaret, 'Representing the Portuguese Empire. Goan Consuls in British East Africa, c.1910–1963' in Eric Morier-Genoud and Michel Cahen (eds.), *Imperial Migrations: Colonial Communities and Diaspora in the Portuguese World*, New York, Palgrave Macmillan, 2012

Frenz, Margaret, 'Swaraj for Kenya, 1949–1965. The Ambiguities of Transnational Politics', *Past and Present*, 218, 2013, pp. 151–77

Gandhi, Mahatma, *India of My Dreams*, Delhi, Rajpal and Sons, 2009, pp. 149–60

Gelfand, Michael, *Tropical Victory: An Account of the Influence of Medicine on the History of Southern Rhodesia, 1890–1923*, Cape Town, Juta, 1953

Gelfand, Michael, *A Service to the Sick: A History of the Health Services for Africans in Southern Rhodesia, 1890–1953*, Gweru, Mambo Press, 1976

Ghai D.P. and Y.P. Ghai (eds.), *Portrait of a Minority: Asians In East Africa*, Oxford University Press, 1970

Ghai Dharam P. and Yash P. Ghai, '*Asians in East Africa*: Problems and Prospects', *Journal of Modern African Studies*, 3, 1965, pp. 35–51

Gilks, John Langton, [Speech to East Africa Branch of BMA], 'The Annual Dinner of the Kenya Branch of the British Medical Association, *Kenya Medical Journal*, 2.11, 1925–26, pp. 321–25

Gilks, John Langton, 'The Medical Department and the Health Organization in Kenya, 1909–1933', *The East African Medical Journal*, 9, 1932–3, pp. 340–54

Gopal, M., D. Balasubramanian, P. Kanagarajah, A. Anirudhan, P. Murugan, 'Madras Medical College, 175 Years of Medical Heritage', *The National Medical Journal of India*, 23.2, 2010, pp. 117–20

Grant, Kevin, Philippa Levine and Frank Trentmann (eds.), *Beyond Sovereignty, 1880–1950: Britain, Empire and Transnationalism*, London, Palgrave, 2007

Green, N., 'African in Indian Ink: Urdu Articulations of Indian Settlement in East Africa', *Journal of African History*, 53, 2012, pp. 131–50

Gregory, J.R., *Under the Sun (A Memoir of Dr RW Burkitt, of Kenya)*, Nairobi, The English Press Ltd., 1951

Gregory, Robert G., *India and East Africa: A History of Race Relations within the British Empire, 1890–1939*, Oxford, Clarendon Press, 1971

Gregory, Robert G., *South Asians in In East Africa: An Economic And Social History 1890–1990*, Oxford, Westview Press, 1993

Guha, Ranajit, 'The Small Voices of History' in Shahid Amin and Dipesh Chakrabarty (eds.), *Subaltern Studies: Writings on South Asian History and Society*, Vol. IX, Oxford, Oxford University Press, 1988

Gunston, H., 'The Planning and Construction of the Uganda Railway', *Transactions of the Newcomen Society*, 74, 2004, pp. 45–71

Hailey, Lord, *An African Survey: A Study of the Problems Arising in Africa South of the Sahara*, Oxford, Oxford University Press, 1938

Hardiman, David (ed.), *Healing Bodies, Saving Souls: Medical Missions in Asia and Africa*, London, Rodopi, 2006

Hardy, Ronald, *The Iron Snake*, London, Collins, 1965

Harrison, Mark, *Public Health in British India: Anglo Indian Preventative Medicine, 1859–1914*, Cambridge, Cambridge University Press, 1994

Harrison, Mark, *Climates and Constitutions: Health, Race and British Imperialism in India, 1600–1850*, Oxford, Oxford University Press, 2002 [first published 1999]

Harrison, Mark, 'Medical Experimentation in British India: The Case of Dr Helenus Scott', in Hormoz Ebrahimnejad (ed.), *The Development of Modern Medicine in Non-Western Countries: Historical Perspectives*, Royal Asiatic Society Books, London and New York, Routledge, 2009, pp. 24–40

Hawley, John C., (ed.), *India in Africa; Africa in India: Indian Ocean Cosmopolitanisms*, Bloomington, IN, Indiana University Press, 2008

Haynes, Douglas M., *Imperial Medicine: Patrick Manson and the Conquest of Tropical Disease*, Philadelphia, University of Pennsylvania Press, 2001

Headrick, Daniel R., *The Tools of Empire: Technology and European Imperialism in the Nineteenth Century*, Oxford University Press, 1981

Heussler, Robert, *Yesterday's Rulers: The Making of the British Colonial Service*, New York, Syracuse University Press, 1963

Hicks, Andrew, 'Forty Years Of the British Medical Association (Kenya Branch)', *East African Medical Journal*, 38.1, 1961, pp. 43–53

Hicks, Andrew, 'Historical Note', *Nairobi Hospital Proceedings*, 1, 1997, p. 69

Hill, M.F., *The Permanent Way: The Story of Kenya and Uganda Railways*, Nairobi, East African Literature Bureau, 1961

Hillary, Richard, *The Last Enemy*, London, Macmillan and Co. Ltd., 1950

Himbara, David, 'The "Asian" Question in East Africa: The Continuing Controversy on the Role of Indian Capitalists in Accumulation and Development in Kenya, Uganda and Tanzania', *African Studies*, 56.1, 1997, pp. 1–18

Hinnells, J.R., *The Zoroastrian Diaspora; Religion and Migration*, London, Oxford University Press, 2005

Hokkanen, Marku, *Medicine and Scottish Missionaries in the Northern Malawi Region, 1875–1930*, Lampeter, Edwin Mellen Press, 2007

Hollingsworth, W. *The Asians of East Africa*, London, Macmillan and Company Limited, 1960

Hughes, Deborah L., 'Kenya, India and the British Empire Exhibition of 1924', *Race & Class*, 47, 2006, pp. 66–85

Hunt, Nancy Rose, *A Colonial Lexicon of Birth, Ritual, Medicalization, and Mobility in the Congo*, Durham. N.C., Duke University Press, 1999

Hunter, R.N., 'Plague in Kenya', *Kenya Medical Journal*, 3.3, 1925, pp. 75–86, 75

Huxley, Elspeth, *White Man's Country: Lord Delamere and The Making of Kenya*, London, Chatto and Windus, 1935

Huxley, Elspeth, *The Sorcers's Apprentice:A Journey Through East Africa*, London, Chatto and Windus, 1949

Huxley, Elspeth, *Out in the Midday Sun: My Kenya*, London, Chatto and Windus, 1958

Huxley, Elspeth, *The Flame Trees of Thika:Memoirs of An African Childhood*, London, Chatto and Windus, 1959

Huxley, Elspeth, *Pioneers' Scrapbook: Reminiscences of Kenya, 1890 to 1968*, London, Evan Brothers Ltd., 1980

Hyam, Ronald, *Britain's Declining Empire: The Road to Decolonisation 1918–1968*, Cambridge, Cambridge University Press, 2006

Iliffe, John, *East African Doctors: A History of the Modern Profession*, Cambridge, Cambridge University Press, 1998

Ingram, Kenneth, 'Medicine in East Africa', *The Journal of African History*, 12, 1971, pp. 162–3

Jeffery, Roger, 'Recognizing India's Doctors: The Institutionalization of Medical Dependency, 1918–39', *Modern Asian Studies*, 13.2, 1979, pp. 301–26

Jeffery, Roger 'Doctors and Congress: The Role of Medical Men and Medical Politics in Indian Nationalism' in Mike Shepperdson and Colin Simmons (eds.), *The Indian National Congress and the Political Economy of India, 1885–1985*, Aldershot, Brookfield, USA, Avebury, 1988, pp. 160–73

Jeffries, Charles, *Partners for Progress: The Men and Women of the Colonial Service*, London, 1949

Jewell, Norman Parsons [ed. Tony Jewell], *On Call*, Privately Published, forthcoming, 2015

Joalahliae, Randolph M.K., *The Indian as an Enemy: An Analysis of the Indian Question in East Africa*, Bloomington, Authorhouse, 2010

Jochelson, Karen, *The Colour of Disease: Syphilis and Racism in South Africa, 1880–1950*, Basingstoke, Palgrave, 2001

Johnson, Ryan and Amna Khalid (eds.), *Public Health in the British Empire: Inter-mediaries, Subordinates, and the Practice of Public Health*, New York & London, Routledge, 2012

Johnson, Ryan, '"An All White Institution": Defending Private Practice and the Formation of the West African Medical Staff', *Medical History*, 54.2, 2010, pp. 237–54

Johnson, Ryan, 'The West African Medical Staff and the Administration of Imperial Tropical Medicine, 1902–1914', *Journal of Imperial and Commonwealth History*, 38.3, 2010, pp. 419–39

Kapila, Neera, *Race, Rail and Society: Roots of Modern Kenya*, Nairobi, East African Educational Publishers, 2011

Karve, S.D., 'An Experiment in Midwifery', *East African Medical Journal*, 10, 1933–34, pp. 358–63

Karve, S.D., 'Some Indian Methods of Midwifery', *East African Medical Journal*, 11, 1934–35, pp. 286–87

Kennedy, Dane, *Islands of White: Settler Society and Culture in Kenya and Southern Rhodesia, 1890–1939*, Durham, Duke University Press, 1990

Kennedy, Dane and Durba Ghosh (eds.), *Decentring Empire: Britain, India, and the Transcolonial World*, Hyderabad, Longman Orient Press, 2006

Kennedy, Dane, 'Constructing the Colonial Myth of Mau Mau', *International Journal of African Historical Studies*, 25.2, 1992, pp. 241–60

Kirk-Greene, Anthony, *On Crown Service: A History of HM Colonial and Overseas Civil Services, 1837–1997*, London, I.B. Tauris, 1999

Kirk-Greene, Anthony, *Symbol of Authority*, London, I.B. Tauris, 2005

Kirk-Greene, Anthony M., 'The Thin White Line: The Size of the British Colonial Service in Africa', *African Affairs*, 79, 1980, pp. 25–44

Kochhar, R., 'European Medical Men in India', *Journal of Bioscience*, 24, 1999, pp. 259–68

Kumar, Deepak, 'Racial Discrimination and Science in Nineteenth Century India', *Indian Economic and Social History Review*, 19, 1982, pp. 63–82

Kumar, Deepak, 'Probing History of Medicine and Public Health in India', *Indian Historical Review*, 37.2, 2010, pp. 259–73

Kyle, Keith, 'Gandhi, Harry Thuku and Early Kenya Nationalism', *Transition*, 27, 1966, pp. 16–22

Laidlaw, Zoe, *Colonial Connections 1815–45; Patronage, the Information Revolution and Colonial Government*, Manchester, Manchester University Press, 2005

Lawrance, Benjamin N., Emily Lynn Osborn and Richard L Roberts (eds.), *Intermediaries, Interpreters, and Clerks: African Employees in the Making of Colonial Africa*, Madison, University of Wisconsin Press, 2006

Lester, Alan, *Imperial Networks: Creating Identities in Nineteenth-Century South Africa and Britain*, London, Routledge, 2001

Lester, Alan, 'Imperial circuits and networks: geographies of the British Empire', *History Compass*, 4, 2006, pp. 124–41

Levine, Philippa, *Prostitution, Race and Politics: Policing Venereal Disease in the British Empire*, London, Routledge, 2003

Lloyd-Jones, W., *Havash! Frontier Adventures in Kenya*, London, Arrowsmith, 1925

Lloyd-Jones, W., *K.A.R.: Being An Unofficial Account of the Origin and Activities of the King's African Rifles*, London, Arrowsmith, 1926

Lonsdale, John, 'The Conquest State, 1895–1904', in William R. Ochieng (ed.), *A Modern History of Kenya 1895–1980 in Honour of B.A. Ogot*, Nairobi, London, Evans Brothers (Kenya), 1989, pp. 6–34

Lonsdale, John, 'Kenya: Home Country and African Frontier', in Robert Bickers (ed.), *Settlers and Expatriates: Britons over the Seas*, Oxford, Oxford University Press, 2014 [First published, 2010], pp. 74–111

Lugard, Frederick D., *Rise of Our East African Empire*, London, William Blackwood and Sons, 1893

Lyons, Maryinez, *The Colonial Disease: A Social History of Sleeping Sickness in Northern Zaire 1900–1940*, Cambridge University Press, 2002 [first published, 1992]

Lyons, Maryinez, 'The Power to Heal: African Medical Auxiliaries in Colonial Belgian Congo and Uganda', in David Engels and Shula Marks (eds.), *Contesting Colonial Hegemony: State and Society in Africa and India*, London, British Academic Press, 1994

MacDonald, J.R.L [with a new introduction by A.T. Matson], *Soldiering and Surveying in British East Africa, 1891–1894*, London, Dawsons, 1973

Macleod, Roy, and Milton Lewis (eds.), *Disease Medicine and Empire*, London, Routledge, 1988

Mahone, Sloan 'The Psychology of Rebellion: Colonial Medical Responses to Dissent in British East Africa', *Journal of African History*, 47.2, 2006, pp. 241–60

Malowany, Maureen, *Medical Pluralism: Disease, Health and Healing in the Coast of Kenya, 1840–1940*, PhD Thesis, McGill University, Canada, 1997

Mamdani, Mahmood, *From Citizen to Refugee: Uganda Asians Come to Britain*, Oxford, Pambazuka Press, 2011 [first pub.1973]

Mangat, J.S., *A History of the Asians in East Africa, c.1886–1945*, Oxford, Oxford University Press, 1969

Manson-Bahr, Philip, *History of the School of Tropical Medicine in London: 1899–1949*, London, HK Lewis & Co. Ltd., 1956

Marett, Paul, *M.P. Shah: His Life and Achievements*, London, Bharatya Vidta Bhavan, 1988

Marks, Shula, 'What is Colonial about Colonial Medicine? And what has Happened to Imperialism and Health?', *Social History of Medicine*, 10, 1997, pp. 205–19

Martin, C.J., 'A Demographic Study of an Immigrant Community: The Indian Population of British East Africa', *Population Studies*, 6.3, 1953, pp. 233–47

Maxon, R.M., *John Ainsworth and the Making of Kenya*, Lanham, MD, University Press of America, 1980

Maxon, Robert M., *Struggle for Kenya: the Loss and Reassertion of Imperial Initiative, 1912–1923*, Rutherford, N.J., Farleigh Dickinson University Press, 1993

Maxon, R.M., 'The Years of Revolutionary Advance: 1920–1929', in William R. Ochieng (ed.), *A Modern History of Kenya 1895–1980*, Nairobi, London, Evans Brothers (Kenya) Ltd., 1989

Mazuri, Ali, 'European Exploration and African Self Discovery', *The Journal of Modern African Studies*, 4, 1969, pp. 661–76

McClintock, Anne, *Imperial Leather: Race, Gender and Sexuality in the Colonial Context*, New York, Routledge, 1995

McCulloch, Jock *Colonial Psychiatry and 'The African Mind*, Cambridge, Cambridge University Press, 1995

McKelvey Jr., J.J., *Man Against Tsetse: Struggle for Africa*, London, Cornell University Press, 1973

Mehta, Makrand, 'Gujarati Business Communities in East African Diaspora: Major Historical Trends', *Economic and Political Weekly*, 36.20, 2001, pp. 1738–47

Mehta, N.K., *Dream Half-Expressed: An Autobiography*, Bombay, Vakils, Feffer, and Simons, [c.1966]

Metcalf, Thomas R., *Ideologies of the Raj*, Cambridge, Cambridge University Press, 1994

Metcalf, Thomas R., *Imperial Connections: India in the Indian Ocean Arena, 1860–1920*, Berkeley and London, University of California Press, 2007

Miller, C., *The Lunatic Express: An Entertainment in Imperialism*, London, Macdonald, 1971

Mills, Stephen and Brian Yonge, *A Railway to Nowhere: The Building of the Lunatic Line, 1896–1901*, Nairobi, Mills Publishing, 2012

Milne, Arthur Dawson, 'The Rise of the Colonial Medical Service', *Kenya and East African Medical Journal*, 5, 1928–29, pp. 50–58

Morgan Philip D., and Sean Hawkins (eds.), *Black Experience and the Empire*, Oxford, Oxford University Press, 2004

Morris, H.F., *Government Publications Relating to Kenya Including Those Relating to the East Africa High Commission and the East African Common Services Organisation, 1897–1963*, Wakefield, EP Microform, 1976

Morris, Stephen, 'Indians in East Africa: A Study in a Plural Society', *The British Journal of Sociology*, 7, 1956, pp. 194–211

Mukharji, Projit, *Nationalizing the Body: The Market, Print and Healing in Colonial Bengal, 1860–1930*, London, Anthem Press, 2009

Mungeam, G.H., *Kenya: Select Historical Documents 1884–1923*, Nairobi, East African Publishing House, 1979

Muraleedharan, V.R., 'Professionalising Medical Practice in Colonial South India', *Economic and Political Weekly*, 27.4, 1992, pp. PE27–30, PE35–37

Nair, T.D., 'A Tana River Yaws Campaign', *Kenya and East Africa Medical Journal*, 4.7, 1927–28, pp. 201–07

Najmudean, G., 'Ocular Manifestations in Syphilis', *East African Medical Journal*, 8, 1931–32, pp. 31–22

Nazareth, John Maximilian, *Brown Man Black Country: A Peep into Kenya's Freedom Struggle*, New Delhi, Tidings Publications, 1981

Ndege, George Odour, *Health, State and Society in Kenya*, Rochester, NY, University of Rochester Press, 2001

Neill, Deborah J., *Networks in Tropical Medicine: Internationalism, Colonialism, and the Rise of a Medical Specialty, 1890–1930*, Stanford, Stanford University Press, 2012

Nelson, Donna, 'Problems of Power in a Plural Society: Asians in Kenya', *Southwestern Journal of Anthropology*, 23, 1972, pp. 255–64

Nicholls, Christine Stephanie, *Red Strangers: The White Tribe of Kenya*, London, Timewell Press, 2005

Nicolls, Christine Stephanie, *Elspeth Huxley: A Biography*, London, Harper & Collins, 2007

Oliver, Roland Anthony, *Sir Harry Johnston and the Scramble for Africa*, London, Chatto and Windus, 1957

Olluumwallah, O.A., *Disease in the Colonial State: Medicine, Society and Social Change Among the Aba Nyole of Western Kenya*, Westport Connecticut, Greenwood Publishing Group, 2002

Oonk, G., *Settled Strangers:Asian Business Elites in East Africa: 1800–2000*, London, Sage Publications, 2013

Oonk, G., 'The Changing Culture of the Hindu Lohana Community in East Africa', *Contemporary Asian Studies*, 13.1, 2004, pp. 7–23

Oza, U.K., *The Rift in the empire's Lute: Being a History of the Indian Struggle in Kenya*, Bombay, Advocate of India Press, 1931

Packard, Randall, *White Plague, Black Labor: Tuberculosis and the Political Economy of Health and Disease in South Africa*, Berkeley, University of California Press, 1989

Page, Malcolm, *A History of the King's African Rifles and East African Forces*, Pen and Sword, Barnsley, 2011

Paice, Edward, *The Lost Lion of Empire: The Life of 'Cape to Cairo' Grogan*, London, Harper Collins, 2001

Paice, Edward, *Tip and Run: The Untold Tragedy of the Great War in Africa*, London, Weidenfeld and Nicolson, 2007

Parker, Mary, 'Race Relations and Political Development in Kenya', *African Affairs*, 50.198, 1951, pp. 41–52

Pashid, Abdur, *History of the King Edward Medical College Lahore, 1860–1960*, Lahore, King Edward Medical College, 1960

Patel, Zarina, *Challenge to Colonialism: The Struggle of Alibhai Mulla Jeevanjee for Equal Rights in Kenya*, Nairobi, Publisher Distributon Services, 1997

Patel, Zarina, *Unquiet: The Life and Times of Makhan Singh*, Nairobi, Zand Graphics, 2006

Patel, Zarina, *Manilal Ambalal Desai: The Stormy Petrel*, Nairobi, Zand Graphics, 2010

Paterson, Arthur Rutherford, *The Book of Civilization: Part 1, On Cleanliness and Health. The Care Of Your Children, Food, and How to Get Rid Of Flies.* London, New York, Toronto, Longmans, Green & Co., 1934

Pati, Biswamoy and Mark Harrison (eds.), *The Social History of Health and Medicine in Colonial India*, London and New York, Routledge, 2009

Patton, Adell, *Physicians, Colonial Racism and Diaspora in West Africa*, Gainesville: University of Florida Press, 1996

Pearson, Michael N., (ed.), *The World of Indian Ocean: 1500–1800*, Aldershot, Ashgate, 2005

Pearson, Michael N., *Port Cities and Intruders: The Swahili Coast, India and Portugal in the Early Modern Area*, Baltimore, John Hopkins University Press, 1998

Perham, Margery, *Race and Politics in Kenya, A Correspondence Between Elspeth Huxley and Margery Perham*, London, Faber and Faber, 1944

Potter, Simon J., 'Webs, Networks and Systems: Globalization and the Mass Media in the Nineteenth- and Twentieth-Century British Empire', *Journal of British Studies*, 46, 2007,pp. 621–646

Pradhan, S.D., *The Indian Army in East Africa, 1914–1918*, New Delhi, National Book Organisation, 1991

Preston, R.O., *Oriental Nairobi*, Nairobi, Preston Publications, 1938

Proctor, R.A.W., 'The Health of Labour on the Kisumu-Yala Railway Construction', *Kenya and East African Medical Journal*, 7.10, 1931, pp. 276–81

Quaiser, Neshat, 'Colonial Politics of Medicine and Popular Unani Resistance', *Indian Horizons*, 47.2, 2000, pp. 29–42

Rafferty, Anne-Marie, 'The Rise and Demise of the Colonial Nursing Service: British Nurses in the Colonies, 1896–1966', *Nursing History Review*, 15, 2007, pp. 147–154

Ramanna, Mridula, *Western Medicine and Public Health in Colonial Bombay, 1845–1895*, Delhi, Orient Longman, 2002

Ramanna, Mridula, 'Indian Doctors, Western Medicine and Social Change, 1845–1885', in Mariam Dossal and Ruby Maloni (eds.), *State Intervention and Popular Response: Western Indian in the Nineteenth Century*, Mumbai, Popular Prakashan Press, 1999, pp. 40–62

Rao, M.S., 'The History of Medicine in India and Burma', *Medical History*, 12.1, 1968, pp. 52–61

Ribeiro, Ayres, 'Multicultural General Practice in Nairobi', *British Medical Journal*, 23 October 1954, pp. 979–81

Roberts, J.I., 'Plague Conditions in an Urban Area of Kenya', *Journal of Hygiene* 36.4, 1936, pp. 467–84

Rodney, Walter, *How Europe Underdeveloped Africa*, London, Bogle-L'Ouverture and Tanzanian Publishing House, 1972

Ross, W.M., *Kenya From Within: A Short Political History*, London, George Allen & Unwin, 1927

Salvadori, Cynthia [Andrew Fedders (ed.)], *Through Open Doors: A View of Asian Cultures in Kenya*, Nairobi, Kenway publications, 1989

Salvadori, Cynthia, *We Came in Dhows*, Nairobi, Paperchase Kenya Ltd, 1996, Vols. 1–3

Salvadori, Cynthia, *Two Indian Travellers: East Africa 1902–1905*, Mombasa, Friends of Fort Jesus, 1997

Sambon, L. Westenra, 'Acclimatization of Europeans in Tropical Lands', *The Geographical Journal*, 12, 1898, pp. 589–98

Sarkar, Sumit, 'The Decline of the Subaltern in Subaltern Studies', in Sumit Sarkar (ed.), *Writing Social History*, 88, Delhi, Oxford University Press, 1997, pp. 82–108

Sastri, V.S. Srinivasa 'The Kenya Question', *The New Age*, 10 May 1923, p. 19

Schram, Ralph, *A History of the Nigerian Health Services*, Ibadan, Ibadan University Press, 1971

Schram, Ralph, *Heroes of Healthcare in Africa*, unpublished folio, Isle of White, 1997

Searle, C., 'The Second Class Doctor and the Medical Assistant in South Africa', *South African Medical Journal*, 31 March 1973, pp. 509–12

Seidenberg, D.A., *Uhuru and the Kenya Indians: The Role of a Minority Community in Kenya Politics, 1939–1963*, New Delhi, Vikas Publishing House, 1983

Seidenberg, D.A., *Mercantile Adventurers*, Delhi, New Age International Ltd., 1996

Sen, S.N., *Scientific and Technical Education in India 1781–1900*, New Delhi, Indian National Science Academy, 1991

Shah, Imam Sultan Muhammad (Aga Kahn III), *India in Transition: A Study in Political Evolution*, Bombay, Bennett, Coleman and Co. Ltd., 1918

Shah, R.D., 'The Kenya Medical Association', *East African Medical Journal*, 53.12, 1976, pp. 724–49

Shapiro, Karen, 'Doctors or Medical Aids—The Debate over the Training of Black Medical Personnel in the Rural Black Population in South Africa in the 1920s and 1930s', *Journal of Southern African Studies*, 13, 1987, pp. 234–55

Sheriff, Abdul, *Slaves, Spices & Ivory in Zanzibar: Integration of an East African Commercial Empire into the World Economy, 1770–1873*, London, James Curry, 1987

Sheriff, Abdul, *Dhow Cultures and the Indian Ocean: Cosmopolitanism, Commerce, and Islam*, New York, NY and Chichester, Columbia University Press, 2010

Simpson, W.J., *Report on the Sanitary Matters in the East Africa Protectorate, Uganda and Zanzibar*, London, Colonial Office, 1915

Singh, Chanan, 'Manilal Ambalal Desai', in Kenneth King and Ahmed Salim (eds.), *Kenya Historical Biographies*, Nairobi, East African Publishing House, 1971, pp. 129–41

Singh, Chanan, 'Later Asian Protest Movements', in B.A. Ogot (ed.), *Hadith 4: Politics and Nationalism in Colonial Kenya*, Nairobi, East Africa Publishing House, 1972

Sivaramakrishnan, Kavita, *Old Potions, New Bottles: Recasting Indigenous Medicine in Colonial Punjab 1850–1940*, New Delhi, Orient Longman, 2006

Smyth, R., 'The Development of British Colonial Film Policy, 1927–1939, With Special Reference to East and Central Africa', *Journal of African History*, 20.3, 1979, pp. 437–50

Sood, Satya V., *Victoria's Tin Dragon: A Railway that built a Nation*, Cambridge, Vanguard Press, 2007

Sorrenson, M.P.K., *Origins of European Settlement in Kenya*, Nairobi, Oxford University Press, 1968

Spurr, David, *The Rhetoric of Empire: Colonial Discourse in Journalism, Travel Writing and Imperial Administration*, Durham and London, Duke University Press, 1993

Stanley, Henry Morton, *Through the Dark Continent*, London, Sampson Low, Marston, Searle & Rivington, 1878

Stepan, Nancy Leys, *Picturing Tropical Nature*, London, Reaktion Books Ltd., 2001

Strayer, Robert W., *The Making of Mission Communities in East Africa: Anglicans and Africans in Colonial Kenya, 1875–1935*, London, Heinemann Educational, 1978

Sutphen, Mary P. and Bridie Andrews (eds.), *Medicine and the Colonial Identity*, London, Routledge, 2003

Swainson, Nicola, *The Development of Corporate Capitalism in Kenya, 1918–77*, London, Heinemann Educational, London, 1980

Theroux, Paul, 'Hating the Asians', *Transition*, 33, 1967, pp. 46–51

Thuku, Harry, *An Autobiography*, Nairobi, Oxford University Press, 1970

Tignor, Robert L., *The Colonial Transformation of Kenya: The Kamba, Kikuyu and Masai from 1900–1939*, Princeton, Guildford, Princeton University Press, 1976

Tilling, H.W., 'Medical Report of the Municipal Council of Nairobi', Nairobi, Nairobi Municipality, 1939

Tinker, Hugh, *A New System of Slavery: The Export of Indian Labour Overseas, 1830–1920*, Oxford, Oxford University Press, 1974

Topiwala, H.T., 'Mulykno'(in Gujarati), *Jyotna Magazine*, c.1940, p. 33

Trowell, H.C., 'Presidential Address of the Conference of Physicians', *East African Medical Journal*, 1955, p. 433

Trowell, H.C., The Medical Training of Africans', *East African Medical Journal*, 11.10, 1935, pp. 338–53

Trowell, Margaret, *African Tapestry*, London, Faber and Faber, 1957

Trzebinski, Errol, *The Kenya Pioneers*, London, Mandarin Paperbacks, 1991

Twaddle, Michael, 'Z.K. Sentongo and the Indian Question in East Africa', *History in Africa*, 24, pp. 309–36

Van Heyningen, E.B., '"Agents of Empire": The Medical Profession in the Cape Colony, 1880–1910', *Medical History*, 33, 1989, pp. 450–71

Vassanji, M.G., *The Gunny Sack*, Oxford, Heinemann International, 1989

Vassanji, M.G., *No New Land*, New Delhi, Penguin, 1992

Vassanji, M.G., *The Book of Secrets*, London, Picador, 1996

Vaughan, Megan, *Curing Their Ills: Colonial Power and African Illness*, Cambridge, Polity Press, 1991

Visram, M.G., *Alidina Visram: The Trail Blazer*, Mombasa, M.G. Visram Publisher, 1990

Watermann, R., 'Medicine and Hospitals along East—African Coasts in 16th Century', in *Actes du XXXIIe Congrès International d'Histoire de la Médecine, Anvers, 3–7 Septembre 1990/Proceedings of the XXXIInd International Congress on the History of Medicine, Antwerp, 3–7 September 1990*, Bruxelles [Belgium], Societas Belgica Historiae Medicinae, pp. 1029–38

Wiggins, Clare Aveling, 'Early Days in British East Africa and Uganda', *The East African Medical Journal*, 37, 1960, pp. 699–708

Wiggins, Clare Aveling, 'Early Days in British East Africa and Uganda: Second Tour— 1904–1907', *The East African Medical Journal*, 37, 1960, pp. 780–93

Wilson, C.J., *Before The Dawn in Kenya*, Nairobi, The English Press, 1952

Wilson, C.J., *Kenya Warning: The Challenge to White Supremacy in Our British Colony*, Nairobi, The English Press, 1954

Worboys, Michael, 'Colonial Medicine', in Roger Cooter and John Pickstone (eds.), *Medicine in the Twentieth Century*, London, Routledge, 2002, pp. 67–80

Worboys, Michael, 'Colonial and Imperial Medicine', in Debra Brunton (ed.), *Medicine Transformed: Health, Disease and Society in Europe, 1800–1930*, Manchester, The Open University and Manchester University Press, 2004, pp. 211–38

Worsley, Peter 'The Anatomy of Mau Mau', *The New Reasoner*, 1, 1957, pp. 13–25

Wylie, Diana, 'Confrontation over Kenya: The Colonial Office and Its Critics, 1918–1940', *Journal of African History*, 18.3, 1977, pp. 427–47

Wylie, Diana, 'Norman Leys and McGregor Ross: A Case Study in the Conscience of African Empire 1900–39', *The Journal of Imperial and Commonwealth History*, 5.3, 1977, pp. 294–309

Youé, Christopher P., 'The Threat of Settler Rebellion and the Imperial Predicament: The Denial of Indian Rights in Kenya, 1923', *Canadian Journal of History*, 12, 1978, pp. 347–60

Younghusband, Ethel, *Glimpses of East Africa and Zanzibar*, London, John Long, 1910

Zwanenburg, R.G., 'Robertson and the Kenya Critic', in K. King and A Salim (eds.), *Kenya Historical Biographies*, Nairobi, East African Publishing House, 1971, p. 152

Archival Sources

British Library (BL), India Office Records (IOR), London
British Library, Colindale (BLC)
British Medical Association Archive, London (BMAA)
National Archive, Kew (NA)
Rhodes House Library, Oxford (RHL)
School of Oriental and African Studies, London (SOAS)
London School of Hygiene and Topical Medicine (LSHTM)
Wellcome Trust Library, London (WTL)
Kenya National Archive, Nairobi (KNA)
Syracuse University (SU), Kenya National Archive Collection (KNA)
Maharashtra State Archives, Mumbai (MSL)

Newspapers

Daily Chronicle
East African Standard
East African Chronicle
The Leader
The Official Gazette
The Times

Index

Note: N stands for note; *f* for figure.